FORCES AND PARTICLES

A Title in the Nature–Macmillan
Physics Series

FORCES and PARTICLES

An Outline of the Principles of Classical Physics

A. B. Pippard, Sc.D., F.R.S.
Cavendish Professor of Physics and President of Clare Hall in the University of Cambridge

A HALSTED PRESS BOOK

JOHN WILEY & SONS
NEW YORK · TORONTO

First Published 1972

THE MACMILLAN PRESS LTD
London and Basingstoke
Associated companies in New York
Toronto Melbourne Dublin
Johannesburg and Madras

Published in the U.S.A. and Canada by
HALSTED PRESS,
a Division of John Wiley & Sons, Inc.,
New York

ISBN 0 470-69015-1

Library of Congress Catalog Card No.: 72-6261

Printed in Great Britain by
WILLIAM CLOWES & SONS, LIMITED
London, Beccles and Colchester

Preface

This book is an expanded version of a lecture course given for several years to first-year Cambridge undergraduates reading physics as one of three experimental sciences for the Natural Sciences Tripos, Part 1A. For reasons of space the course had to be duplicated, with two lecturers covering the same material, though at different levels of difficulty. It was my task to lecture to those who wanted the more searching treatment; surveys afterwards showed a substantial fraction to be sympathetic, though there were always those who found the approach pointlessly unconventional. Recognizing the impossibility of pleasing everyone, I have now addressed myself, as I did when giving the course, to the more adventurous who are prepared to work out for themselves congenial patterns of thought that will make sense of what can so easily remain a disorderly assembly of observations and ideas.

Every practising physicist develops his own favourite modes of thought, and no one can, or should wish to, pass on to others all the details of his private system. Indeed, the more original his mode, the harder it is to transmit. Nothing in this book, however, is original; in so far as I reveal my own ways of thinking about physics, it is my hope that they will seem perfectly ordinary—for this offers the best chance that they will prove acceptable to others and will play some part in leading other imaginations into paths where genuine intellectual satisfaction is to be found.

A.B.P.

Preface

Contents

Note: Certain rather advanced sections of the discussion are distinguished by square brackets [ ].

Acknowledgements

The author acknowledges with thanks the courtesy of all who have kindly given permission for the reproduction of the following illustrations.

Page

Weather Map, 16 Oct. 1970. Reproduced with the permission of the controller of Her Majesty's Stationery Office and *The Times* 41

Figure from Roll, Krotkov and Dicke, *Ann. Phys.* **26** (1964), p. 442. By kind permission of the author and editor. © Academic Press. 58

Figure 2.7.2 from Batchelor, *Introduction to Fluid Dynamics*, C.U.P., p. 106. By kind permission of the publishers. 101

Figure from Lamb, *Hydrodynamics*, C.U.P., p. 228. By kind permission of the publishers. (*upper figure*) 106

Figure 125 from Prandtl, *Fluid Dynamics*, Vieweg, p. 149. By kind permission of the publishers. 141

Figure 18 from Jeans, *Electricity and Magnetism*, C.U.P., p. 51. By kind permission of the publishers. (*lower figure*) 146

Figure 17 from Jeans, *Electricity and Magnetism*, C.U.P., p 50. By kind permission of the publishers. (*left-hand figure*) 151

Figure 64 from Jeans, *Electricity and Magnetism*, C.U.P., p. 193. By kind permission of the publishers. (*upper figure*) 152

Figure 14 from Gomer, *Field Emission and Field Ionization*, Harvard, p. 33. By kind permission of the publishers. 153

Figure 3 from *The Fermi Surface* (ed. Harrison and Webb), Wiley, p. 149. By kind permission of the publishers. (*lower figure*) 231

Figure from *The Physics of Metals* (ed. Ziman), C.U.P., p. 65. By kind permission of the publishers. 236

'Descartes put forward his conjectures as verities, almost as if they could be proved by his affirming them on oath. He ought to have presented his system of physics as an attempt to show what might be anticipated as probable in this science, when no principles but those of mechanics were admitted: this would indeed have been praiseworthy; but he went further, and claimed to have revealed the precise truth, thereby greatly impeding the discovery of genuine knowledge.'

CHRISTIAAN HUYGENS (1629–1695)

1

Introduction

This book is devoted to the exposition of certain concepts which form the basis of what has come to be called classical physics. At the time they were developed, over a period of some hundreds of years up to about 1900, there were, of course, no classical physicists, only physicists (or natural philosophers). The ideas became classical when later discoveries showed that, however exact they may have seemed to be, they were inadequate to describe the whole range of observations on physical systems, especially the very small (atoms) and the very large (galaxies), not to mention bodies moving at speeds approaching the speed of light. Classical physics could then be recognized, precisely because its failings were so clearly delineated, as a closed system of thought which works extraordinarily well within its proper sphere; roughly, that sphere comprises phenomena on the human scale that are accessible to observation without the need for such elaborate extensions of the human senses as particle detectors or radio telescopes. Within its own sphere, it provides answers to problems much more economically than the more general quantum and relativistic theories which contain classical physics as a special case, and the answers are experimentally indistinguishable from those given by the more advanced theories. There is no question, therefore, of studying classical physics as a pious duty or merely as a preliminary to real modern ideas—it has enormous value in its own right, sufficient indeed to carry many physicists, and still more engineers, through the whole of their professional careers, and is still capable of further development and refinement. Moreover, it has great intellectual beauty, such as is well worth the pursuit by anyone who finds the exercise of exact thought a source of joy rather than discomfort.

The main stream of the argument is the exposition of the laws of classical physics which are to be found displayed at the head of the appropriate chapters—3 (Newton's laws of motion), 4 (inverse-square laws), 10 (electromagnetism), 11 (force on a moving charge), 12 (electromagnetic induction) and 14 (displacement). Each statement of a law is followed by an outline account of simple laboratory experiments and elementary arguments such as should persuade the willing student to give conditional assent to the law; as the development throws up further experimental tests he should find that he understands its significance ever more deeply, and has reason to trust its application over a far wider range of phenomena than he may have suspected at first. By the end, he will appreciate why these laws have attained special prominence as examples of *fundamental laws*, laws from which it is believed could (in principle) be developed exact mathematical theories describing fully the behaviour of material objects. Some of the other chapters show elementary examples of this process in operation, and should be considered as no less important than those concerned with the laws themselves. There is a tendency, especially among those so engaged, to regard the investigation of fundamental laws as having a peculiar importance denied to other scientific activities, but this attitude will not bear critical examination. It is not the sole, or even an especially important, aim of scientists to reduce all human experience and observation to a few fundamental laws, though it is clearly of great interest to discover whether this can be done. It seems to be intrinsic to Western civilization, with its diverse Jewish, Hellenistic and Christian philosophical background, to believe at heart in the perfection of the order of our universe—even if we have no explicit religious faith, we subscribe to the basic monotheistic belief that all things are ultimately in harmony; the recognition of the patterns that we call fundamental physical laws, and the way in which deeper probing has not overthrown this concept, but only served to replace an imperfect law by one less imperfect, have contributed to a strengthening of this primary article of faith.

On the whole, then, the present consensus of scientific opinion is that the fundamental laws of physics are adequate to describe the mechanisms at work in physical, chemical and biochemical systems, and that detailed analysis of living creatures is unlikely to reveal any new fundamental principle characteristic of life itself—Vital Forces, Animal Magnetism and suchlike concepts seem nowadays to belong to a world of Gothic fantasy. But if it now seems only a romantic dream to have hoped that life should be subject to its own special laws, transcending the mundane laws that govern the inanimate world, surely the alternative, which is to discover how the severely economic first principles underlying the behaviour of matter can proliferate into the wealth of phenomena that we experience every day in our lives, is as exciting a challenge to the imagination as one could ever hope to meet.

This challenge is not to be met by concentrating solely on the fundamental laws. It may be true that the rules of chemical combination are derivable from the laws of quantum mechanics, but no watertight derivation has yet been given; they are derivable in principle but not in practice. If chemists had not discovered how to set up simple models of valence forces which could be refined by experiment and analysis until they represented a coherent and useful system of thought, it is certain that no physicist, starting with quantum mechanics, would have deduced therefrom any set of working rules. Indeed we do not need to seek even so mildly recondite an example; if nature had not presented us with liquids as a matter of everyday experience, it is very doubtful whether their existence would have been predicted. The gas and the solid are obvious forms of aggregation of attracting molecules—the liquid is anything but obvious. The fact is that for all their simplicity the fundamental laws, when they are applied to anything more complex than a small number of particles, can generate a fantastic wealth of behaviour, making it almost impossible for the pure theorist to predict what will actually happen; every conceivable consequence of the basic laws would have to be analysed, at enormous expense of time and intellect, to see whether it could or could not occur, and one need not doubt that the vast majority of guesses would turn out to be erroneous.

Thus in practice it would describe the activity of the theorist more accurately to say that he has to accept the facts as revealed by experiment, and explain if he can their relation to his fundamental laws. But he must not expect other scientists to be particularly interested in his efforts. The chemist, the biologist, the engineer, and even most physicists are far more concerned to develop the working rules and models that they find useful in their own concerns, than to understand in every detail how these rules are related to the laws of physics. And they are right to do so, for the local working rules are a fuller expression of the truth of experience. There is no possibility of writing down a complete theory of the behaviour of matter, starting from first principles; the equations would be so complicated as to be outside the reach of any computer. At every stage, then, approximations are essential, and the art of model-building in any branch of science is the art of choosing appropriate simplifications that yield helpful models; the refinement of a model usually involves restoring certain concepts which were previously rejected as irrelevant, or inconvenient and relatively unimportant. A reasonable ideal for science is not a single unified activity, but a large number of fields of study, each described (or understood) in terms of its own local models and theories, and each related to its neighbours and to fundamental conceptions by plausible approximate arguments. Seen in this light, each branch of science is to be judged on its own merits, and it is not necessarily mere partisanship if those who are concerned with understanding non-physical matters regard their local set of rules as much more valuable than the arid abstractions which are the fundamental laws.

Our concern, nevertheless, in this book is with fundamental laws, and with their use in justifying various rules and techniques of thought that are helpful in solving real problems. But before we begin we should try to be clear about the status of these laws. Such terse and lapidary statements as Newton's laws of motion possess an aura of authority hardly less than the Tables of the Law handed down on Mount Sinai, and one might be tempted to ask what are the consequences of breaking these laws. Of course the question is ill-conceived; they are not laws that can be broken by man—they can only be used or disregarded; in so far as they are rules of conduct, it is in the sense that they indicate the way to proceed in analysing a mechanical problem. It is open to any physicist, without incurring odium, to set up different guide-lines; but if he does so in analysing a problem which lies within the general area of classical mechanics, he must expect to find some difficulty in persuading his colleagues that his result is correct; he will be regarded as a crank rather than a sinner. The justification for Newton's laws is not the authority of a great man, but the experience of countless scientists who have found the logical application of the laws to yield answers which are demonstrably in accord with experiment.

There are certain tests, such as Cavendish's test of the inverse-square law and Newton's explanation of Kepler's laws of planetary motion, which explore the basic ideas very searchingly. The description of crucial experiments such as these has been particularly emphasized in this book, partly for their own sake, for it is possible to learn at least as much about the real nature of science from an imaginatively conceived and meticulously conducted exact measurement that puts the seal on an idea, as from the often simple and even naive exploration that first generated that idea. But of course the principal purpose behind these experiments remains the primary reason for treating them seriously—they provide quantitative measures of the extraordinarily fine limits of precision to which the laws may be trusted to hold good.

It is something almost unique to physics among all the sciences that it has been found possible to reduce the description of its characteristic phenomena to a small number of working rules which are strong enough to form the premises from which long logical chains of argument can be derived. In no other science, except possibly genetics, can mathematical development be safely indulged in without frequent recourse to experiment, to verify that the premises were sufficiently sound to carry the weight of logic. The continued success of this characteristic deductive method of physics (which has often misled philosophers into supposing that physics is the ideal towards which other sciences should aspire) has indeed inspired great confidence in the fundamental laws; and, as was pointed out at the beginning, this confidence is enhanced rather than diminished by the discovery and precise mapping of their limitations.

We must always bear in mind, however, that behind our present certainty there lie endeavours, doubts and mistakes whose reconstruction by historians of science adds much to appreciation of the process of discovery but little to understanding of the theory in its final form. It may be that an expurgated version of the search, with the principal developments presented in orderly array, will bring the student most readily to the point of understanding; but it is mischievous to call this the historical method of presentation, as if the scientists of the past, their coats and minds immaculately white, proceeded inexorably from one great discovery to the next. If we are to eschew a truly historical presentation, surely it is best to avoid the pretence of historicity altogether. The treatment adopted here, therefore, will assume in the reader the mental furniture appropriate to the twentieth century. We all believe in atoms, even though we might find it hard to explain why. We even believe that an atom consists of a massive, positively-charged nucleus surrounded by a cloud of light, negatively-charged electrons. We take for granted, what shocked our quite recent ancestors to discover, that the most solid material object is mainly empty space, contaminated by a light sprinkling of charged particles. It seems sensible, then, to direct attention first to the behaviour of individual particles—how they interact with each other and how they respond to the forces exerted on them—and then to see whether we can understand the behaviour of assemblies of atoms such as constitute the samples of real matter we investigate experimentally.

It would be too much, in the compass of a single book, to develop the fundamental laws in their own right as well as to apply them to building up a coherent picture of the many-sided behaviour of matter. This book is more like a guide than a systematic course of instruction. In Chapter 6, for example, we take a tour of the physics of rigid and elastic bodies, throwing in a little hydrodynamics in passing. No one will become an expert in these matters by sedulously learning all that is included here and no more; on the other hand, when he turns to the books listed at the end of the chapters for something more detailed, the reader will find, I hope, that he begins to see the relation of the parts to the whole. He will then have begun the educational process that will, if he is lucky, continue throughout his lifetime and still not exhaust the possibilities for seeing the complex yet meaningful interrelation of ideas originally developed separately.

But already we are assuming a degree of sophistication on the part of the reader. One cannot start with a description of the whole and proceed therefrom to show the details; a few details are needed before the synoptic vision is meaningful. It is assumed, then, that the reader is already familiar with certain elementary ideas and techniques. If he cannot solve the problem of the simple pendulum, or apply Ohm's law to an electrical network, if he is unfamiliar with the gas laws and the meaning of Boltzmann's constant, if he could not set up a simple potentiometer circuit or Wheatstone's bridge, this book is too advanced for him as yet. A list of more elementary texts is given at the end of this chapter; it is no more than a personal choice from an enormous field, as indeed can be said of all the texts listed after each chapter.

Finally, a word about problems. There are a good many, together with worked examples, scattered throughout the text, but not the generous collection at the end of each chapter that seems to be demanded by convention of all modern textbooks. Instead, reference is made to the appropriate Cavendish Problems,* which are not intended to be a systematic set of drill questions, but which should serve to search the reader's understanding and help him develop his physical insight. If these prove difficult or insufficient, I must refer him to a more conventional text to meet his needs.

READING LIST

D. HALLIDAY and R. RESNICK, *Physics* (2 Vols.), Wiley.
E. M. ROGERS, *Physics for the Inquiring Mind*, Princeton U. P.
PHYSICAL SCIENCE STUDY COMMITTEE, U.S.A. *Physics Text*, Heath.
G. A. G. BENNET, *Electricity and Modern Physics*, Arnold.

* *Cavendish Problems in Classical Physics*, edited by A. B. Pippard, 2nd edition revised by W. O. Saxton (Cambridge University Press, 1971).

2

Kinematics

Kinematics is concerned purely with the description of motion without enquiring into the causes of the motion. It will be assumed that the reader is familiar with the concepts of velocity and acceleration in one-dimensional motion and can derive the following standard results:

If the position of a particle at time t is $x(t)$, its velocity v is defined as $\mathrm{d}x/\mathrm{d}t$ and its acceleration a is defined as $\mathrm{d}^2x/\mathrm{d}t^2$.

A particle starting with velocity u and subject to constant acceleration a has at time t a velocity $u + at$; it has travelled a distance

$$s = ut + \tfrac{1}{2}at^2;$$

and

$$v^2 = u^2 + 2as.$$

PROBLEM

A point on a moving trolley is observed to have the following coordinates at equal intervals of 1 second: 5·1, 1·6, 1·5, 4·8, 11·5, 21·6 cm. Show that the acceleration is uniform and determine its magnitude.

[Answer: 3·4 cm s^{-2}]

We shall now discuss the motion of a particle in three dimensions, using the position vector $\mathbf{r}(t)$ to describe the trajectory. The elementary 313 manipulations of vector algebra are summarized in an Appendix, and from now on we shall refer to any one of these results by the appropriate roman numeral.

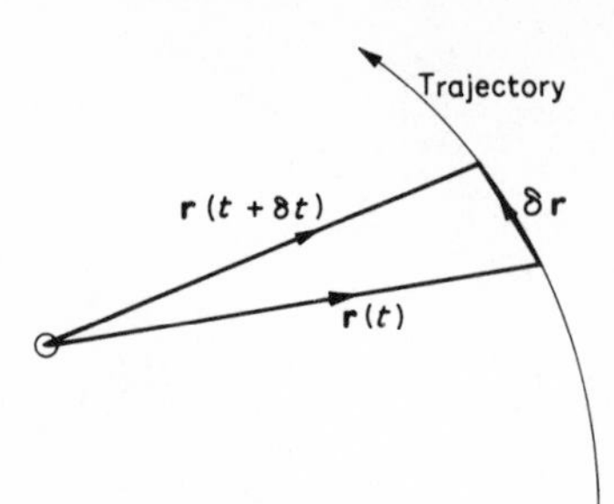

The derivative of the position vector $\mathbf{r}$ with respect to time defines the velocity of the particle, and is formed in the usual way:

$$\mathbf{v} = \mathrm{d}\mathbf{r}/\mathrm{d}t = \lim_{\delta t \to 0} \left[\frac{\mathbf{r}(t + \delta t) - \mathbf{r}(t)}{\delta t} \right]$$

The numerator is the vector $\delta\mathbf{r}$ and clearly $\mathrm{d}\mathbf{r}/\mathrm{d}t$ (or $\dot{\mathbf{r}}$) is directed along the tangent to the trajectory and has magnitude equal to the speed of the particle's tangential motion. The velocity is a vector quantity and velocities may be compounded by vector addition. If, for example, a particle is moving with velocity $\mathbf{v}_1$, relative to a platform which itself moves with velocity $\mathbf{v}_2$ relative to an observer standing on the floor, the observer sees the particle moving with velocity $\mathbf{v}_1 + \mathbf{v}_2$.* Similarly, if a stationary observer sees a particle moving with velocity $\mathbf{v}$, an observer moving with velocity $\mathbf{u}$ relative to him sees the particle moving with velocity $\mathbf{v}' = \mathbf{v} - \mathbf{u}$.

EXAMPLE

A liner moves in a straight line, at a speed of 20 knots, some distance out to sea from a port where a tender is waiting to intercept it. If the maximum speed of the tender is 12 knots, find how late it can afford to postpone leaving, and determine the direction in which it should go.

SOLUTION

(1) *A straightforward solution.* Let X be the port and XN the normal from X to the liner's path. If the tender sets out at full speed when the liner is at P and catches it at Q, $XQ = a \sec \theta$, and therefore PQ, the distance travelled by the

* Although it may seem obvious that velocities can be compounded vectorially in this way, it is implicitly assumed that the two observers are able to agree on their measures of length and time. Thus the argument implies that in time δt the particle moves $\mathbf{v}_1 \delta t$ relative to the platform, which itself moves $\mathbf{v}_2 \delta t$ relative to the floor, and we are to add the two distances vectorially. The observer on the floor has to take the word of the observer on the platform that he has measured $\mathbf{v}_1$ and δt correctly or, better, to compare measuring-sticks and clocks with him. When the velocities involved become comparable to the velocity of light, agreement between the observers proves not to be possible, and the simple law of addition of velocities breaks down. 306 The extreme example of this is the famous Michelson–Morley experiment which shows that a light signal in free space passes all observers at the same speed, even though they are moving relative to one another. This is one of the paradoxical observations that gives support to the Special Theory of Relativity and its corrections to Newtonian mechanics so as to describe correctly the motion of very fast particles.

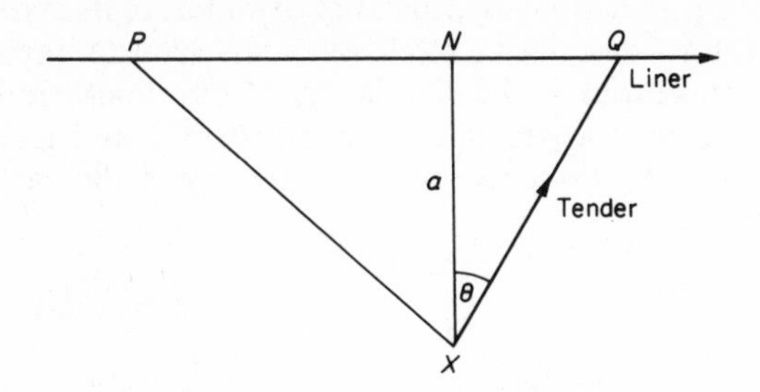

liner while the tender moves along XQ, is $\frac{5}{3}a \sec\theta$. Then PN is $a(\frac{5}{3}\sec\theta - \tan\theta)$. The latest time of setting out corresponds to the minimum value of PN, which by differentiation is found to occur when $\sin\theta = \frac{3}{5}$, i.e.,

$$NQ = \tfrac{3}{4}a, \quad XQ = \tfrac{5}{4}a, \quad PQ = \tfrac{25}{12}a, \quad PN = \tfrac{4}{3}a.$$

From this it is seen that PXQ is a right angle.

(2) *A trick solution using a vector diagram of velocities.* An observer on the liner sees the tender leave X when he is at P, and move directly towards him along a line parallel to XP. If OR represents the velocity of the tender as seen from land, R must lie on a circle of radius 12. OO' represents the liner's velocity relative to land, and has length 20, and $O'R$ is the apparent velocity of the tender as seen from the liner. $O'R$ is parallel to XP, and is as near as possible to the vertical when it is tangential to the circle. Then clearly $\sin\theta = \frac{3}{5}$ as before, and OR is perpendicular to $O'R$.

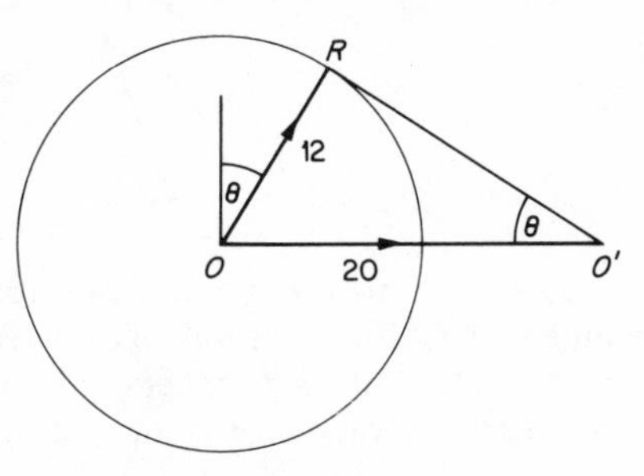

Note. The second solution is elegant and the reader may always be permitted to feel proud when he devises an elegant solution to a problem. He should never forget, however, that it is more important to get a solution than to be elegant, and that the straightforward method is in general the one to aim for. The pursuit of elegance is justified if it encourages deeper understanding (as in this case it illustrates the use of relative velocity), or if it leads to facility in recognizing the essentials of a problem; but if it does neither, its value is aesthetic only.

The second time derivative of the position vector, $\mathrm{d}^2\mathbf{r}/\mathrm{d}t^2$ or $\ddot{\mathbf{r}}$ or $\mathrm{d}\mathbf{v}/\mathrm{d}t$, is defined by the same limiting process applied to $\mathbf{v}$ now rather than to $\mathbf{r}$, and higher derivatives can be defined, as required, in the same way. It should be noted that in general the directions in which the successive vector derivatives point are all different. For example, a particle moving at constant speed in a circle, not centred on the origin, has vectors $\mathbf{r}$, $\dot{\mathbf{r}}$ and $\ddot{\mathbf{r}}$ as shown in the diagram. In this particular case, if the particle has angular velocity ω in a circular trajectory of radius b, its speed is ωb and the velocity vector $\dot{\mathbf{r}}$ describes a circle of radius ωb with angular velocity ω. Just as $\dot{\mathbf{r}}$ is tangential to the circular trajectory described by $\mathbf{r}$, so $\ddot{\mathbf{r}}$ is tangential to the circular trajectory described by $\dot{\mathbf{r}}$ and has magnitude $\omega^2 b$ pointing towards the centre of the trajectory of the particle (*centripetal acceleration*).

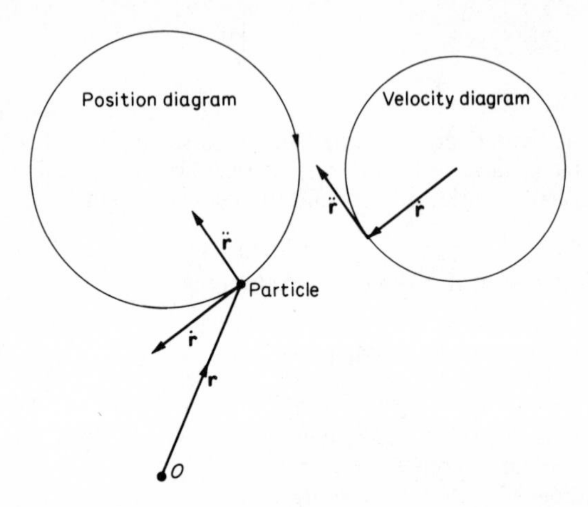

It is convenient to use this example as an introduction to an alternative, and very useful, method of representing vectors in two dimensions by means of complex numbers. The rules for this application of complex numbers are very shortly summarized:

1. A vector $\boldsymbol{\rho}$ with components (A, B) is written as $A + \mathrm{i}B$, and i is to be regarded as the operation of turning through $\pi/2$ in an anticlockwise direction.

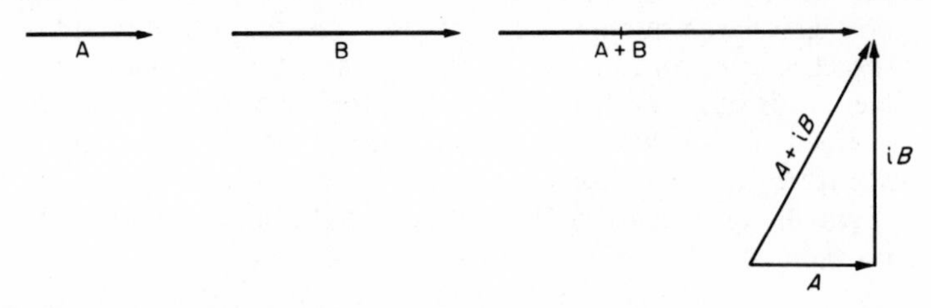

2. The operation i^2 signifies two rotations through $\pi/2$, i.e., reversal of

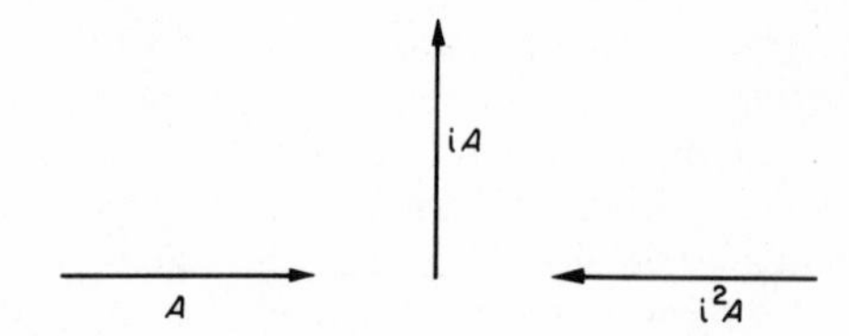

sign: thus $i^2 = -1$. There is no meaning to be attached to the statement $i = \sqrt{(-1)}$ except that $i^2 = -1$.

3. i is to be treated as an algebraic symbol for all purposes of manipulation.

4. Any vector $A + iB$ operated on by i is turned through $\pi/2$; $i(A + iB) = -B + iA$.

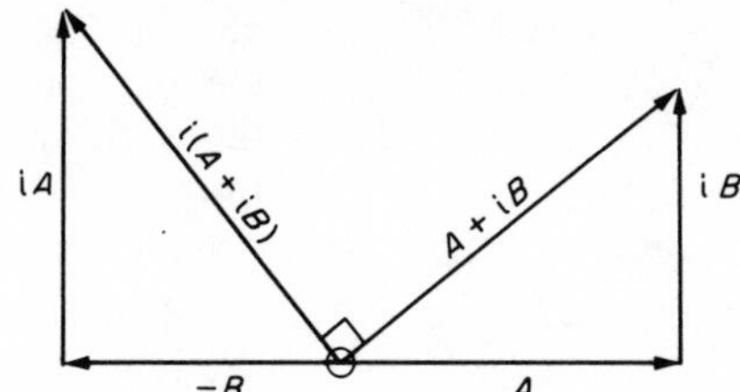

5. $e^{i\theta} = \cos\theta + i\sin\theta$; $e^{i\theta}$ is a unit vector at an angle θ to the real axis.

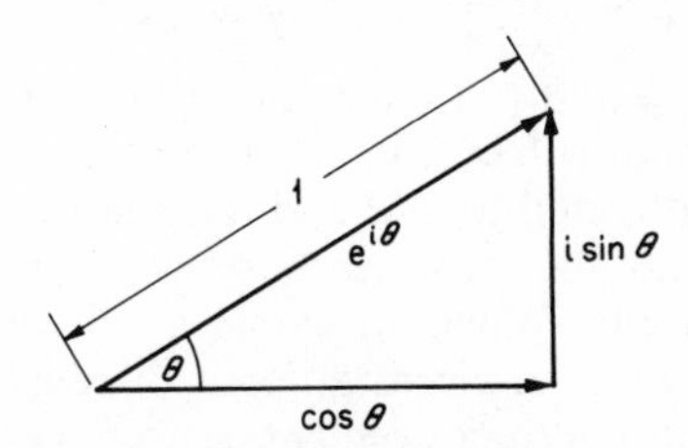

6. Any vector $\boldsymbol{\rho}$ may be represented by $\rho\, e^{i\theta}$, ρ being real and positive; ρ is the *Amplitude* and θ the *Argument*. This is a vector of length ρ pointing in a direction θ. Multiplication by $e^{i\phi}$ turns the vector through ϕ, since $\rho\, e^{i\theta} \times e^{i\phi} = \rho\, e^{i(\theta+\phi)}$.

7. The complex conjugate of a vector $\mathbf{X}$, written as $\mathbf{X}^*$, is obtained by replacing every i by $-i$. The complex conjugate of $\rho\, e^{i\theta}$ is $\rho\, e^{-i\theta}$. $\mathbf{X}$ and $\mathbf{X}^*$ are mirror images with respect to the real axis.

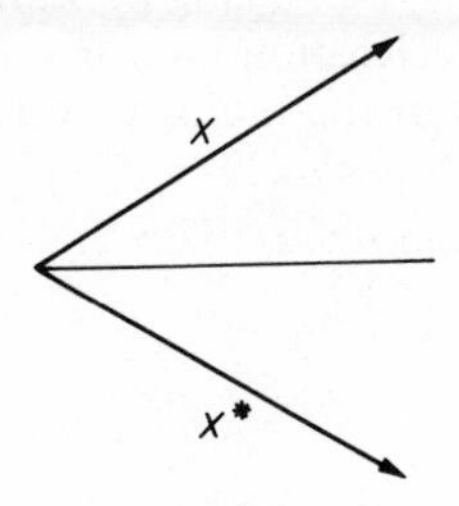

8. To form the scalar product of two vectors $\mathbf{X}$ and $\mathbf{Y}$, construct $\mathbf{XY}^*$ or $\mathbf{X}^*\mathbf{Y}$ and take the real part. If $\mathbf{X} = \rho_1\, e^{i\theta_1}$ and $\mathbf{Y} = \rho_2\, e^{i\theta_2}$, with ρ_1 and ρ_2 real,

$$\text{Re}\,(\mathbf{XY}^*) = \text{Re}\,(\mathbf{X}^*\mathbf{Y}) = \rho_1\rho_2 \cos(\theta_1 - \theta_2),$$

which is by definition the scalar product.

9. The amplitude of a vector is the square root of the product of the vector with its own complex conjugate: $|\mathbf{X}|^2 = \mathbf{X}\mathbf{X}^*$.

Let us apply this representation to the uniform motion of a particle in a circle. If the centre of the circle is O', represented by a complex number $\boldsymbol{\rho}_0$ and if we choose the origin of time as the moment when the particle passes through N, the position P at a subsequent time t is represented by the complex number

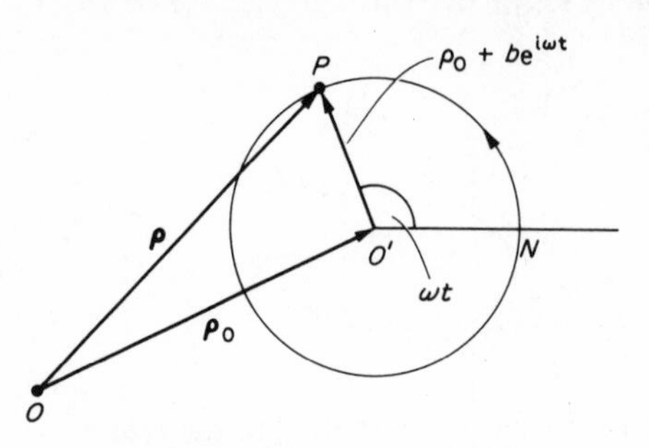

$$\boldsymbol{\rho} = \boldsymbol{\rho}_0 + b\,\mathrm{e}^{\mathrm{i}\omega t}.$$

We now determine the velocity and acceleration by differentiation,

$$\dot{\boldsymbol{\rho}} = \mathrm{i}\omega b\,\mathrm{e}^{\mathrm{i}\omega t}, \qquad \ddot{\boldsymbol{\rho}} = (\mathrm{i}\omega)^2 b\,\mathrm{e}^{\mathrm{i}\omega t} = -\omega^2 b\,\mathrm{e}^{\mathrm{i}\omega t}.$$

The magnitudes of $\dot{\boldsymbol{\rho}}$ and $\ddot{\boldsymbol{\rho}}$ are ωb and $\omega^2 b$ as found previously; the direction of $\dot{\boldsymbol{\rho}}$ is normal to $b\,\mathrm{e}^{\mathrm{i}\omega t}$, i.e., $O'P$ because it is multiplied by i, while $\ddot{\boldsymbol{\rho}}$ is oppositely directed to $O'P$ and is a centripetal acceleration.

The same complex number representation is useful for analysing other two-dimensional motions. When we come to discuss the motion of a particle under the influence of a central force, directed from a fixed origin to the particle, we shall find it convenient to use polar rather than Cartesian coordinates, and to seek to determine from the time-variation of ρ and θ the components of acceleration directed radially and tangentially. The example of circular motion just discussed shows that the radial component of acceleration is not to be found simply by placing the origin at the centre of the circle and evaluating $\ddot{\rho}$, for in this special case with constant ρ the answer would be zero. The error of this procedure lies in neglecting the fact that the radial direction, along which the component of acceleration is required, is itself changing with time. We must therefore first determine the total acceleration and then find its component along the instantaneous direction of the radius vector; this is very simply done with complex numbers.

61
259

If $\boldsymbol{\rho} = \rho\,\mathrm{e}^{\mathrm{i}\theta}$ and both ρ and θ vary with time,

$$\dot{\boldsymbol{\rho}} = (\dot{\rho} + \mathrm{i}\rho\dot{\theta})\,\mathrm{e}^{\mathrm{i}\theta}$$

and

$$\ddot{\boldsymbol{\rho}} = (\ddot{\rho} + 2\mathrm{i}\dot{\rho}\dot{\theta} + \mathrm{i}\rho\ddot{\theta} - \rho\dot{\theta}^2)\,\mathrm{e}^{\mathrm{i}\theta}.$$

The real part of the coefficient of $\mathrm{e}^{\mathrm{i}\theta}$ in this expression denotes that part of $\ddot{\boldsymbol{\rho}}$ which lies parallel to $\boldsymbol{\rho}$, and this is the radial component of acceleration.

Thus

$$\ddot{\rho}_{\mathbf{radial}} = \ddot{\rho} - \rho\dot{\theta}^2,$$

which reduces to $-\rho\dot{\theta}^2$, i.e., $-\omega^2\rho$ if ρ is constant, the same as we found before. The tangential component of acceleration, normal to the radius vector, is given by the imaginary terms in the coefficient of $e^{i\theta}$:

$$\ddot{\rho}_{\text{tang}} = 2\dot{\rho}\dot{\theta} + \rho\ddot{\theta} = \frac{1}{\rho}\frac{d}{dt}(\rho^2\dot{\theta}).$$

This result has a simple geometrical representation. If the particle moves from P to P' in time δt, the radius vector sweeps out an area (the triangle OPP') whose magnitude, to first order in δt, is $\frac{1}{2}\rho^2\dot{\theta}\delta t$. Thus the rate at which the radius vector sweeps out area is $\frac{1}{2}\rho^2\dot{\theta}$, and if this is constant, as Kepler observed to be the case for the radius vector joining any one planet to the Sun, it implies that there is no tangential component of acceleration.

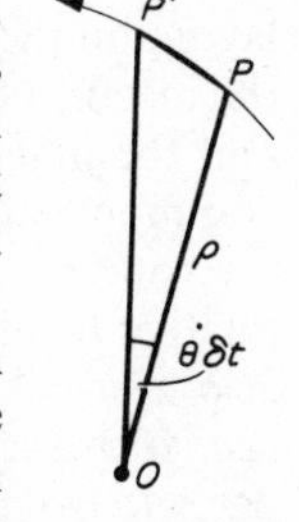

As a further example, let us examine the way in which observers moving relative to one another will describe the same motion of a given particle. If the observers are in uniform relative translational motion at velocity $\mathbf{u}$, without rotation, they will record at any instant velocity vectors for the particle which differ by the constant vector $\mathbf{u}$, and therefore will record the same vector for the acceleration of the particle. This does not hold when the observers are on rotating frames of reference. For instance, if we imagine the Sun to be fixed in space, the effect of the Earth's rotation is to make it appear as if the Sun were performing a circular orbit around us, and we should be inclined to ascribe a centripetal acceleration to it which an observer fixed far out in space would deny. Let us analyse this problem by imagining a particle moving relative to a fixed observer in an arbitrary plane trajectory $\rho\, e^{i\theta}$, where ρ and θ vary with time, and observed by another on a rotating reference frame who plots out its trajectory on a piece of paper attached to his frame. We shall choose as origin of coordinates in both frames the axis of rotation. If the angular velocity is ω and we measure time from a moment when the real axes of both observers coincide, the angle between the frames at time t is ωt, and the angular coordinate of the particle as seen by the rotating observer will be not θ but $\theta - \omega t$; ρ on the other hand will be the same for both.

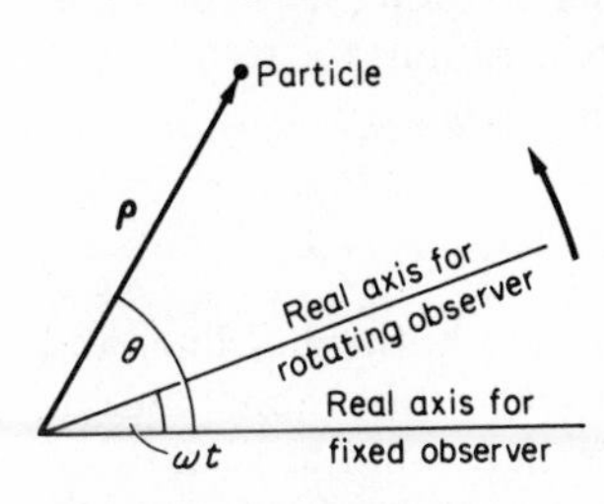

Thus if $\boldsymbol{\rho}$ and $\boldsymbol{\rho}'$ are the position vectors as seen by the fixed and rotating observers respectively,

$$\boldsymbol{\rho} = \boldsymbol{\rho}'\, e^{i\omega t}.$$

Hence by differentiation,

$$\dot{\boldsymbol{\rho}} = (\dot{\boldsymbol{\rho}}' + i\omega\boldsymbol{\rho}')\, e^{i\omega t}, \tag{2.1}$$

and

$$\ddot{\boldsymbol{\rho}} = (\ddot{\boldsymbol{\rho}}' + 2i\omega\dot{\boldsymbol{\rho}}' - \omega^2\boldsymbol{\rho}')\, e^{i\omega t}$$

i.e.,

$$\ddot{\boldsymbol{\rho}}' = \ddot{\boldsymbol{\rho}}\, e^{-i\omega t} - 2i\omega\dot{\boldsymbol{\rho}}' + \omega^2\boldsymbol{\rho}'. \tag{2.2}$$

To see what this result implies, let us note that if at any moment the fixed observer wished to communicate to the rotating observer the actual acceleration of the particle in the fixed frame of reference, he could do so by laying on the ground an arrow of length and orientation describing $\ddot{\boldsymbol{\rho}}$; at the instant t this would be seen by the rotating observer as a vector $\ddot{\boldsymbol{\rho}}e^{-i\omega t}$, and the first term in the equation is therefore the actual acceleration as communicated by the fixed observer. But the other two terms show that the rotating observer does not see this as the acceleration of the particle relative to his frame. It must be supplemented by a term of magnitude $2\omega\dot{\boldsymbol{\rho}}'$ normal to the velocity of the particle in the rotating frame, and a *centrifugal* acceleration of magnitude $\omega^2\boldsymbol{\rho}'$ governed by the distance of the particle from the axis of rotation. The term $2\omega\ddot{\boldsymbol{\rho}}'$, which depends for its existence on the particle being in motion relative to the rotating observer, is called the *Coriolis* acceleration.

It may seem surprising that the radial acceleration is centrifugal rather than centripetal, but a simple example should show that there is no paradox here. Suppose the particle to be at rest in the fixed frame, and therefore moving clockwise in a circle relative to the rotating observer. Then $\dot{\boldsymbol{\rho}}' = -i\omega\boldsymbol{\rho}'$ and the Coriolis acceleration is $-2\omega^2\boldsymbol{\rho}'$, a centripetal acceleration twice as great as the centrifugal acceleration; the resultant of the two is the expected centripetal acceleration $-\omega^2\boldsymbol{\rho}'$.

PROBLEM

A particle moves at constant velocity in a straight horizontal line, and leaves a record of its trajectory on a turntable which rotates about an axis lying on the trajectory. This is the curved path that would be seen by an observer rotating with the turntable. Plot out the shape of this path by constructing about 10 points at equal time-intervals between the centre and rim of the turntable, for the case when the table turns through something like 180° during the particle's progress from centre to rim. Hence estimate (e.g., by joining neighbouring points) how the velocity vector varies and verify, especially near the centre and near the rim, that its variation is in accord with the Coriolis and centrifugal accelerations. Pay particular attention to signs. It is advantageous to do this construction on a large sheet of paper.

We shall now translate (2.2) into conventional vector notation, but to do so we need to devise a vector representation for rotation. This is not quite so straightforward as it may appear, since finite angular displacements are not described by vectors. To justify this statement it is sufficient to show that rotations are not commutative, that is, the resultant of two rotations in sequence depends on the order in which they are performed. Consider, for example, the die shown below, which is twice turned

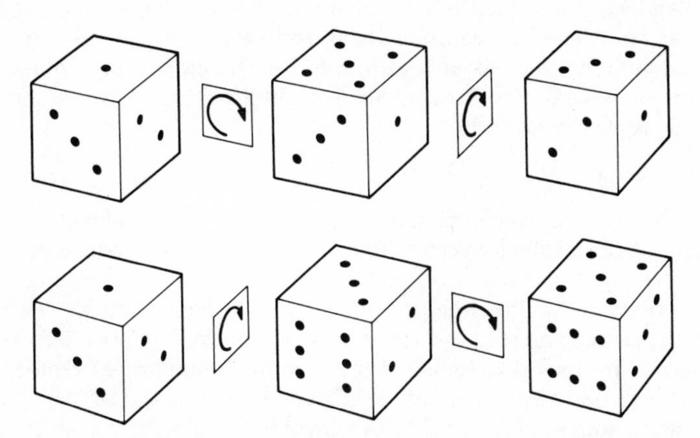

through $\pi/2$ about different axes. In the two rows the operations are reversed in order, and it is seen that a different final result is obtained. If the rotations were capable of being described by vectors, the sum of two rotations would be formed by head-to-tail addition and would be commutative.

Nevertheless, it is possible to represent infinitesimal rotations by vectors, since the difference in resultant, when the order of operations is reversed, is of second order in the angles of rotation, and can therefore be made as small as desired by going to the limit of infinitesimal angles. The

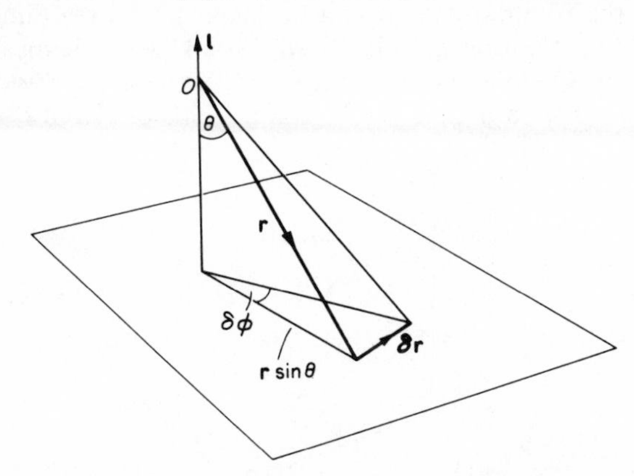

reader may care to verify this statement by direct calculation. The appropriate vector to describe a small rotation through $\delta\phi$ is $\mathbf{l}\ \delta\phi$, where $\mathbf{l}$ is a unit vector parallel to the axis of rotation. If $\mathbf{r}$ is the position vector of a point on the body, measured from a point on the axis of rotation, the movement $\delta\mathbf{r}$ of this point is through a distance $r \sin \theta . \delta\phi$, and normal to both $\mathbf{r}$ and $\mathbf{l}$; in vector notation, $\delta\mathbf{r} = (\mathbf{l} \wedge \mathbf{r})\, \delta\phi$. Now when a body is rotating continuously, its momentary angular velocity is defined by a limiting process $\omega = \lim_{\delta t \to 0} (\delta\phi/\delta t)$, and the fact that infinitesimal rotations can be compounded as vectors implies that angular velocities behave likewise. It is therefore legitimate to introduce an angular velocity vector $\boldsymbol{\omega}$ directed along the axis of rotation, $\mathbf{l}$. The velocity of any point on the rotating body whose position vector is $\mathbf{r}$, relative to an origin situated on the axis, is clearly given by

$$\mathbf{v} = \boldsymbol{\omega} \wedge \mathbf{r}.$$

The direction in which the arrow of $\boldsymbol{\omega}$ is drawn is not defined by inspection as it is for a velocity vector. Instead it is a matter of convention, and we shall adopt the convention that when right-handed axes are used, the arrow points in the direction of travel of a right-handed corkscrew rotating in the sense expressed by $\boldsymbol{\omega}$; for left-handed axes we use a left-handed corkscrew. We may distinguish in fact two different sorts of vector, *polar vectors* like $\mathbf{r}$ and $\mathbf{v}$ which are not sensitive to the right- or left-handedness of the axes, and *axial vectors* like $\boldsymbol{\omega}$ which are. The vector product of two polar vectors is an axial vector by virtue of the convention which determines its sense. This is not a matter which we shall pursue further, for it does not play any significant part in the discussions involved in this book, but it is worth remembering that the two sorts of vectors are really quite different in kind, and that we shall never find any equation in which one sort is equated to the other.

With this preliminary, we are in a position to write the Coriolis and centrifugal accelerations in vector form, and we shall do this by inspection. Although we derived the solution for a particle moving in a plane normal to the rotation axis, it can be taken into three dimensions immediately, since motion parallel to the axis is seen identically by both observers. The Coriolis and centrifugal accelerations expressed in (2.2) are thus associated with $\boldsymbol{\rho}'$ and $\dot{\boldsymbol{\rho}}'$, the components or $\mathbf{r}'$ and $\dot{\mathbf{r}}'$ in the plane

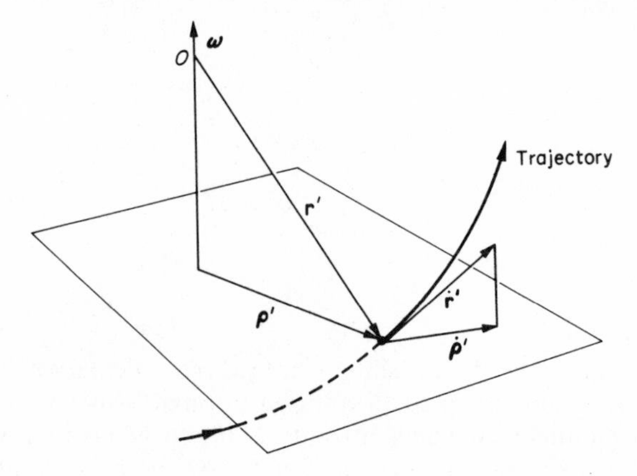

normal to the axis. Now when we form the vector product $\boldsymbol{\omega} \wedge \mathbf{r}'$ the component of $\mathbf{r}'$ parallel to $\boldsymbol{\omega}$ makes no contribution, and the product is the same as $\boldsymbol{\omega} \wedge \boldsymbol{\rho}'$; the same holds for any other vector, that only its component in the plane normal to $\boldsymbol{\omega}$ need be considered. Thus $\boldsymbol{\omega} \wedge \mathbf{r}'$ has magnitude $\omega\rho'$ and lies in the plane normal to $\boldsymbol{\omega}$, making a right angle with $\boldsymbol{\rho}'$. If we now form the vector product of this with $\boldsymbol{\omega}$, $\boldsymbol{\omega}$ and $\boldsymbol{\omega} \wedge \mathbf{r}'$ being normal to one another means that the resulting vector $\boldsymbol{\omega} \wedge (\boldsymbol{\omega} \wedge \mathbf{r}')$ has magnitude $\omega^2\rho'$ and points inwards*. Thus $-\boldsymbol{\omega} \wedge (\boldsymbol{\omega} \wedge \mathbf{r}')$ correctly represents the centrifugal acceleration. Similarly it will be seen that the Coriolis acceleration is correctly represented by $-2\boldsymbol{\omega} \wedge \dot{\mathbf{r}}'$, so that in vector notation the acceleration seen by the rotating observer is related to that communicated to him by the stationary observer through the equation

$$\ddot{\mathbf{r}}' = [\ddot{\mathbf{r}}]_{\text{stat}} - 2\boldsymbol{\omega} \wedge \dot{\mathbf{r}}' - \boldsymbol{\omega} \wedge (\boldsymbol{\omega} \wedge \mathbf{r}'). \tag{2.3}$$

It may also be noted that (2.1), describing the velocity in the moving frame in terms of the velocity seen and communicated by a stationary observer, may be written in vector form:

$$\dot{\mathbf{r}}' = [\dot{\mathbf{r}}]_{\text{stat}} - \boldsymbol{\omega} \wedge \mathbf{r}'. \tag{2.4}$$

* Note that the order of multiplication is very important; $(\boldsymbol{\omega} \wedge \boldsymbol{\omega}) \wedge \mathbf{r}' = 0$ since $\boldsymbol{\omega} \wedge \boldsymbol{\omega}$ vanishes identically.

3

Newton's laws of motion

N1 Every body continues in its state of rest, or of uniform motion in a straight line, unless it is compelled to change that state by forces impressed upon it.

N2 The change of momentum is proportional to the motive force impressed, and is made in the same direction as that of the impressed force.

N3 Action and reaction are equal and opposite; or, the mutual actions of two bodies upon each other are always equal and oppositely directed.

These three statements* must rank among the most condensed utterances in science. They are so terse as to have led some commentators to suggest that they have very little actual content, and yet one must accept that they form the basis of the very successful science of dynamics. The fact is that they are primarily meaningful as reminders of the techniques to be used in analysing mechanical systems—their real content is revealed by practice in their application. We shall therefore not trouble to discuss them as isolated statements, but shall point to the principles which they enshrine

* For Newton's own treatment see Newton's *Principia*, Motte's translation revised by F. Cajori (University of California Press, 1947), which has valuable explanatory and historical notes.

and the simple observations which make them seem a plausible starting-point for systematic application. Here, as in the other parts of this book, the reader is presumed to have some understanding of simple problems in dynamics, so that the following analysis may be taken as a brief resumé of what he should already be aware of, even if not quite explicitly.

The first law, N1, was probably far more essential in Newton's day than in ours, when rival theories that he had to reject are forgotten except by historians. We may take it to idealize the sort of observation that is commonplace nowadays in teaching laboratories—brass discs floating freely on cushions produced by dry ice or other means are seen to run over flat surfaces with very little change of speed. We are prepared to believe that if we could observe a body moving in an otherwise empty universe, it would not come to rest of its own accord, nor would it perform a circular path, but would travel in a straight line at constant speed for ever. Obviously this is outside the bounds of observation, and indeed modern cosmologists would vehemently reject almost everything contained in the last sentence, but we must not pedantically insist on telling nothing but the truth. It is enough to be persuaded that Newton was not over-credulous in accepting a view of the universe that sees uniform rectilinear motion as the characteristic of an isolated body, and interprets any departure from such rectilinear motion as due to the influence of other bodies. In other words, we establish in the statement of N1 a criterion for recognizing what we call a *force*, even though we cannot assign a magnitude to it, and we shall assert that forces are not occult mysteries but, once recognized, are to be studied quantitatively so that their origins and modes of operation can be described as exactly as possible. In making the hypothesis that we now call the First Law of Motion, we commit ourselves to the belief that when bodies are in motion we may separate the causes of the motion (or rather, changes of motion) from the bodies undergoing motion and regard them, the forces, as distinguishable quantities capable of being studied and described in isolation from the bodies on which they act. Whether we are justified depends on the results of analysis and comparison with experiment. This first law by itself yields no consequences that can be made the subject of critical experiment.

Between leaving N1 and reaching N2 we find ourselves making an implicit assumption. We need not go so far as to accept immediately that we can measure a force quantitatively, though the statement of N2 may seem to imply this. It does, however, seem to be necessary to believe that we can recognize the constancy of a force. It is a consequence of N2 that a body subjected to a constant force suffers constant acceleration (momentum is here used as the product of a certain as yet undefined property of a body, called *mass*, which is believed to be unchanged by its motion, and the velocity of the body) and this is something that we can test without being able to measure force, if only we can recognize when it is remaining constant. Now one can perform simple experiments in which constant acceleration is observed under conditions when it may be accepted without serious misgivings that the force may remain constant. The constant acceleration, g, of a massive falling body is such an observation, and provided one adopts the philosophy implicit in N1 that the body falls because it is pulled by the Earth, rather than because it is in the nature of bodies to fall, it is easy to be persuaded that the Earth's pull may well remain constant however fast the body may be falling. Even more convincing, perhaps, is the sort of experiment that one may make with a light and smoothly-running trolley which can carry various loads. As the weight W

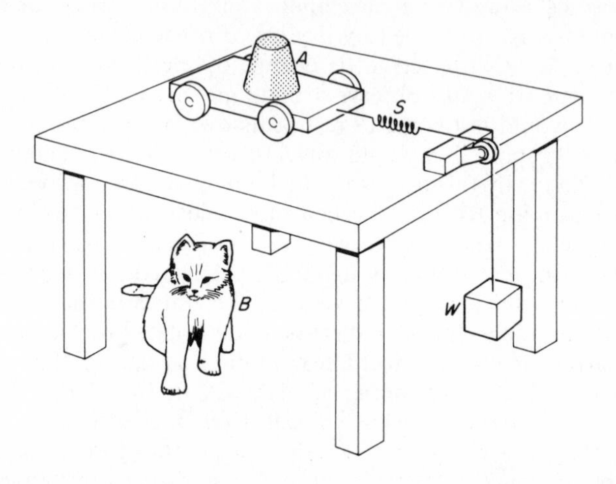

draws the trolley along, its position at equal intervals may be recorded (with old-fashioned apparatus by means of a vibrating stylus leaving a wiggly trace on a smoked plate attached to the trolley; or at greater expense with a flashing light photographing a mark on the trolley at equal intervals) and from the record the acceleration may be deduced. If all goes well, and the trolley does not stick, the acceleration will be found to be constant; moreover, a spring-balance S in the cord will remain at constant extension, encouraging the belief that the force is constant. By experiments such as this, the observer who wishes to assent to belief in Newton's laws will be persuaded that he may accept that a body moving under the influence of a constant force suffers constant acceleration. This is a very important result, one of the several central experimental results hidden in the statement of N2. It may indeed be claimed to be the most important step in appreciating Newton's laws, since the rest of the analysis involves only experiment, not the development of a new philosophical point of view.

We now proceed to investigate how the acceleration of the trolley depends on the load it carries. Suppose we choose two different bodies, A and B, to mount on the trolley, measure their accelerations, a_A and a_B, when a given weight W is attached to the cord, and calculate the ratio a_A/a_B. Then we repeat the experiment with another weight W', which yields different values a'_A and a'_B. But when we calculate the ratio a'_A/a'_B we discover it to be the same as a_A/a_B. And we can repeat the experiment many times with various different weights on the cord to find (if the accelerations are kept much less than g)* that the ratio a_A/a_B is constant. It follows, then, that the ratio is a property of the two bodies and does not depend on the nature of the force. Let us display this important experimental result:

Two bodies subjected to the same force experience accelerations whose ratio is not determined by the force.

In the light of this, we may proceed to assign a number to every body, in inverse proportion to the acceleration it suffers under the influence of a given force. For example, let us arbitrarily choose one body and assign to it the number unity, and compare its acceleration with that of a second body when both are acted upon by the same force; if the second body is accelerated less than the first by a factor m, we assign a number m to the second body. Clearly we can assign such a number, represented by the symbol m, to every body in a consistent fashion such that for every pair subjected to a given force the ratio a_1/a_2 is equal to m_2/m_1. Otherwise stated, the product m_1a_1 is equal to m_2a_2 and to the product ma for every body subjected to this particular force.

From this result we see that the numerical value of ma, being the same for every body subjected to a given force, cannot be a property of any special body, but is determined by the force alone. We may therefore assign this number to the force, as a measure of its magnitude, and represent it by the symbol F. By means of this experiment, we have assured ourselves that we may assign numerical magnitudes to forces and to the inertial properties of bodies (i.e., their disinclination to be accelerated), and that these magnitudes are related to the measured acceleration by the equation

$$F = ma = m\,\mathrm{d}v/\mathrm{d}t.$$

This is the result that is summed up in the first part of N2, and it is seen to be a reasonable interpretation of laboratory observations. There is nothing in the way that m, which we call the *inertial mass*, is here introduced to tell us of any other properties of a body with which it may be correlated, but a few extra experiments show that two bodies piled on the trolley behave as if endowed with a mass equal to the sum of the individual masses—*mass is additive*. Further, the observation attributed to Galileo that all bodies fall with the same acceleration shows that the pull of the Earth on any body is strictly proportional to its inertial mass. The

* The reader who is familiar with dynamical problems will appreciate the reason for this proviso and may object to the statement that a_A/a_B is constant. But it is so nearly constant as to justify the statement as a first approximation; when we have used the result to develop the rules for solving problems, we may return to this example and find the departures from constancy fully explained, thereby confirming our belief.

chemical experiments of Lavoisier and others, that demonstrated the conservation of mass during substantial changes of structure of the participants in chemical change, give added strength to the contention of Newton that the mass of a body is a measure of the amount of matter in it; that is, mass is dependent on the number of basic structural elements, atoms or whatever they may be, which from their very fundamental nature might be expected to be eternally durable. We may extend the trolley experiment by attaching one or more extra cords to it, either at the front or back, and hanging weights on them, to assure ourselves that when all the forces are parallel their resultant effect is the sum of their individual effects. And finally we may carry out the experiment with the trolley pointing in different directions to verify that the mass of a body is the same for all directions of acceleration. This last point, a rough test of the isotropy of space, can be very highly refined, e.g., by observing that the oscillation frequency of a tuning-fork or, better, a quartz crystal is not affected by its orientation.

Let us continue for the moment to confine our analysis to one-dimensional motion, and see the implication of N3 in such a situation. This may be said to be the least controversial of the three laws, if only because by the time we reach it we are conditioned to accept the concepts upon which it depends. The experiments we conducted to investigate N2 were performed with macroscopic bodies of elaborate and varied atomic constitution. The question arises how it is that so complex a body can automatically show such simple behaviour, and Newton's answer is that the equality of action and reaction ensures this. If any particle A exerts a force F on any other particle B, we are to suppose that B exerts a force $-F$ on A. Let us now write down the equations of motion of a swarm of particles, each exerting a force on all the rest, and each subject to forces exerted by other influences outside the swarm. Thus the total force on the first particle, being the sum of individual forces, may be written as $F_1 + (F_{21} + F_{31} + \cdots)$, in which F_1 is the force exerted from outside, and F_{n1} is the force exerted by the nth particle on the first. In general, we can write for the force on the ith particle $F_i + \sum_j F_{ji}$, and equate this to the product of its mass and acceleration:

$$m_i\ddot{x}_i = F_i + \sum_j F_{ji}.$$

This expression summarizes a set of N differential equations, if there are N particles in the swarm. Clearly, if the forces depend on the positions of all the particles, the complete integration of these equations to determine the details of the motion of each particle will normally be out of the question, but we can get enough for present purposes simply by adding all these equations together:

$$\sum_i m_i\ddot{x}_i = \sum_i F_i + \sum_i \sum_j F_{ji}.$$

The double summation means, as the reader may verify by writing out a few of the equations *in extenso*, that we allow the two suffixes, i and j, in F_{ji} to run through all possible values from 1 to N, so that there are N^2 terms* in the sum implied by the shorthand notation. Now within this set we shall find for every term F_{mn} a term F_{nm} with the subscripts interchanged, and according to N3 these are equal and opposite, being the action of the mth particle on the nth and the corresponding reaction of the nth on the mth. The double summation therefore vanishes, and we may rewrite the resulting equation in the form

$$M\ddot{X} = F; \tag{3.1}$$

here, M is the total mass of all particles, $\sum_i m_i$; F is the total force exerted by outside influences, $\sum_i F_i$; and X is the coordinate of the centre of mass (centroid) of the particles, such that

$$X = \sum_i m_i x_i / \sum_i m_i.$$

The special meaning of (3.1), relevant to the present discussion, is that N3 ensures that the centroid of a rigid body (in one-dimensional motion, in fact, any point in a rigid body) reacts to an applied force as though it were a particle of mass M. The third law is an expression of the simplest way of ensuring this result, and there is a wealth of evidence on the laboratory scale that provides direct confirmation of the principle; the firing of bullets into suspended sandbags and the detailed study of the collision of billiard balls show that when the interaction occurs, however violently, the centroid of the whole system continues to move uniformly in accordance with N1. As (3.1) shows, if the system is isolated so that $F = 0$, $M\dot{X}$ does not change with time. This is the principle of *conservation of linear momentum*.

It is very important to recognize that the equality of action and reaction does not only apply to systems in equilibrium. Probably no student would fall into error in this matter but for the unfortunate habit of referring to the 'reaction' of a table, say, to a body resting on it. The body is subject to two vertical forces, one downwards, its weight W which is the force exerted by the Earth on it, and one upwards, the 'reaction' R of the table. If the body is at rest, W and R are equal and opposite, but they are *not* the action and reaction referred to in N3. The Newtonian reaction to W is the upward force

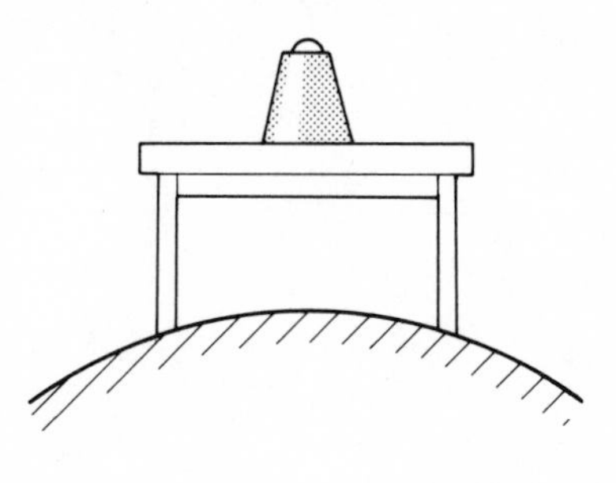

* F_{ii} is always zero, since by N1 an isolated particle does not accelerate. Of the N^2 terms in the double summation, then, the N diagonal terms, with two identical subscripts, automatically vanish.

exerted by the body on the Earth. If we suppose for simplicity that the table is weightless, the forces acting on each body are like this:

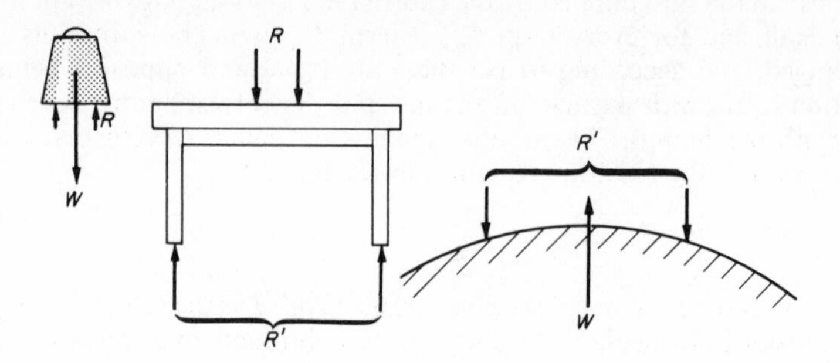

Each pair of forces labelled by the same symbols is a Newtonian pair of action and reaction. The members of each pair are always equal and opposite. Equilibrium is ensured by the equality of R, R' and W. If we kick away the table, R and R' disappear while W remains; the body then falls down while the Earth rises up, their accelerations being in inverse proportion to their masses so as to leave the centroid of the combined (Earth + body) system at rest, or moving as before.

We now return to the second clause of N2, which is important when motion in more than one dimension is being considered. The additive property of two parallel forces may be considered (and was so considered by Newton) to be sound evidence for the acceleration due to any two forces being the sum of the accelerations that each would produce if it acted alone. And since the rule for compounding accelerations is the rule for vector addition, we may expect to find that forces too are to be treated as vectors. The simplest direct test of this point is to hang strings with

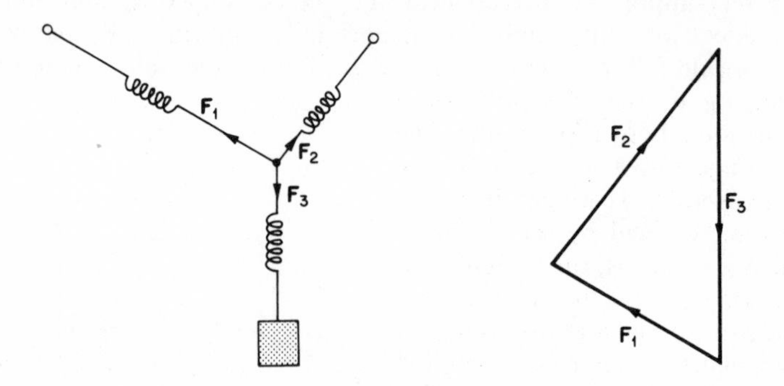

calibrated spring balances as shown in the diagram, and to verify by drawing a triangle of forces that the three forces compound vectorially to zero resultant. The zero resultant is of course what one needs if the knot where the three strings join is to remain at rest.

Linear and angular momentum

This concludes our exposition of Newton's laws as plausible articles of faith on the laboratory scale. From now on we shall see what deductions can be made on the assumption that they are strictly true, not only to show how they can be checked more closely, but also to indicate how they can be developed so as to shed light on an ever-widening variety of phenomena. As a start we shall write the equation of motion of a particle in vector form,

$$m\ddot{\mathbf{r}} = \mathbf{F}, \tag{3.2}$$

and extend the one-dimensional analysis of a swarm of particles to yield certain general results for motion in three dimensions. Using the same notation as before, we have for one member of the swarm

$$m_i\ddot{\mathbf{r}}_i = \mathbf{F}_i + \sum_i \mathbf{F}_{ji}, \tag{3.3}$$

and if we add the equations for all particles, remembering that, by N3, $\mathbf{F}_{ij} = -\mathbf{F}_{ji}$, we have that

$$M\ddot{\mathbf{R}} = \mathbf{F}, \tag{3.4}$$

where $\mathbf{R}$ is the vector coordinate of the centroid, defined as $\sum_i m_i\mathbf{r}_i/\sum_i m_i$. It may be noted that this definition gives a unique position for the centroid, since if we shift the zero of coordinates so as to add to each $\mathbf{r}_i$ a constant $\mathbf{r}_0$, the coordinate of the centroid is also altered by $\mathbf{r}_0$, which is the same as saying that its position is not changed relative to the positions of the particles. For an isolated system $\mathbf{F} = 0$ and (3.4) shows that $M\dot{\mathbf{R}}$, the total linear momentum, is invariant.

Another result of equal importance is obtained by a simple manipulation of (3.2), involving vector multiplication of each side by $\mathbf{r}$ so that the equation reads

$$m\mathbf{r}\wedge\ddot{\mathbf{r}} = \mathbf{r}\wedge\mathbf{F}. \tag{3.5}$$

We then note that by the rule for differentiating products,

$$\frac{\mathrm{d}}{\mathrm{d}t}(\mathbf{r}\wedge\dot{\mathbf{r}}) = r\wedge\ddot{r} + \dot{\mathbf{r}}\wedge\dot{\mathbf{r}},$$

and that the second term vanishes. Thus (3.5) may be written in the form

$$m\frac{\mathrm{d}}{\mathrm{d}t}(\mathbf{r}\wedge\dot{\mathbf{r}}) = \mathbf{r}\wedge\mathbf{F}$$

or

$$\dot{\mathbf{L}} = \mathbf{G}, \tag{3.6}$$

in which $\mathbf{L}$ is $\mathbf{r}\wedge m\dot{\mathbf{r}}$, the moment of momentum (or *angular momentum*) of the particle about the origin, and the torque $\mathbf{G}$ is $\mathbf{r}\wedge\mathbf{F}$, the moment of the force $\mathbf{F}$ about the origin. If we look at the plane defined by the vector $\mathbf{F}$

and the radius vector $\mathbf{r}$, the vector product $\mathbf{r} \wedge \mathbf{F}$ is directed normal to the plane and indicates the axis about which the torque $\mathbf{G}$ exerts its turning action; for example, a lamina lying in the plane and supported from a fixed point O, when acted upon by $\mathbf{G}$ would start to turn about an axis through O parallel to $\mathbf{G}$. The magnitude of the torque is the product of $\mathbf{F}$ and the shortest distance ON from O to the line of action of the force, i.e., the line through the particle drawn parallel to $\mathbf{F}$. Similarly the angular momentum $\mathbf{L}$ has a magnitude given by the product of the linear momentum of the particle, $m\dot{\mathbf{r}}$, and the shortest distance from O to the line parallel to $m\dot{\mathbf{r}}$ passing through the particle. At any instant the vectors $\mathbf{G}$ and $\mathbf{L}$ are not necessarily parallel, but (3.6) shows that $\dot{\mathbf{L}}$ is always parallel to $\mathbf{G}$. The angular momentum is closely related to the rate of sweeping out of area by the radius vector, which is $\frac{1}{2}\, \mathbf{r} \wedge \dot{\mathbf{r}}$ and therefore $L/(2m)$. A particle moving under the influence of a central force directed always towards or away from the origin experiences no torque $\mathbf{G}$ and therefore moves so as to sweep out area at a constant rate.

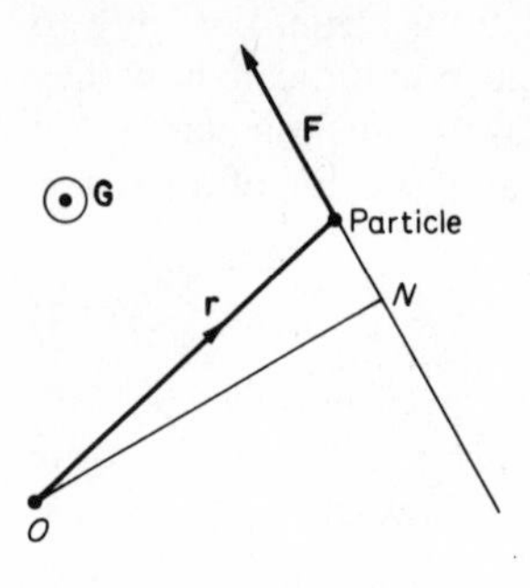

We now apply this analysis to a swarm of particles, for one of which (3.6) typically takes the form

$$\dot{\mathbf{L}}_i = \mathbf{G}_i + \sum_j \mathbf{r}_i \wedge \mathbf{F}_{ji} \tag{3.7}$$

in which $\mathbf{G}_i$ is the moment of forces exerted from outside and the $\mathbf{F}_{ji}$ are, as before, mutual forces between particles. It should be observed that when we add all the equations together we now no longer find the double sum vanishing, since corresponding terms are $\mathbf{r}_i \wedge \mathbf{F}_{ji}$ and $\mathbf{r}_j \wedge \mathbf{F}_{ij}$, which add to give $(\mathbf{r}_i - \mathbf{r}_j) \wedge \mathbf{F}_{ji}$ if $\mathbf{F}_{ij} = -\mathbf{F}_{ji}$. Unless $\mathbf{F}_{ij}$ is a central force, directed along the line joining the ith and jth particles, the pair of oppositely directed forces exerts a net torque. To be sure, the mutual forces which will principally concern us are the gravitational and electrostatic inverse-square forces which are central, and if we could be certain that all forces were of this kind we could with a clear conscience forget about their contribution to the torque, and sum over all particles to obtain the equation

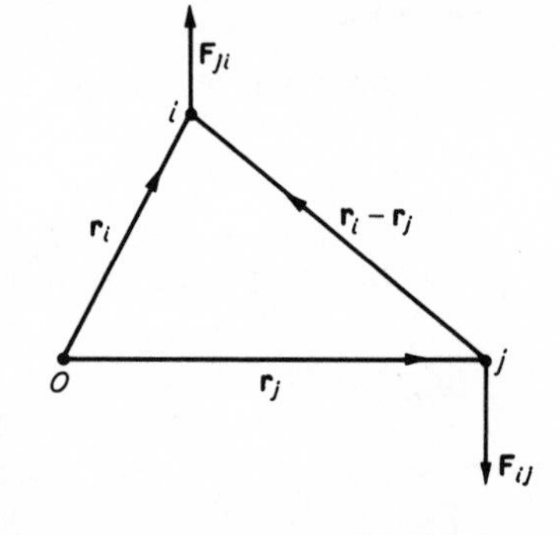

$$\dot{\mathbf{L}} = \mathbf{G}, \tag{3.8}$$

the same as (3.6) except that now $\mathbf{G}$ is the total torque exerted from outside and $\mathbf{L}$ is the total angular momentum. In particular, an isolated system of particles according to this result would conserve constant angular momentum about any fixed point.

At this stage, however, we have to admit that the situation is unsatisfactory. We have no certainty that atoms interact through central forces only; indeed the magnetic interactions, though weak, are non-central. Yet experience shows that bodies left to themselves do not spontaneously start spinning. Is this because all forces are basically central if only we analyse the atoms into their most fundamental constituents, or is it because the statistical average of the couples due to non-central forces is zero, and cannot be made otherwise by any trick of lining up the atoms to point in preferred directions? Here we reach a point where we must defer the enquiry until we have learnt much more. There is no virtue in a deep analysis at the sub-atomic level where we know Newtonian mechanics to be quite inadequate, and further discussion is therefore outside the scope of this book. It is no bad thing, however, to recognize at this comparatively elementary stage that we are faced with a rather fundamental problem, and that our only sensible course in such a situation is to pass on, without putting it entirely out of our minds; we shall indeed come back to it again later. We must, however, take the law of angular momentum in the form (3.8) as ultimately justified by experiment and therefore as additional to (though of course not in conflict with) Newton's laws.

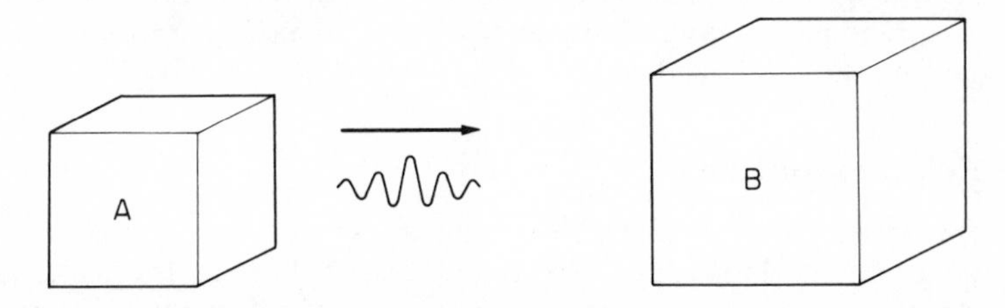

Another difficulty arises with N3 and conservation of linear momentum if we probe deeply enough, as a consequence of the finite speed of propagation of forces. An example of the problem is presented by the radiation of a light signal by a body A, the signal being absorbed, on arrival, by another body B. Now light exerts a small, but detectable, pressure on reflexion or absorption, and correspondingly when A emits the signal it recoils to the left. Thus if A and B were initially at rest, while the signal is on its way to B the system A + B has a non-zero linear momentum. When the signal is absorbed by B, however, B begins to move to the right with momentum equal and opposite to that of A, and the initial state of zero total momentum is restored. It appears, then, that action and reaction are here equal and opposite, except for a time lag, but in the long run momentum is conserved. Are we to abandon N3 because of this? Surely not. We are here dealing with an effect that requires extreme skill and sophistication to detect, and we should discuss its importance in the context of an understanding of physics that is adequate to devise and conduct the experiment. In this case, by the time we understand enough about light to see where its pressure comes from, we have sufficient control of our ideas not to worry over a little thing like ascribing momentum to a light signal. Then at all times the total momentum is zero, since the momentum of the emitted signal is equal and opposite to the recoil momentum of A. The picture that emerges is hardly different from that of A ejecting a particle which is absorbed by B, and in this situation we do not suppose that the time lag between ejection and absorption constitutes a violation of N3. Rather, we deny that the recoil of A and the later opposite recoil of B are the effects of a Newtonian action–reaction pair. The true pairs of forces are those between A and the particle, which lead to the latter's ejection, and (later) those between the particle and B when it is absorbed.

This example illustrates the difficulties which can arise if we adopt a legalistic attitude to the laws of motion, trying to formulate them in such a way as to enable them to apply rigorously in all sorts of situations far outside the context in which they were first developed, and then treating them as little short of Holy Writ. It is healthier to expect that they will require re-interpretation as we become more learned and experienced; to put the matter crudely, we need not hesitate to twist the meaning of the original words, provided that we do so in a logical and unambiguous fashion, so that the resulting reformulation is capable of critical experimental test. It is in this sense that we referred at the start to Newton's laws as a reminder of the techniques of dynamical analysis rather than an explicit and self-contained code.

Relative motion

The experiments cited as plausible demonstrations of Newton's laws were performed in the laboratory, which our observation of the stars and planets leads us to suppose is pursuing a rather complicated trajectory in space. It is proper to ask whether our conclusions may have been influenced by this fact, though we should probably be surprised if laws so simple should turn out to be the accidental outcome of the complex conditions of the experiments. We shall show, by analysis, that any effects due to the Earth's motion are too subtle to be detected in crude laboratory experiments, but that refined measurements do indeed reveal just what we might expect. Our belief is thereby strengthened that Newton's laws have more than merely local validity.

We may begin by assuming, with Newton, that there is such a thing as a stationary frame of reference, though with our present knowledge of relativity theory we know better than to regard this as likely. It is, however, convenient to pretend that there is an ideal observer at rest with respect to this frame, who will see the laws of motion exemplified in their simplest form. We shall suppose that he finds Newton's laws to be exactly true, and shall deduce what an observer in a moving frame will think the laws of motion to be. First, let our moving frame be *inertial*, i.e., have constant linear velocity $\mathbf{u}$ with respect to the fixed frame, and no rotation. Then the position, velocity, and acceleration of a particle seen by the moving observer (represented by primed quantities) are related to the corresponding quantities (unprimed) seen by the fixed observer, by the equations:

$$\mathbf{r}' = \mathbf{r} - \mathbf{u}t; \quad \dot{\mathbf{r}}' = \dot{\mathbf{r}} - \mathbf{u}; \quad \ddot{\mathbf{r}}' = \ddot{\mathbf{r}}.$$

From these equations it follows that if the fixed observer notes a free particle travelling at a uniform velocity in a straight line ($\dot{\mathbf{r}}$ = constant), so also does the moving observer, though the velocities they observe differ by $\mathbf{u}$. Thus, if N1 is true for the fixed observer, it is also true for the moving observer. Next we conduct the trolley experiments, or their equivalent, and watch them from both frames. Since both observers see the same acceleration, they will agree that constant force produced constant acceleration, and that the ratio of accelerations produced in two bodies by a given force is the same for all bodies, etc. In other words, if one agrees that N2 is verified, so does the other; moreover, they will agree on the same measure for the mass of a given body and the force producing a given acceleration. So, finally, if one finds action and reaction to be equal and opposite, so does the other, and both agree as to the truth or otherwise of N3.

For the moment we shall be content with this overall view of the agreement between different observers, but at a later stage, when we have discussed the detailed character of some of the most important fundamental forces, the electric and magnetic forces, we shall ask how it comes about that the observers can agree; it will then turn out that comparison of the observations made by different observers gives great insight into the relationship between electric and magnetic phenomena. The following paragraphs, which touch superficially on the cosmological significance of a study of relative motion, should not therefore mislead the reader into supposing that this is the sole reason for introducing the topic; not only the philosopher, but the everyday physicist, has much to learn from the comparison of different observers, and this should become clear in due course.

242
305

To return, however, to Newton's laws, it is clear that uniform motion, without rotation, relative to a fixed system of axes does not alter an observer's view of the truth of Newton's laws, and does not allow him to determine, if he finds them obeyed, how fast he is moving relative to the fixed axes. If we are to determine absolute velocity in the fixed frame we must seek other laws which are not invariant in going from one inertial frame to another. But such have not yet been found, and the Principle of Relativity asserts that we shall not find them. If then we believe we have no means of determining our absolute velocity in the fixed frame of the universe, have we any reason to suppose that the frame exists? There are experimental facts that prevent us from giving a categorical 'no' to this question, and these are concerned with questions of acceleration rather than velocity. For example, an observer shut in a box which was being accelerated by tugging on a rope attached to it would have no difficulty, without looking out, in recognizing that something was happening; a ball rolled across the table would move in a parabola rather than a straight line, and if the pull on the rope altered, the ball would respond to the change. Such an observer would not recognize the truth of Newton's laws in his own environment, and so we must conclude, having recognized these laws, that to a good approximation our Earth is an inertial frame; there are discernible effects due to its rotation, which we shall analyse in a moment, but if we neglect them for the present we have to conclude that our own framework is inertial. If we believe in an absolute frame, we can say that, whatever velocity we may have with respect to it, that velocity is sensibly constant. But how do we define an inertial frame, if we do not believe in an absolute frame?

To deal with this question, it is helpful to take a broader look at the problem from a point of view which has proved enormously stimulating and fruitful. We cannot do more than remark on this in passing, for the concepts of General Relativity are deep indeed and far outside our domain. It must suffice to point out that it is not necessary for an observer to be moving at constant velocity relative to us for him to agree that Newton's laws (and the other laws of physics) are valid within his own observations. An astronaut in free orbit round the Earth experiences weightlessness, as does all his equipment. The constancy of the gravitational acceleration for all bodies at a given point implies that they suffer no relative accelerations and do not perform any motions that can hint to the astronaut, shut up in his capsule, that he is not simply moving at constant velocity in a gravity-free space, rather than continually accelerating under the influence of gravity. It is possible to develop fundamental cosmological theories on the basis of the idea that a body falling freely in a gravitational field behaves as if in an inertial frame, and that this indeed is what an inertial frame truly is, and not a frame moving at constant speed with respect to an absolute frame of reference. Within this scheme, the coordinate system with respect to which motions must be measured is no longer a hypothetical system that would be present even in a totally empty universe, but is one determined by the very presence of matter, which itself creates gravitational fields and provides locally the appropriate coordinates. In so far as anything can be said to be fixed, it is the centre of mass of all matter, and the axes with respect to which the angular momentum of all matter is zero.

The mention of angular momentum leads us to enquire about the effect of rotation on an observer's view of the laws of motion, and here, as with linear acceleration, we find no difficulty in recognizing the existence of rotation. Two bodies, joined by a string, and resting on a smooth table in a rotating cabin, will spontaneously draw apart until the string is taut. If we fail to observe this centrifugal effect in a very delicate experiment, and then look out of our cabin, we shall find that it is not rotating with respect to the distant stars. The Foucault pendulum, which we discuss soon, is an example of such an experiment, but it must be admitted that it is, for all its delicacy in human terms, a very crude experiment from which to draw any firm conclusions that one could embody in a cosmological theory. The size of the universe is so vast, that any suggestion of an observation that showed it rotating as a whole relative to some equipment that we believed detected the absolute non-rotating framework, would be met with considerable scepticism. Most of the matter in the universe is millions of light-years away, and even if it were rushing at the speed of light with respect to the fixed axes would still take millions of years to perform one revolution.

19

From the point of view of the ordinary physicist, these general ideas of cosmology are still without immediate consequence, however fascinating they may be for their imaginative and intellectual content. One day they will be found to be related to other ideas which are now central to physics, but till then we must admit their irrelevance and pass on to humbler matters, simply noting that the distant stars provide us with axes relative to which we may measure rotational motion. The kinematics of uniform rotation has already been developed in Chapter 2, and we can proceed from this point. The acceleration $[\ddot{r}]_{\text{stat}}$ seen by the stationary observer and, as we now realize, interpreted by him as the effect of a real force, may be interpreted also by the rotating observer as the effect of a real force—he has, after all, the same ability as the stationary observer to discern what it is due to, a string or gravitation or something else. He will therefore be in a position to write the equation of motion for the particle in the form

$$m\ddot{\mathbf{r}}' = \mathbf{F} + \mathbf{F}_1 + \mathbf{F}_2,$$

in which $\mathbf{F}$ is the real force, $\mathbf{F}_1$ is the Coriolis force, $-2m\boldsymbol{\omega}\wedge\mathbf{r}'$, and $\mathbf{F}_2$ is the centrifugal force, $-m\boldsymbol{\omega}\wedge(\boldsymbol{\omega}\wedge\mathbf{r}')$. These last two forces are fictitious, in the sense that they have to be introduced by an observer in a non-inertial frame if he is to pretend that his frame is inertial and that Newton's laws hold in it.

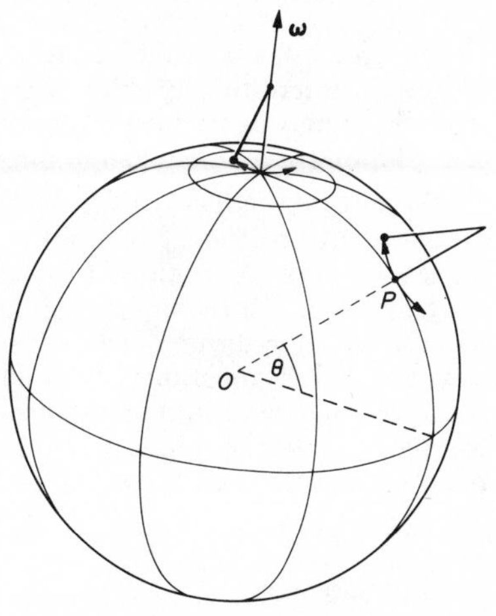

If the stars define the fixed axes, an observer on the Earth is in a frame rotating once in 24 hours, with an angular velocity vector directed from South to North and of magnitude $7{\cdot}3\times10^{-5}$ rad s^{-1}. Any body whose motion is recorded with respect to a set of axes fixed to the Earth should appear to be influenced by Coriolis and centrifugal forces. The most famous test of this prediction is Foucault's pendulum; a simple pendulum consisting of a heavy bob on the end of a long wire, and set swinging in a certain direction, is found to change its direction of swing steadily. Under ideal conditions, which we shall discuss, the time taken for the plane of oscillation to rotate through one whole revolution is cosec θ days, θ being the latitude at which the experiment is conducted. It is easy to see why this should be by imagining the experiment performed first at the North Pole. The pendulum should remain swinging in the same plane relative to the stars, and, as the Earth rotates, the plane of swing, observed by someone on the Earth, should apparently swing round once a day, in a clockwise sense viewed from on top. If the earthbound observer wished to explain the precession as due to the Coriolis force, he would note that this force, $-2m\boldsymbol{\omega}\wedge\dot{\mathbf{r}}'$, acts from left to right as the pendulum swings away from him, and reverses as it swings towards him, as shown in the diagram, so that the clockwise precession is naturally accounted for. A pendulum at latitude θ is also subject to a Coriolis force, the effective part of which is less by a factor $\sin\theta$ than at the pole. For if we resolve $\boldsymbol{\omega}$ into a radial component $\omega\sin\theta$ along OP and a tangential component $\omega\cos\theta$ parallel to the surface, the latter contributes to vertical Coriolis forces, which only serve to make the bob lighter or heavier by negligible amounts, and the radial component $\omega\sin\theta$ alone provides the horizontal force needed for precession. The precession rate is thus multiplied by $\sin\theta$ and the time for one revolution by $\operatorname{cosec}\theta$. There is no precession at the equator, and precession is anticlockwise in the southern hemisphere. It is not very difficult to construct a Foucault pendulum that precesses in the expected sense at something like the expected rate (say within 10%), but anyone who has tried to exhibit the effect reliably knows that the pendulum is capable of embarrassing irregularities, unless extraordinary care is taken in the design.

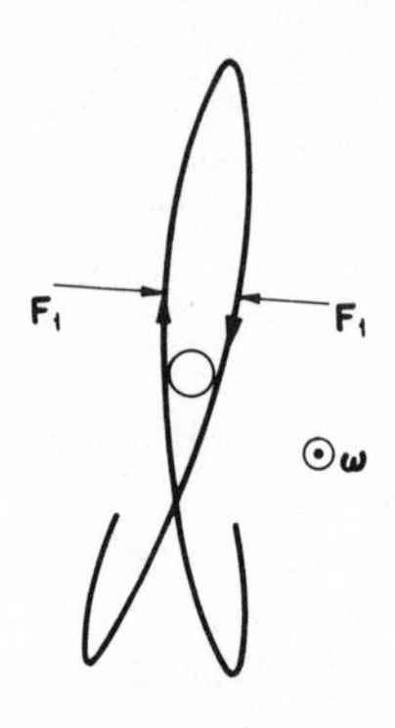

For this reason, and also because it provides an excuse to display the solution of certain interesting problems, we shall look into the possible disturbing effects that may cause the precession to go wrong. But first let us write down the solution for an ideal pendulum, with and without Coriolis forces.

1. *The ideal pendulum without Coriolis force.* The reader is assumed to be familiar with the theory of a simple pendulum, and to know that for small angles of swing the motion of the bob may be treated as motion in the horizontal plane under the influence of a restoring force proportional to displacement from equilibrium. The ideal simple pendulum has the same force constant for all directions of displacement in the horizontal plane, i.e., the restoring force is central and isotropic. If we describe the displacement by a complex number, $\boldsymbol{\rho}$, representing direction and magnitude, the equation of motion takes the form

$$m\ddot{\boldsymbol{\rho}} + \alpha\boldsymbol{\rho} = 0,$$

or

$$\ddot{\boldsymbol{\rho}} + \omega_0^2\boldsymbol{\rho} = 0,$$

where ω_0 is written for $(\alpha/m)^{1/2}$. The general solution of this equation is conveniently written as

$$\boldsymbol{\rho} = A\exp(\mathrm{i}\omega_0 t) + B\exp(-\mathrm{i}\omega_0 t).$$

If $A = 0$, the solution $B \exp(-i\omega_0 t)$ represents clockwise circular motion with radius B (conical pendulum); if $B = 0$, $A \exp(i\omega_0 t)$ represents anticlockwise circular motion with radius A. If A is real and equal to B, $\boldsymbol{\rho} = 2A \cos \omega_0 t$, which is a linear harmonic oscillation along the real axis. If $B = -A$, $\boldsymbol{\rho} = 2iA \sin \omega_0 t$, which is a linear harmonic oscillation along the imaginary axis. In general, if A and B are complex numbers, the motion is an ellipse, whose axes remain fixed. It is left as an exercise for the reader to determine in this general case the lengths and directions of the principal axes of the ellipse, and the sense of rotation. (This is achieved with most clarity by considering the two contributions $A \exp(i\omega_0 t)$ and $B \exp(-i\omega_0 t)$ as rotating vectors). Under all conditions the motion is strictly periodic with angular frequency ω_0.

It is worth comment here that the complex notation is also of great value in representing the simple harmonic oscillation of a single variable, x say. The trick is to interpret the statement $x = A\, e^{i\omega t}$ as implying that x is the real part only of $A\, e^{i\omega t}$ (e.g., if A is real, $A \cos \omega t$) which represents simple harmonic motion. The convenience of this device stems from the property of an exponential that taking its derivative simply multiplies it by a constant: $(d/dt)\, e^{i\omega t} = i\omega\, e^{i\omega t}$, and this allows a differential equation of motion to be replaced by an algebraic equation, as the following examples will illustrate.

2. *The ideal pendulum with Coriolis force.* If we were to take complete account of the Earth's rotation we should include the centrifugal force in the calculation. This has the effect of displacing the equilibrium position of the bob very slightly, and its variation with the position of the bob makes a contribution to the forces appearing in the equation of motion. Since, however, the variations of centrifugal force are of the order of $m\omega^2\rho$, if ω is the Earth's angular velocity, while the Coriolis force is of the order of $m\omega\omega_0\rho$ (since $\dot{\rho} \sim \omega_0\rho$), the latter is larger by a factor ω_0/ω, enough to allow us to neglect centrifugal effects. We therefore concentrate on the horizontal component of the Coriolis force, which has the form $-2im\omega'\dot{\boldsymbol{\rho}}$, where ω' is the vertical component of the Earth's angular velocity. The new equation of motion is then

$$\ddot{\boldsymbol{\rho}} + 2i\omega'\dot{\boldsymbol{\rho}} + \omega_0^2\boldsymbol{\rho} = 0,$$

and if we seek a solution of the form $e^{i\omega t}$ we find the differential equation is replaced by an algebraic equation for ω:

$$-\omega^2 - 2\omega\omega' + \omega_0^2 = 0.$$

Adding the (negligible) second-order term ω'^2 to the left-hand side, we see that

$$\omega = \pm\,\omega_0 - \omega',$$

and the general solution is a combination of the two solutions,

$$\boldsymbol{\rho} = A\, e^{i(\omega_0 - \omega')t} + B\, e^{-i(\omega_0 + \omega')t}.$$

The two conical pendulum motions proceed at angular frequencies differing from the unperturbed ω_0 by $\pm\omega'$, the vertical component of the Earth's angular frequency. The significance of this is obvious at the North Pole, where a stationary observer sees that a conical pendulum can rotate with angular frequency $\pm\omega_0$, but the observer on the Earth sees the anticlockwise sense of rotation slowed down by ω' and the clockwise sense equally speeded up.

The special case $A = B$ leads to the precessing linear oscillation

$$\rho = 2A \cos \omega_0 t . e^{-i\omega' t},$$

which represents the usual mode in which Foucault's pendulum is demonstrated.

3. *Non-ideal pendulum without Coriolis force.* The requirement that the restoring force be the same for all directions of swing is rather difficult to achieve in practice. The wire supporting the heavy bob is necessarily strong and not therefore perfectly flexible. If it is soldered into a bushing at the top, or held in a pin-chuck, it may not bend in exactly the same way in all directions, and thus the effective point of support, and effective length of pendulum, may vary with direction. The same result is produced if the pendulum is hung from a roof-beam which is less rigid against lateral forces in some directions than in others. The swaying of the support as the pendulum swings increases its effective length, and any variation of this increase with direction makes the restoring force anisotropic.

PROBLEM

A simple pendulum with bob of weight W is hung from a beam which is elastically deflected through a horizontal distance βF when a force F is applied laterally to it. Show that the pendulum is effectively longer than its real length by βW, i.e., the amount by which the beam would be deflected by a lateral force equal to the weight of the bob. A simple demonstration may be set up by clamping a wooden lath or metre-stick to the edge of a table and swinging a pendulum from the free end; the period will be longer for swings in the plane normal to the stick that cause it to bend.

A harmonic oscillator with different natural frequencies $\omega_0 \pm \delta\omega_0$ for linear swings in two directions at right angles, such as is the consequence of the anisotropy just referred to, can only maintain linear motion unchanged if swung in one or other of these two principal directions—for only in these directions is the restoring force strictly radial. If swung in any other direction, the bob executes the well-known *Lissajous figures*, as can be easily shown by means of the simple demonstration apparatus just described. The orbit varies slowly from linear through elliptical to a differently oriented linear, and then back again to the original linear orbit after a time $\pi/\delta\omega_0$. Clearly if $\delta\omega_0$ begins to approach ω' in magnitude the orbit variations of the Lissajous figures will completely mask the Coriolis precession.

Suppose the effective length of the pendulum is $R \pm \delta R$ for the two principal directions; then the corresponding frequencies are $[g/(R \pm \delta R)]^{1/2}$. That is, $\omega_0 = (g/R)^{1/2}$ and $\delta\omega_0 \approx \frac{1}{2}g^{1/2}\delta R/R^{3/2}$. For a pendulum 20 m long, δR need be only $\frac{1}{3}$ mm for the pendulum to swing from one linear orbit to the other in the course of a day. Since this would clearly be enough to spoil quantitative observations of the Coriolis effect, great efforts must be made to keep the effective length of the pendulum, even one as long as this, the same to better than 0·1 mm for all directions of swing.

A further effect, which will be discussed later, arises from allowing the pendulum to swing with too large an amplitude and in a not quite linear orbit. With larger amplitudes, the restoring force is not exactly proportional to displacement, and an elliptical orbit is then not stable as regards the direction of its axes; the axes themselves precess steadily in the same sense as the bob moves in its elliptical orbit. Obviously this can also mask the Coriolis effect, and it is essential to use small amplitudes and avoid ellipticity if good results are to be obtained. This analysis should make clear why the Foucault pendulum is good enough, when well set up, to exhibit the Earth's rotation but too sensitive to errors to be of any use in detecting much slower rotations.

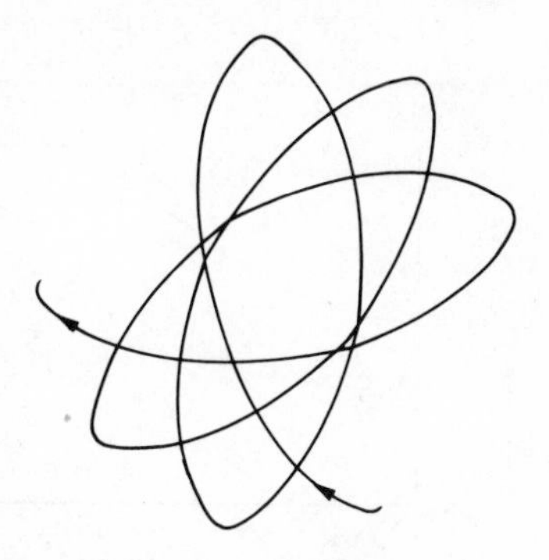

4. *Non-ideal pendulum with Coriolis force.* We make the foregoing discussion an excuse to exhibit the solution of the instructive problem presented by the non-ideal pendulum when it is affected by Coriolis forces. It is no longer convenient to use complex notation for the displacement, as the reader will find if he tries to work out even the simple theory of Lissajous figures by this technique, and we shall use ordinary Cartesian coordinates, choosing as axes those directions in which the restoring force is strictly radial. Then the equations of motion take the form,

$$\left.\begin{aligned} &\ddot{x} + (\omega_0 + \delta\omega_0)^2 x - 2\omega'\dot{y} = 0 \\ \text{and} \quad & \\ &\ddot{y} + (\omega_0 - \delta\omega_0)^2 y + 2\omega'\dot{x} = 0, \end{aligned}\right\} \tag{3.9}$$

the last term in each equation being the Coriolis acceleration. We now use complex notation in the alternative sense mentioned on p. 35, representing the simple harmonic oscillations of x and y by $X\,\mathrm{e}^{\mathrm{i}\omega t}$ and $Y\,\mathrm{e}^{\mathrm{i}\omega t}$ respectively (in which X and Y may be complex numbers) while understanding by this that only the real parts describe the actual variations with time of x and y. We must find values of X, Y and ω which satisfy the simultaneous differential equations (3.9). Substituting the exponential forms, we find the equations are reduced to algebraic equations—this, we repeat, is the peculiar merit of the exponential representation of simple harmonic motion—

$$\left.\begin{aligned} -\omega^2 X + (\omega_0 + \delta\omega_0)^2 X - 2\mathrm{i}\omega\omega' Y = 0, \\ -\omega^2 Y + (\omega_0 - \delta\omega_0)^2 Y + 2\mathrm{i}\omega\omega' X = 0. \end{aligned}\right\} \tag{3.10}$$

Now we are concerned with cases where $\delta\omega_0$ and ω' are very much smaller than ω_0, and where the actual oscillation frequency ω is expected to differ from ω_0 only by a quantity of the order of $\delta\omega_0$ or ω'. It is therefore justifiable to replace ω by ω_0 in the last term of each equation (3.10) and to make other comparable second-order approximations which reduce these equations to the form

$$\left.\begin{aligned} (\Omega - \delta\omega_0)X &= -\mathrm{i}\omega' Y \quad \text{(a)} \\ (\Omega + \delta\omega_0)Y &= \mathrm{i}\omega' X \quad \text{(b)} \end{aligned}\right\} \tag{3.11}$$

where Ω is written for $\omega - \omega_0$. For these two equations to be compatible it is necessary that

$$(\Omega - \delta\omega_0)(\Omega + \delta\omega_0) = \omega'^2$$

i.e., (3.12)

$$\Omega = [\omega'^2 + (\delta\omega_0)^2]^{1/2}.$$

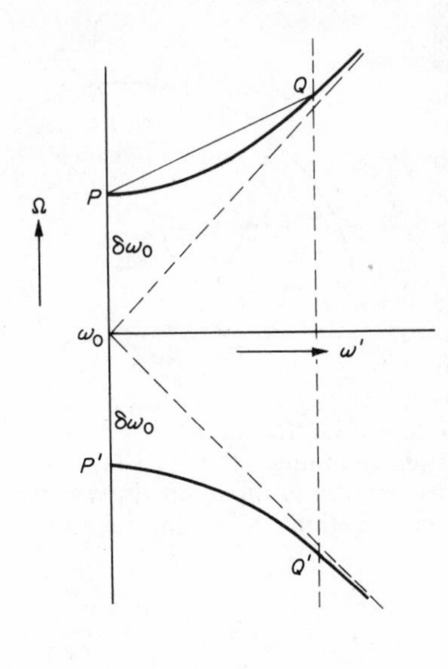

If we imagine the Coriolis force, represented by ω', to be increased steadily from zero, initially (as represented by P and P') the pendulum has stable patterns of linear oscillation (*normal modes*) of frequency $\omega_0 \pm \delta\omega_0$, which are what its non-ideality allows. When the Coriolis term is strong compared to the non-ideality ($\omega' \gg \delta\omega_0$), the normal modes are circular and of frequency $\omega_0 \pm \omega'$; this is the ideal Foucault pendulum. The hyperbolic curves show how one form of behaviour merges into the other. To verify the statements about the patterns of oscillation and to see which are the normal modes in the intermediate cases when ω' and $\delta\omega_0$ are comparable, we refer back to 3.11(a). If we take X to be a real number, Y is purely imaginary, and the stable pattern is determined by the real parts of $X\,\mathrm{e}^{\mathrm{i}\omega t}$ and $Y\,\mathrm{e}^{\mathrm{i}\omega t}$, i.e.,

$$x = X\cos\omega t$$

$$y = -\mathrm{Re}\,\{(\Omega - \delta\omega_0)X\,\mathrm{e}^{\mathrm{i}\omega t}/(\mathrm{i}\omega')\} = -\frac{\Omega - \delta\omega_0}{\omega'}X\sin\omega t.$$

The oscillations of x and y in phase-quadrature combine to form a stable elliptical motion. That represented by the upper branch (Q) of the curve

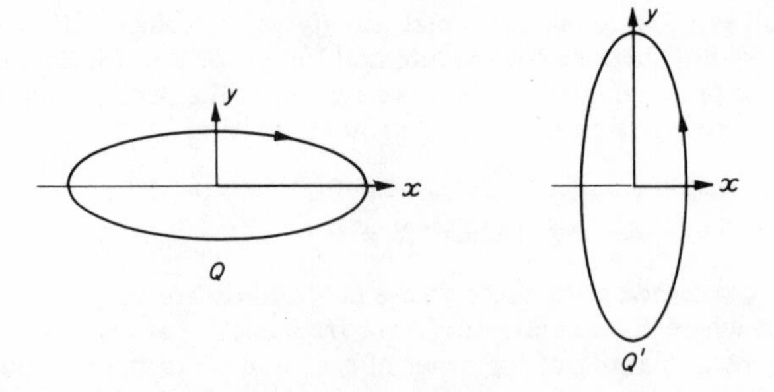

($\Omega > 0$) has its major axis along the x-direction, and the sense of rotation is clockwise. The minor/major axial ratio is $(\Omega - \delta\omega_0)/\omega'$, represented by the slope of the line PQ and varying from zero when $\omega' = 0$ (linear motion) to unity when ω' is large (circular motion). For the lower branch (Q') the ellipse has the same axial ratio but with the axes interchanged, and has an anticlockwise sense of rotation, as shown directly by the use of 3.11(b). If the pendulum is set swinging arbitrarily, its initial motion may be resolved into a mixture of these two stable modes which then beat against one another to produce a fluctuating pattern analogous to a Lissajous figure but rather more complicated.

The Coriolis force achieves prominence from time to time when the belief is aired, in the correspondence columns of newspapers and elsewhere, that the vortex formed as the bathwater runs away spins in one sense in the northern hemisphere and the opposite sense in the southern hemisphere. If the reader has access to a bath which forms a vortex he may carry out some simple experiments with advantage. First, let the bath remain undisturbed for at least five minutes, and then withdraw the plug as gently as possible; there is a very good chance that no vortex will form. Then repeat the experiment, but after withdrawing the plug stir the water very gently; it should be easy to induce either sense of vortex rotation with only very small movements. Finally, after a vortex has begun to form and its sense become clear, stir the other end of the bath very vigorously in the same sense and then watch developments; with luck, the vortex will stop and begin again in the opposite sense. These experiments may not work with all designs of bath, but certainly do with the author's. The observation that gentle stirring can set up a powerful vortex illustrates the instability of outflow without vorticity. Just as a ball resting on a hill between two valleys can remain there so long as it is undisturbed, but when disturbed even slightly rolls into one or other of the valleys where it finds a position of stable equilibrium, so the outflow may remain vortex-free for a while, but is only waiting for a chance disturbance to precipitate it into a new, and stable, pattern of flow. Left- and right-handed vortices are both preferred to no vortex at all, and it is therefore easy to establish either, but rather hard to switch from one to the other; equally, it requires more effort to transfer the ball from one valley to the other than to roll it into either from the hilltop.

These simple experiments show how dependent on chance motions is the sense in which the vortex is established, and therefore how cautious we must be in supposing that it is only the Earth's rotation that controls the situation. Consider, for example, a large circular vessel at the North Pole. From the point of view of an observer fixed in space, the vessel and its contents are rotating once a day. When water is allowed to escape through a hole at the centre, any small element of water starting at the edge and moving towards the centre possesses angular momentum at the outset, because of the Earth's rotation, and, since the forces causing it to flow are radial, there is no change in this angular momentum as it approaches the centre. It must therefore move in a direction which is not strictly radial, and it is its tangential component of velocity that sets up the vortex. To see things in proportion, consider a (domestically speaking) rather large circular bathtub of 1 m radius with its plug-hole in the middle. At the Pole, where the effect is strongest, the Earth's rotation gives the water a tangential speed, at the edge, of only $\frac{1}{14}$ mm per second; it is necessary to wait for a very long time, in a very still room, before accidental motions of the water are reduced to less than this. In other words, on the domestic scale, vortex motion is determined by influences other than the Earth's rotation.

However, with vessels 1 km or more in radius, the speed due to the Earth's rotation is 1000 times greater, more than 70 mm s^{-1}. This may well be the dominant movement, and it has indeed been observed that vortices at the outflow of reservoirs feeding hydro-electric turbines do spin in the expected sense, anticlockwise in the northern hemisphere, clockwise in the southern.

On a larger scale again, Coriolis forces unmistakably affect atmospheric movement. Where a low pressure region (cyclone) develops, the force on an element of air is radially inwards, since the pressure on its outward surface is greater than that on its inward surface. This radial force does not, however, cause radial movement as it would on a non-rotating Earth. Instead, vortex motion is set up, with the wind blowing almost tangentially, i.e., along the lines of constant pressure (*isobars*). It is helpful at this stage to consider orders of magnitude. Only rarely in temperate climates is the wind-speed much greater than 20 m s^{-1} (on the Beaufort scale a gale is defined as 17–21 m s^{-1}), while the centre of the cyclone, around which the air is supposed to be circulating, may be 500 km or, usually, further away. The time needed for an element of air to make one revolution round the isobar is, in these circumstances, nearly 2 days; this, being more than the Earth's period, indicates that the centripetal force needed to keep the air moving in such a circular trajectory on a stationary Earth is correspondingly less than the Coriolis force due to the movement on a rotating Earth. For anything short of a full gale, it is a reasonable assumption that the force due to the pressure gradient and the Coriolis force balance each other, to allow virtually unaccelerated motion of the air. Since in the northern hemisphere the Coriolis force is from left to right for an observer looking along the wind direction, the force due to the pressure gradient must run from right to left, and the low pressure region must be on the observer's left. We expect the wind in the northern hemisphere to blow in an anticlockwise sense round a low-pressure region, and clockwise in the southern hemisphere.

It must be appreciated that the above analysis is highly idealized and neglects many important effects, notably the frictional forces acting on the air as it travels close to the ground, spinning off eddies that carry

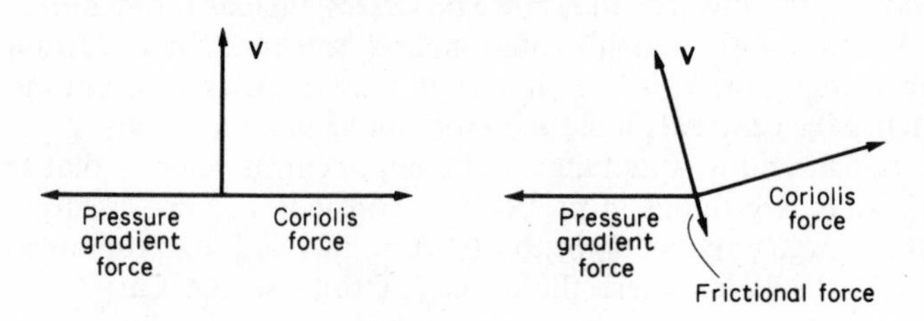

away linear momentum and transfer it to the ground. This swings the direction of the wind towards the low-pressure region and diminishes its speed, as the vector diagrams illustrate. The diagram on the left shows the pressure gradient and Coriolis force balancing, in the absence of friction, with **v** making a right angle with the pressure gradient. On the right, a frictional force opposite in direction to **v** produces a situation in which the three forces now present can only balance if **v** is tilted towards the direction of the pressure gradient.

EXAMPLE

The weather map for the Atlantic and W. Europe on October 16, 1970, the day this paragraph was written, shows isobars at intervals of 8 mbar ($800\ \mathrm{N\ m^{-2}}$) (the other lines, which are warm and cold fronts, do not concern

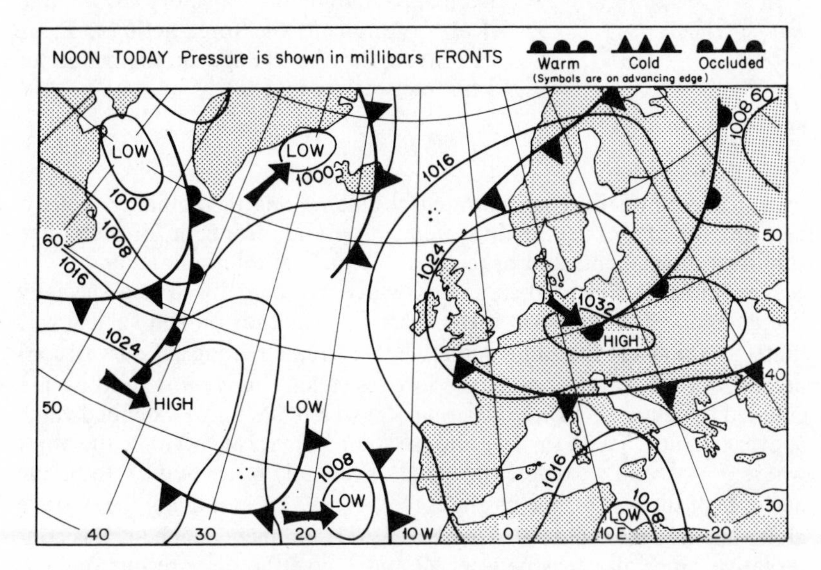

us). The pressure gradient over the centre of the British Isles is about $10^{-3}\ \mathrm{N\ m^{-3}}$ and this is the force acting on $1\ \mathrm{m^3}$ of air, of mass about 1·2 kg, which must be balanced by the Coriolis force, $2\,m\omega v \sin\theta$. The air-speed needed for this is about $7\ \mathrm{m\ s^{-1}}$ and the wind direction, if it follows the isobar, should be nearly due south. In fact the weather forecast was that the wind would be south and light or moderate in strength ($5\text{–}6\ \mathrm{m\ s^{-1}}$). Since one would expect the wind-speed at ground level to be rather less than higher up, where the simple calculation should apply more closely, the agreement may be regarded as good.

99

Work and energy

When a force **F** acts at a point on a body, and that point moves a distance $\delta\mathbf{r}$, the work δW done by the force is defined as the product of F and the component of δr parallel to $\mathbf{F}$, i.e., $\delta W = \mathbf{F}.\delta\mathbf{r}$. If the point is moving at a velocity $\mathbf{v}$, the rate at which the force is doing work, $\dot{W} = \mathbf{F}.\mathbf{v}$. The work done by a force on a moving particle can be expressed as a change in the kinetic energy, $\frac{1}{2}mv^2$, of the particle by the following simple transformation.

According to N2, $\mathbf{F} = m\dot{\mathbf{v}}$, so that

$$\mathbf{F}.\mathbf{v} = m\dot{\mathbf{v}}.\mathbf{v} = \frac{\mathrm{d}}{\mathrm{d}t}\left(\frac{1}{2}m\mathbf{v}.\mathbf{v}\right) = \frac{\mathrm{d}}{\mathrm{d}t}\left(\frac{1}{2}mv^2\right).$$

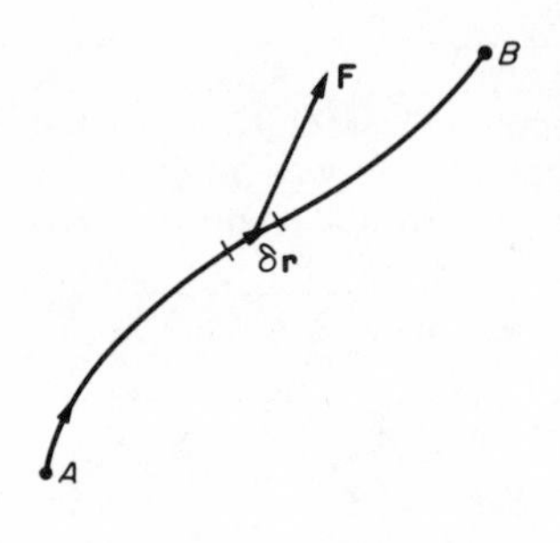

Let us now integrate with respect to time, remembering that in any time interval δt, the meaning of $\mathbf{v}\,\delta t$ is the distance $\delta\mathbf{r}$ moved by the particle. If, then, the particle moves from A to B along the trajectory shown, and if at any moment the force acting is $\mathbf{F}$, we can write for the change in the value of $\frac{1}{2}mv^2$ between A and B

$$[\tfrac{1}{2}mv^2]_A^B = \int_A^B \mathbf{F}.\mathrm{d}\mathbf{r}, \tag{3.13}$$

in which the integral is a line integral taken along the actual trajectory, i.e., we divide the trajectory into elements $\delta\mathbf{r}$ and form $\mathbf{F}.\delta\mathbf{r}$ for each, then sum over all elements and proceed to the limit of infinitesimal $\delta\mathbf{r}$.

Now (3.13) is a general statement which can be reduced to a specially simple and important form if the force $\mathbf{F}(\mathbf{r})$ depends on $\mathbf{r}$ in such a way that $\int_A^B \mathbf{F}.\mathrm{d}\mathbf{r}$ takes the same value when evaluated along all curves connecting A to B. If this is true, the force is called *conservative*, and we are enabled to define a *potential function* $\phi(\mathbf{r})$ which takes a well-defined value at every point $\mathbf{r}$. To do this, we choose some point O at will to be the point where $\phi = 0$, and we define the value of ϕ at any other point P to be the negative of the work done by the force, $-\int_O^P \mathbf{F}.\mathrm{d}\mathbf{r}$, in moving the particle along any line from O to P. Clearly ϕ at P is uniquely defined if $\mathbf{F}$ is conservative, since all paths between O and P give the same result; further, the difference in ϕ between any two points A and B is simply the value of $-\int_A^B \mathbf{F}.\mathrm{d}\mathbf{r}$ evaluated along any line joining them.

Thus for conservative forces, (3.13) may be integrated:

$$[\tfrac{1}{2}mv^2]_A^B = \phi_A - \phi_B,$$

or (3.14)

$$\tfrac{1}{2}mv_A^2 + \phi_A = \tfrac{1}{2}mv_B^2 + \phi_B.$$

If, as (3.14) states, $\frac{1}{2}mv^2 + \phi$ takes the same value at two arbitrarily chosen points on the trajectory of the particle, it must take the same value at all points on the trajectory. This quantity, $\frac{1}{2}mv^2 + \phi$, is called the *total energy*, E, of the particle, and is made up of the *kinetic energy* $T = \frac{1}{2}mv^2$ and the *potential energy* ϕ. It may be noted that ϕ is not unique, in the sense that a change in origin adds a constant to $\phi(\mathbf{r})$, but this is not significant. The essential property of the trajectory is that E is constant along it, not that E takes a particular value, and therefore additive constants are irrelevant.

The components of the vector $\mathbf{F}$ are readily related at any point to the variation of $\phi(\mathbf{r})$ around this point. If we move a small distance δx, the work done by $\mathbf{F}$ is $F_x\,\delta x$ and is also equal to $-(\partial\phi/\partial x)\,\delta x$. Hence

$$\mathbf{F} = -(\partial\phi/\partial x,\ \partial\phi/\partial y,\ \partial\phi/\partial z).$$

The differential function of ϕ defined by the brackets is written grad ϕ or $\nabla\phi$, ∇ being a vector differential operator with Cartesian components $(\partial/\partial x, \partial/\partial y, \partial/\partial z)$. The reason for the symbol grad (short for gradient) is seen by considering an undulating surface, such as is represented by the contour map. A portion round P is shown magnified on the right to a

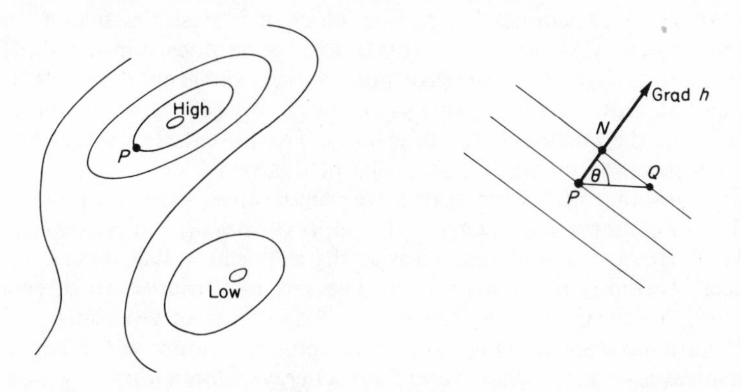

sufficient degree for the contours to be effectively straight and evenly spaced (the diagram on the right could equally well be thought of as the contours of the plane tangent to the surface at P). We can define the local gradient of the surface at P by drawing a two-dimensional vector, lying in the horizontal plane and pointing along the direction of steepest slope, i.e., normal to the contours, with magnitude defined by the rate of change of height, h, with horizontal coordinate along PN. It is clear that the rate of change of height with position along PQ is less than along PN by a factor $\cos\theta$, since to achieve the same change requires a movement along PN only $\cos\theta$ times as far as along PQ. Thus if we call the vector grad h, and PQ is parallel to the x-axis, $\partial h/\partial x$ is just grad $h\cos\theta$, from which it is obvious that grad h, defined in this way in terms of the magnitude and direction of steepest slope, has components $(\partial h/\partial x, \partial h/\partial y)$. This then is a two-dimensional example of the vector introduced as grad ϕ.

Now imagine the surface to be mirror-smooth and let a ball roll on it. As it passes P it will experience a vertical gravitational force and a 'reaction' force normal to the surface. The two combine to give a resultant pointing downhill parallel to PN, and if the undulations are very gentle so that the surface never departs far from horizontal, the horizontal force on the ball is very close to $-mg$ grad h. Thus the force is defined by a potential $\phi = mgh$, and the ball always rolls with such horizontal velocity as keeps $mgh + \frac{1}{2}mv^2$ constant.

PROBLEM

The above argument was developed approximately for very gently undulating surfaces. Show that if v is the total velocity, $mgh + \frac{1}{2}mv^2$ is constant even for steeply corrugated surfaces, and discuss the form of grad ϕ under these conditions.

This example has been used simply as an illustration of a force derivable from a potential. The forces we meet in practice on a macroscopic scale are sometimes of this form, e.g., gravitational attraction, and sometimes not, e.g., frictional forces, for which it is possible to perform a closed trajectory in which the total work done does not vanish. The Coriolis force depends on velocity, not position, and is not derivable from a potential. However, being always normal to $\mathbf{v}$ it does no work and does 223 not change the kinetic energy of a body. The Lorentz force acting on a charged particle moving in a magnetic field is very similar.

The constancy of E when forces are conservative is not a statement of the law of conservation of energy; it simply defines certain circumstances in which energy is conserved, without any implication that these circumstances invariably arise in practice. The law of conservation of energy springs from the conviction, based on a wide variety of experiments, that on the atomic scale either all forces are conservative or the fundamental dynamical laws are such as to enforce energy conservation. By way of illustration, consider a highly simplified model of matter considered as an assembly of particles interacting by means of central forces, of which on the atomic scale the inverse square Coulomb force of electrostatic interaction is the most important. First we note that any isotropic central force is conservative; a force directed outwards from a point O and of magnitude $F(r)$ is generated as $-$grad ϕ by a potential $\phi = -\int F(r)\,\mathrm{d}r$. The surfaces of constant ϕ (*equipotentials*) are spheres centred on O, and the force at any point is directed normal to the equipotential at that point, i.e., radially. To extend the argument consider two particles at $\mathbf{r}_1$ and $\mathbf{r}_2$ which interact with a force $\pm\mathbf{F}_{12}(\rho)$ along the line joining the particles, ρ being the separation of the particles. If they are acted on by forces $\mathbf{F}_1$ and $\mathbf{F}_2$ from outside, their equations of motion can be written in the form

$$\dot{T}_1 = \dot{\mathbf{r}}_1 . (\mathbf{F}_1 + \mathbf{F}_{12})$$

and

$$\dot{T}_2 = \dot{\mathbf{r}}_2 . (\mathbf{F}_2 - \mathbf{F}_{12});$$

i.e.,

$$\dot{T} = \dot{T}_1 + \dot{T}_2 = \dot{\mathbf{r}}_1 . \mathbf{F}_1 + \dot{\mathbf{r}}_2 . \mathbf{F}_2 + \mathbf{F}_{12} . \frac{\mathrm{d}}{\mathrm{d}t}(\mathbf{r}_1 - \mathbf{r}_2) = \dot{W} + F_{12}\dot{\rho},$$

in which $\dot{W}$ is the rate of work on the part of external forces. The second term can be written as $-\dot{\phi}_{12}$ by introducing a potential function ϕ_{12} defined as $-\int F_{12}\,\mathrm{d}\rho$, and thus for the two particles taken together

$$\dot{E} = \dot{T} + \dot{\phi}_{12} = \dot{W}.$$

If the two particles are isolated from other influences, $\dot{W} = 0$ and E is constant.

PROBLEMS

(1) Show, by extension of the foregoing arguments, that for a swarm of particles interacting with each other in pairs by central forces, $F_{ij}(\rho_{ij})$, a potential function ϕ_i may be defined for each particle, in the form $-\frac{1}{2}\sum_j \int F_{ij}\,\mathrm{d}\rho_{ij}$, such that the total kinetic energy of the particles plus the sum of ϕ_i over all particles changes by the amount of work done by external forces. Note the factor $\frac{1}{2}$ in the definition of ϕ_i, which arises from the division of the potential energy of interaction of any two particles evenly between them.

87 166

(2) Show that the kinetic energy of a swarm of particles whose centroid has velocity $\mathbf{v}_0$ is equal to the sum of T_0, the kinetic energy of a particle of mass $\sum_i m_i$ moving with velocity $\mathbf{v}_0$, and T', the sum of individual kinetic energies as seen by an observer moving with the centroid.

(3) Show that if a number of bodies interact inelastically (i.e., without conserving kinetic energy) the change in total kinetic energy is seen to be the same by all inertial observers.

88

The results contained in these problems are basic to the development of a kinetic theory of matter, in that they provide plausible explanations of the first law of thermodynamics as a general statement of the conservation of energy, provided that heat is included as one of the forms of energy. Heat is here interpreted as that part of the energy which may be added to or taken from a body but which does not appear as a change in the potential or kinetic energy of the body as a whole. It is to be imagined as transformed into random motions of the constituent atoms and to result in changes in their individual kinetic and potential energies without any change in the gross dynamical variables describing the body. When, for example, two equal lumps of putty, moving at equal speed in opposite directions, collide and coalesce the following statements may be made:

(1) The momentum of the system is unchanged; the centroid was initially at rest and remains so after coalescence.

(2) The kinetic energy of the lumps considered as mass points is decreased from a non-vanishing value to zero.

(3) If the particles within the putty interact by conservative forces, this kinetic energy is not lost, but is distributed among the particles partly as kinetic and partly as potential energy.

(4) The increase of random kinetic energy is apparent to a sensitive thermometer as a slight temperature rise of the putty.

So far, so good; but it happens that the proviso in (3) is not strictly true—we know that some forces between particles (e.g., those due to magnetic interactions) are not central. We shall see, however, that the non-central forces which are not derivable from a potential are still such as to conserve energy. Nevertheless, we are not justified at this point in asserting a law of conservation of energy except as an additional postulate, as we were forced to do with angular momentum. We shall refrain, however, from making this additional postulate, since it is more instructive to show how each new type of force that we come to consider strengthens our faith in the conservation of energy; but to maintain this principle involves rethinking the definition of energy at every stage. Thus, to take a historical example, it was necessary for Joule to define *heat* firmly as a form of energy in order to account quantitatively for the disappearance of kinetic energy and the consequent rise of temperature of the bodies concerned. And we shall find that when we introduce electrostatic and magnetic forces, we can associate energy with their fields in a consistent manner that allows the conservation law to hold. Ultimately, in the development of relativistic mechanics, it turns out that the two cherished conservation laws, of energy and mass, are only tenable if conflated into a single conservation law. The development of these conceptions shows us that the law of conservation of energy is not something so well understood right from the start that we may take for granted that consequences deduced from it will be confirmed by experiment. Until we have analysed each new phenomenon, we do not know how to express its contribution to the energy which is to be conserved. The law appears indeed to be a very deep principle which asserts that every new discovery will be found not to violate energy conservation, in the sense that it will always prove possible to invent a new form of energy which can be consistently introduced into the energy balance so as to maintain it. One of the threads running through this book is the substantiation of this statement by a succession of examples analysed in detail.

Cavendish Problems: 1, 21, 34, 40, 41, 42, 111.

READING LIST

General Relativity: H. Bondi, *Cosmology*, Cambridge U.P.

Weather Maps: A. Watts, *Weather Forecasting Ashore and Afloat*, Adlard Coles.

Kinetic Theory: F. Reif, *Statistical Physics* (Berkeley Physics Course, Vol. 5), McGraw-Hill.

4

The inverse-square laws (Newton, Michell, Coulomb)

Laws of Force

Two point masses m_1 and m_2 attract each other with a force along the line joining them of magnitude proportional to m_1m_2/r^2 (Newton).

Two point charges q_1 and q_2 (which may be positive or negative) attract each other with a force along the line joining them of magnitude proportional to $-q_1q_2/r^2$; if q_1 and q_2 are opposite in sign the charges attract; if the same, they repel (Coulomb).

Conservation Law

The total mass and the total charge of a closed, isolated system remain constant.

Superposition Law

The force exerted by a number of masses or charges on another mass or charge is the vector sum of the forces each would exert if it alone were present.

Among the forces with which bodies interact, the fundamental gravitational and electrostatic forces are of such supreme importance as to deserve a preliminary discussion before we begin to consider the application of mechanics to real problems. Historically, Newton's law of gravitation came first and, developed with the aid of his laws of motion, is the cornerstone of his system of the world as expounded in *Philosophiae Naturalis Principia Mathematica* (1687). It was propounded by him in order to account quantitatively for the motions of planets and their satellites, and he collected a wealth of evidence to support his theory. 198 Some of this will be discussed in detail in the next chapter. In 1750 Michell, the inventor of the torsion balance, studied the forces between magnets and concluded that they could be explained quantitatively on the assumption of the existence of north and south poles which interacted by means of an inverse-square central force. In 1766 Priestley, having satisfied himself of the truth of Franklin's observation that an electrically charged cork ball experienced no force when hung inside a charged metal vessel, deduced from this that electric charges also obey an inverse-square law. A refined version of this experiment was carried out a few years later by Cavendish, and has since been improved greatly to fix the exponent in the law of electric attraction as 2 within very fine limits. This will be discussed later in the chapter. In 1785, Coulomb used a torsion balance (independently invented) to demonstrate the electrostatic law directly, and to show that when bodies exchanged charge on contact the total charge present was unaltered.

There is thus excellent reason for investigating the consequences of an inverse-square central force law in detail with a view to careful comparison with experiment. We shall also see how the unique correlation of inertial mass and gravitational force implied in Newton's law, and substantiated by the identical acceleration shown by all falling bodies, can be subjected to very stringent test. We shall express the gravitational law in an unconventional fashion so as to give it the same form as the conventional expression of Coulomb's law.

$$\text{Newton's law} \qquad \mathbf{F} = -\frac{m_1 m_2 \mathbf{r}}{4\pi\gamma r^3} \tag{4.1}$$

$$\text{Coulomb's law} \qquad \mathbf{F} = \frac{q_1 q_2 \mathbf{r}}{4\pi\varepsilon_0 r^3}. \tag{4.2}$$

Both formulae give vector form to the laws; the coefficient $1/(4\pi\gamma)$ is written instead of the conventional Newtonian constant of gravitation G, 298 in line with the SI formulation of Coulomb's law. The constants γ and ε_0 are to be found experimentally, and their values depend on the choice of other units. This is a matter we return to in Chapter 14. For the moment, we are mainly concerned with the form of the equations, not numerical magnitudes, though it is interesting to note in passing the relative sizes of the electrostatic and gravitational forces between two elementary particles, for example, electrons. The ratio of these two forces is clearly independent of the distance between the particles:

$$\frac{F_q}{F_m} = \frac{\gamma}{\varepsilon_0}\left(\frac{q}{m}\right)^2.$$

In the SI system of units,

$$\gamma = 1{\cdot}2 \times 10^9 \text{ kg s}^2 \text{ m}^{-3}, \quad \varepsilon_0 = 8{\cdot}9 \times 10^{-12} \text{ C}^2 \text{ kg}^{-1} \text{ s}^2 \text{ m}^{-3},$$

and for an electron $q = 1{\cdot}6 \times 10^{-19}$ C, $m = 9 \times 10^{-31}$ kg. It is now seen that the electric force between electrons is 4×10^{42} times stronger than the gravitational force.

If it were not for the fact that charges of both sign exist, but all masses attract each other, gravitation would always be completely masked by electric forces. On the atomic and molecular scale, this is true, but gravitation begins to be potentially significant in a body containing 10^{18} particles and of mass about 1 μg; one electron in excess of the number of protons in two such bodies leads them to interact equally strongly by Coulomb and gravitational forces. With bodies of planetary size, the excess charge needed to give a Coulomb force comparable with the gravitational force would speedily be neutralized by currents flowing through the rarefied ionized interplanetary gas. The Coulomb force is dominant on the atomic scale, gravitation on the cosmic scale, and Man stands between at a point where both are significant.

The influence of an assembly of charges (or masses) on another test charge moved among them may be plotted out in detail as a vector $\mathbf{E}(\mathbf{r})$ which varies with position and determines the force acting on unit test charge placed at $\mathbf{r}$, on the assumption that the presence of the test charge does not displace any members of the assembly. The vector function $\mathbf{E}(\mathbf{r})$ is the *field of force*, or electric field, and $\mathbf{E}q$ is the force experienced by a charge q in the field $\mathbf{E}$.

PROBLEM

Show that if $\mathbf{E}(\mathbf{r})$ is constant (*uniform field*), a dipole consisting of charges $\pm q$ separated by a distance $\mathbf{l}$ experiences no resultant force but a torque $\mathbf{p} \wedge \mathbf{E}$, where $\mathbf{p}$ is the dipole moment $q\mathbf{l}$.

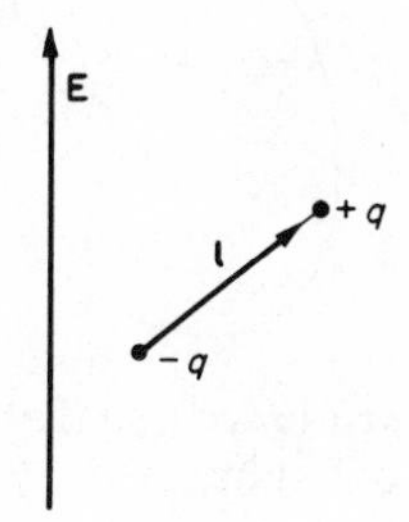

According to the superposition law $\mathbf{E}(\mathbf{r})$ is made up of contributions from every charge in the vicinity. Thus a charge q_i at $\mathbf{r}_i$ exerts, according to (4.2), a force $q_i(\mathbf{r} - \mathbf{r}_i)/[4\pi\varepsilon_0|\mathbf{r} - \mathbf{r}_i|^3]$ on unit charge at $\mathbf{r}$, and $\mathbf{E}(\mathbf{r})$ is the vector sum of all such terms. To form a vector sum of many terms is usually an awkward affair, and often the field is better expressed by means of its potential. The potential energy of unit charge in the conservative field of a charge q at a distance R is $q/(4\pi\varepsilon_0 R)$, so that the potential function defining $\mathbf{E}(\mathbf{r})$ is the scalar sum of such terms

$$\phi(\mathbf{r}) = \sum_i \frac{q_i}{4\pi\varepsilon_0|\mathbf{r} - \mathbf{r}_i|}; \quad -\text{grad}\,\phi = \mathbf{E}(\mathbf{r}). \tag{4.3}$$

The potential, as a scalar function of position, may be represented by means of equipotential surfaces. Around a single charge q, the equipotentials are spheres, and the electric field, being determined by $-\text{grad}\,\phi$, is directed normal to the equipotentials, i.e., radially. If we calculate or measure the equipotentials for any system of charges and then draw their orthogonal trajectories, these *field lines* (or *lines of force*), being by construction everywhere normal to the equipotentials, point always in the local direction of $\mathbf{E}$. As such, they are useful in indicating the general pattern of the field, but it happens that for inverse-square laws of force they do more than this, and by their closeness indicate also the strength of the field. To show this, we shall first prove Gauss' theorem, a general result for inverse square fields which is of inestimable value in discussing and solving special problems.

Gauss' theorem

The total flux of electric field outwards through a closed surface is equal to $1/\varepsilon_0$ times the charge enclosed.

The flux of a vector $\mathbf{A}$ through an element of area $\mathrm{d}S$ is defined as the product of A and the projected area of $\mathrm{d}S$ on a plane normal to $\mathbf{A}$; if $\mathrm{d}\mathbf{S}$ is a vector drawn normal to the surface element, the flux is $\mathbf{A}.\mathrm{d}\mathbf{S}$. Gauss' theorem states that if S is a closed surface, $\int_S \mathbf{E}.\mathrm{d}\mathbf{S} = \sum q/\varepsilon_0$, if $\sum q$ is the total charge enclosed in S.

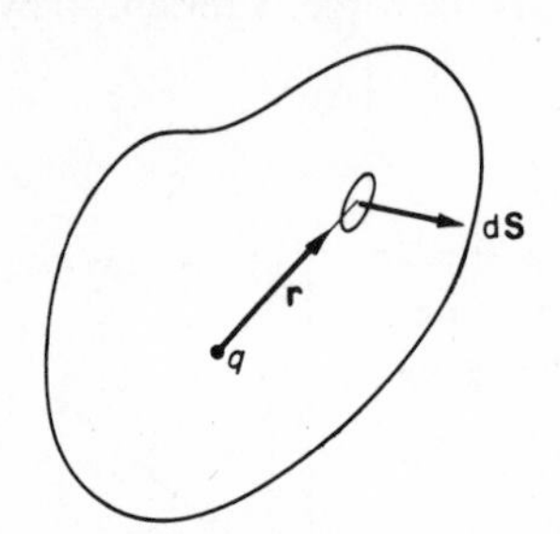

The proof is simple. Consider a charge q and an element $\mathrm{d}\mathbf{S}$ of the surrounding surface, represented by an outward-pointing vector. The field strength at $\mathrm{d}\mathbf{S}$ due to q is $q\mathbf{r}/(4\pi\varepsilon_0 r^3)$, so that $\mathbf{E}.\mathrm{d}\mathbf{S} = q\mathbf{r}.\mathrm{d}\mathbf{S}/(4\pi\varepsilon_0 r^3)$. Now $\mathbf{r}.\mathrm{d}\mathbf{S}/r^3$ is the solid angle $\mathrm{d}\Omega$ subtended by $\mathrm{d}\mathbf{S}$ at the position occupied by the charge, so that $\mathbf{E}.\mathrm{d}\mathbf{S} = q\,\mathrm{d}\Omega/(4\pi\varepsilon_0)$. When the integral is taken over the whole surface, $\int_S \mathbf{E}.\mathrm{d}\mathbf{S} = (q/4\pi\varepsilon_0)\int_S \mathrm{d}\Omega = q/\varepsilon_0$, since the solid angle subtended at any interior point by a closed surface is 4π. The contribution of q to the total flux is independent of its position within the surface and therefore all interior charges may be summed to give the result required.

Gauss' theorem serves to demonstrate the useful properties of field lines. These were drawn as orthogonal trajectories of the equipotentials, and we may use them to construct an element of volume bounded at the sides by field lines and at the ends by equipotentials. If there are no charges inside the element, the electric field within is everywhere smooth and finite and, since field lines cannot cross, any entering one end must leave at the other end. Now according to Gauss' theorem, if there is no charge within, the same flux enters and leaves the two ends; therefore constancy of flux and constancy of the number of field lines go hand in hand through this element of a *tube of force*. We may extend this argument from element to element to show how in a charge-free region the field lines in the neighbourhood of a given line have a density (number per unit normal area) that everywhere matches the strength of the field. This permits us to fill the whole of a charge-free space with lines whose direction at any point gives the direction of **E** and whose density gives the strength of **E**; e.g., we may choose a constant C arbitrarily so that the number of lines crossing unit normal area is everywhere CE. So far we have treated only a charge-free space, where the lines are continuous, but the argument is readily extended to include charges, which serve as sources of new field lines. If the number of lines leaving any closed surface S exceeds by N the number entering, the total flux of **E** through the surface is N/C, and this must be equated to q/ε_0, by Gauss' theorem. Hence $N = Cq/\varepsilon_0$, and we see that every charge q generates Cq/ε_0 new lines.

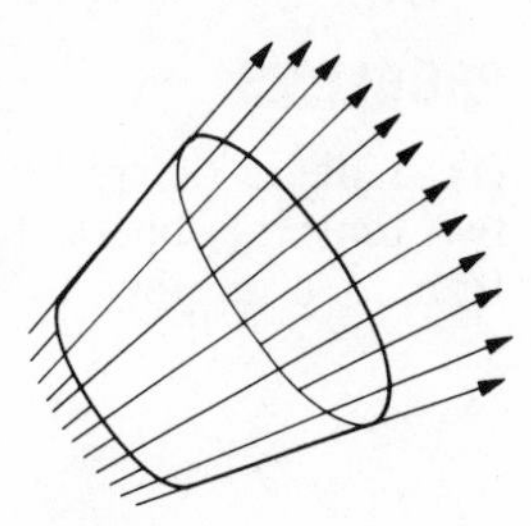

It should be noted that the field lines, drawn as orthogonal trajectories of the equipotentials, only describe the strength of the field if the inverse-square law is obeyed. For other laws, it would be necessary to start and stop lines continuously to keep their density proportional to field strength, and the representation would be far less valuable.

We shall make considerable use of the concept of field lines in later chapters, but for the present we concentrate on certain corollaries of Gauss' theorem which are immediately useful.

(1) There is no field inside a uniform spherical shell of charge; for any such field must by symmetry be radial and isotropic, and in particular constant at all points on a concentric inner sphere. It must therefore vanish since no charge is enclosed in the sphere.

(2) The field outside a uniform spherical shell of charge is the same as if all the charge were concentrated at the centre; for the shell and the concentrated charge must both, by symmetry, give radial isotropic fields, and by Gauss' theorem the fields must have equal magnitude.

(3) Two spherically symmetric (non-intersecting) charge distributions, A and B, act upon one another as if their charges were concentrated at their centres. For the field produced by A in the vicinity of B is the same as if A were concentrated to a point charge, and therefore the force on B is

unchanged by concentrating A. Since the forces obey N3, B must then act on A with the same force whether or not A is concentrated. By the same argument, B may now be concentrated to a point at its centre without affecting its influence on A, and the required result is proved.

PROBLEMS

(1) A dielectric sphere when uniformly polarized may be thought of as two uniform spherical charge densities, $\pm\rho$ per unit volume, displaced from one another by a small distance $\mathbf{a}$. By considering the vectors

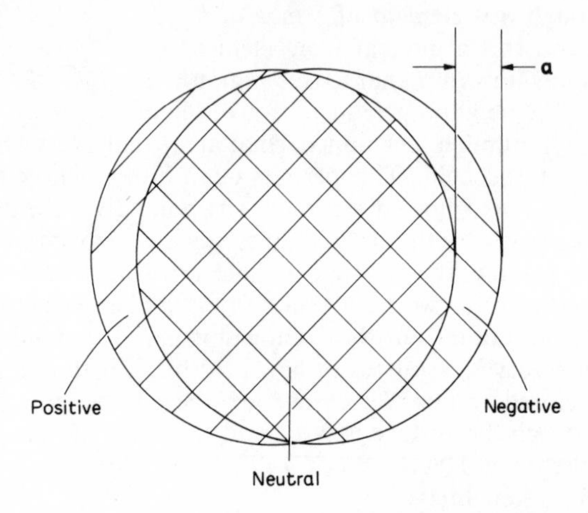

191

representing at any internal point the fields due to these two charge distributions, show that their resultant is constant throughout the sphere, and has magnitude $-\mathbf{P}/(3\varepsilon_0)$ where $\mathbf{P}$ is $\rho\mathbf{a}$, the dipole moment per unit volume.

(2) A spherical cavity is cut in an otherwise uniform sphere. By treating the cavity as an additional sphere of negative mass, show that the gravitational field inside the cavity is uniform and of a magnitude and direction determined by the line joining the centres of sphere and cavity.

The fact that we can regard charged or massive spheres as concentrated into points, whilst they still interact according to inverse-square-law forces greatly simplifies analysis of the motion of planets and other celestial bodies. To a good approximation the problem is reduced to one of particle dynamics involving a small number of interacting particles. We shall turn to this in the next chapter. Meanwhile we examine two important experiments designed to test with very high sensitivity two basic points in these laws. The first is Cavendish's test of the inverse-square law of electrostatics, which was repeated 100 years later by Maxwell, and more recently with far greater sensitivity by Plimpton and

Lawton. The second is the test of the equivalence of inertial and gravitational mass by a method due to Eötvös but considerably refined by Roll, Krotkov and Dicke.

The Cavendish experiment is designed to discover if there is any field within a uniform sphere of charge. If the inverse-square law holds, we have seen that there is none, but Gauss' theorem does not apply to any other law of force and in general a spherical distribution must be expected to generate an internal field. Let us work this out for the case of a nearly inverse-square law, governed by a potential varying as $1/r^{1+h}$ rather than $1/r$, with $h \ll 1$. To be precise, let the potential due to a charge q be Aq/r^{1+h}. Then if we set up a spherical shell of charge with surface density σ per unit area, we can evaluate the potential at B by direct integration. That part of the shell lying between the planes x and $x + \mathrm{d}x$ has area $2\pi a\,\mathrm{d}x$ and carries a charge $2\pi a\sigma\,\mathrm{d}x$, so that its contribution to the potential at B is given by

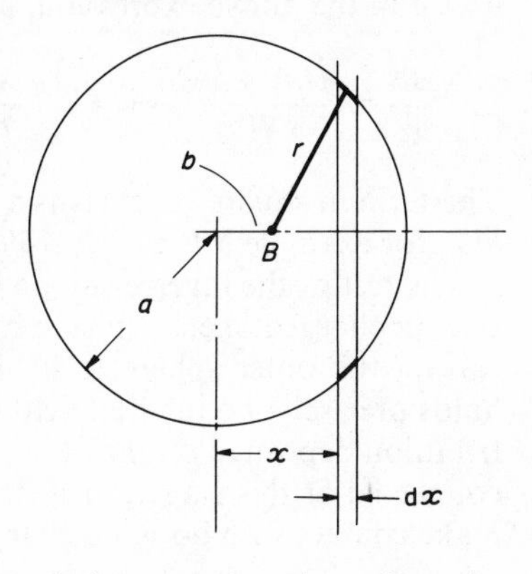

$$\mathrm{d}\phi_b = \frac{2\pi A a\sigma\,\mathrm{d}x}{r^{1+h}} = \frac{2\pi A a\sigma\,\mathrm{d}x}{(a^2 + b^2 - 2bx)^{1/2(1+h)}}.$$

Integrating over the whole sphere, we have

$$\phi_b = 2\pi A a\sigma \int_{-a}^{a} \frac{\mathrm{d}x}{(a^2 + b^2 - 2bx)^{1/2(1+h)}}$$

$$= \frac{2\pi A a\sigma}{b(1-h)}\,[(a+b)^{1-h} - (a-b)^{1-h}].$$

Since h is very small, we may write

$$\begin{aligned}(a+b)^{1-h} &= (a+b)(a+b)^{-h}\\ &= (a+b)\exp\,[-h\ln(a+b)]\\ &= (a+b)[1 - h\ln(a+b) + \cdots],\end{aligned}$$

and

$$(a-b)^{1-h} = (a-b)[1 - h\ln(a-b) + \cdots].$$

Therefore

$$\phi_b \approx K\left\{1 - \tfrac{1}{2}h\left[\frac{a+b}{b}\ln(a+b) - \frac{a-b}{b}\ln(a-b)\right]\right\}$$

where

$$K = 4\pi A a\sigma/(1-h).$$

Clearly if $h = 0$, $\phi_b = K$ and does not vary inside the sphere, in accordance with Gauss' theorem. But if $h \neq 0$ there is a radial variation of ϕ inside. It is convenient to compare the potential difference between the shell and an interior point B with the potential difference between the shell and infinity ($\phi = 0$). To find the potential of the shell we put $b = a$ in the above expression, and then we have that

$$\frac{\Delta\phi}{\phi} \equiv \frac{\phi(a) - \phi(b)}{\phi(a)} = \tfrac{1}{2}h\left[\frac{a+b}{b}\ln\left(\frac{a+b}{2a}\right) - \frac{a-b}{b}\ln\left(\frac{a-b}{2a}\right)\right].$$

The term in square brackets is a numerical multiplier depending only on b/a; for example if $b/a = \frac{4}{5}$, $\Delta\phi/\phi = 0{\cdot}17h$.

The test of the inverse-square law based on this result involves placing one uncharged metal sphere concentrically inside another, and then raising the outer sphere to a high potential. If the inverse-square law holds precisely, no internal field is produced by the spherical charge distribution deposited on the outer sphere. If, therefore, the two spheres are connected at this stage, for instance by dropping a metal chain so as to make contact with both, no charge will flow; and if the connection is then removed and the outer sphere taken apart so as to leave the inner sphere standing alone, any charge it possesses can be revealed by connecting an electroscope to it, to observe its potential. This in outline is what Cavendish did, and Maxwell after him, to show by the absence of a discernible charge on the inner sphere that h was less than 10^{-4}. In 1936, Plimpton and Lawton achieved much higher sensitivity by actually incorporating an amplifier inside the inner sphere, to detect any potential variations between the two spheres. The schematic diagram shows the amplifier A connected across the spheres, with its output connected to a galvanometer G having a natural period of $\frac{1}{2}$ second. The outer sphere is connected to a 3000 V a.c. generator of 2 Hz (cycle/second) frequency, and any signal picked up by the amplifier will set the galvanometer swinging at its resonant frequency. The galvanometer is observed through a window consisting of a bath of salt solution which conducts well enough to complete the conducting outer sphere.

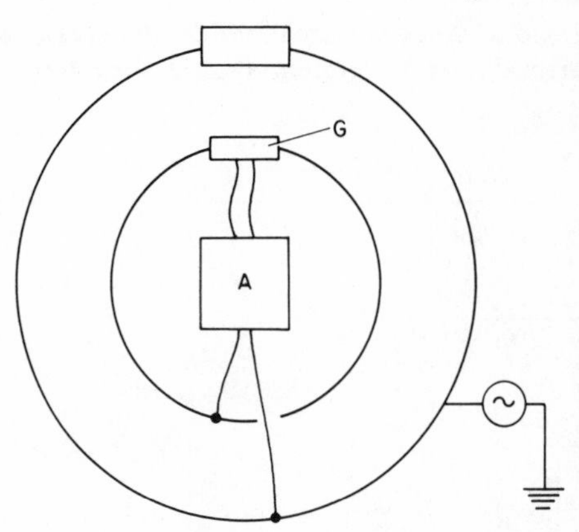

256

It is worth remarking, though the relevant theory will not be discussed until Chapter 12, that the resistance of the input stage of the amplifier must be very high. If a potential difference is set up between the two spheres, and they are then connected by a resistance R, the time required for sufficient charge to flow to eliminate the potential difference is a few times CR, if C is the capacitance between the spheres. R must be large enough for this characteristic time to be rather greater than $\frac{1}{2}$ second, so that the amplifier does not by its own conductivity eliminate the effect it is intended to detect. The spheres used by Plimpton and Lawton had radii of about 0·75 and 0·60 m and the capacitance between them was about 300 pF, while the input resistance of the amplifier was 10^{10} Ω; thus CR was about 3 s and the criterion was well satisfied.

The result of the experiment was null, within the limits of sensitivity of the detector, i.e., 3000 V variations of ϕ produced $\Delta\phi$ less than 10^{-6} V. This indicates that h is less than 2×10^{-9}, so that the inverse-square law may be taken to be substantiated to a very high degree of accuracy.

At this point the critical reader may well ask how it is possible to produce a sphere so perfect that one is justified in carrying out an experiment designed to detect a discrepancy of only one part in 10^9. Surely even a tiny bump will throw the whole experiment into confusion? The answer is simple—only the theory, not the experiment, is thrown into confusion. As will be shown in Chapter 8, if the inverse-square law is obeyed there is no field inside a conducting shell of any shape, and therefore dents and bumps cannot turn a null result, arising from an inverse-square law, into a positive result. The reason for using a sphere is that it allows one to calculate, as we have done, what effect would be observed if the inverse-square law were not true. If, therefore, Plimpton and Lawton used a badly dented sphere we should not be so sure that h was less than 2×10^{-9}—it might be as much as twice this value!

149

The experiment of Roll, Krotkov and Dicke on the equivalence of gravitational and inertial mass was designed to test more sensitively what can be inferred from the fact that all bodies fall with the same acceleration, if air resistance is negligible. A better simple verification is provided by experiments with simple pendulums whose bobs are made of different materials. Let us fix our ideas by working out the theory of a simple pendulum without assuming the equivalence of gravitational and inertial mass. In other words, we take the bob to have inertial mass m_{I} and to be attracted towards the earth with a force αm_{G} where α is a constant and m_{G} the gravitational mass. Then the equation of motion of the bob takes the form $m_{\mathrm{I}} l \ddot{\phi} = -\alpha m_{\mathrm{G}} \sin\phi \approx \alpha m_{\mathrm{G}} \phi$ if $\phi \ll 1$. Consequently the frequency ω is $(l/\alpha)^{1/2}(m_{\mathrm{I}}/m_{\mathrm{G}})^{1/2}$. With a simple pendulum it is not too difficult to determine l and ω within 1 part in 10^4 and hence to verify to this degree that $m_{\mathrm{I}}/m_{\mathrm{G}}$ is constant. The ingenious experiment of Roll, Krotkov and Dicke provides a far more sensitive test by a quite different method, making use of the Earth's rotation round the Sun as well as on its own axis.

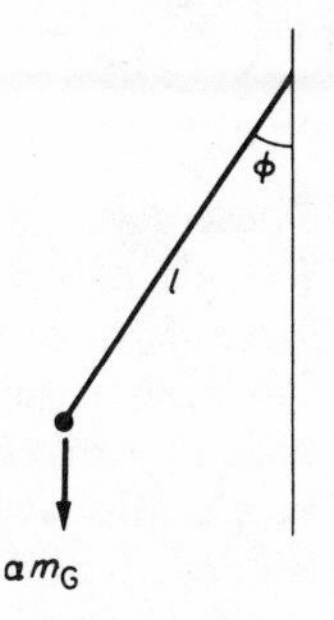

It is convenient to imagine the experiment carried out at the North Pole, and to forget that the Earth's axis is not exactly normal to the plane of its orbit. Let us introduce the gravitational mass defined in accordance with (4.1), and let the Sun's gravitational mass be $m_G^{(S)}$ and the Earth's $m_G^{(E)}$, while their inertial masses will be represented by subscripts I. Then the motion of the Earth in its solar orbit, supposed circular and of radius R, with angular frequency Ω, involves the gravitational attraction $m_G^{(S)}m_G^{(E)}/(4\pi\gamma R^2)$ producing acceleration $R\Omega^2$; i.e.,

$$\frac{m_G^{(S)}m_G^{(E)}}{4\pi\gamma R^2} = m_I^{(E)}R\Omega^2,$$

or

$$\frac{m_G^{(E)}}{m_I^{(E)}} = \frac{4\pi\gamma R^3\Omega^2}{m_G^{(S)}}.$$

Any body resting on a smooth table at the Pole will stay at rest if it has the same ratio m_G/m_I as the Earth, since it too will find the Sun's attraction just right to keep it in orbit with the Earth. But a body with any other ratio will tend to slide, and can only be kept at rest by application of an extra force. Let us then take two bodies having the same gravitational mass m_G, but with different inertial masses $\beta_1 m_G$ and $\beta_2 m_G$, β_1 and β_2 being of course very nearly the same as the value β for the Earth as a whole; and let us hang them at opposite ends of a bar supported by a torsion wire. This is a very sensitive way of detecting small forces. When the system is in equilibrium, each, as a result of the Earth's orbit, has acceleration $R\Omega^2$. Therefore body 1 is subjected to a force $\beta_1 m_G R\Omega^2$ of which $\beta m_G R\Omega^2$ is provided by the Sun's attraction and $(\beta_1 - \beta)m_G R\Omega^2$ must be provided by the bar. Similarly body 2 experiences a force $(\beta_2 - \beta)m_G R\Omega^2$ from the bar.

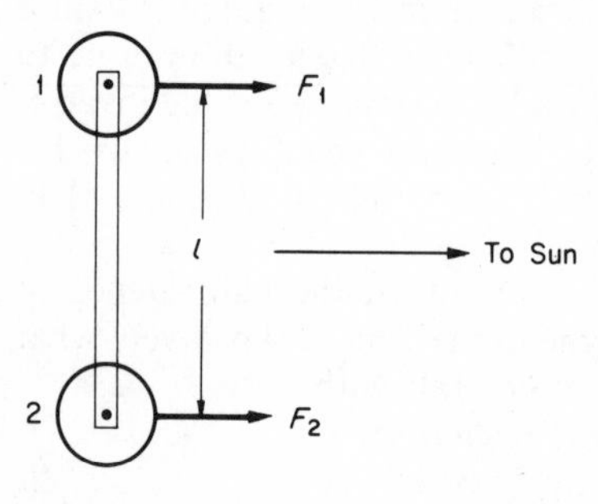

These two forces are shown as F_1 and F_2, and can be rewritten as $\bar{F} + \Delta F$ and $\bar{F} - \Delta F$, where $\bar{F} = \frac{1}{2}(F_1 + F_2)$ and $\Delta F = \frac{1}{2}(F_1 - F_2)$. The force $\bar{F}$ is provided by the tilting of the whole apparatus, while the force ΔF is provided by the twisting of the suspension. Remembering that F_1 and F_2 are the extra forces needed to provide the centripetal acceleration, we can see that if $F_1 > F_2$ in the diagram, a clockwise torque is needed, and this is provided by twisting the arm in an anticlockwise sense. If we write μ for the torsion constant of the suspension, i.e., the torque needed to twist it through one radian, the actual angle of twist, ϕ, needed to balance the variations of β is given by

$$\mu\phi = l\Delta F = \tfrac{1}{2}l(\beta_1 - \beta_2)m_G R\Omega^2.$$

Now it is clearly out of the question simply to set up the torsion arm and observe directly the twist ϕ, since we have no zero position to refer it to. Here the Earth's rotation on its own axis comes in. After 12 hours the situation in the diagram is replaced by one in which body 2 is above body 1, and the torque needed for equilibrium is reversed. If therefore we set up the experiment and watch it over a long period, we may expect any effect due to $(\beta_1 - \beta_2)$ to show up as a to-and-fro oscillation of the torsion arm with a period of 24 hours. The magnitude of the swing determines $(\beta_1 - \beta_2)$, and if no swing is observed an upper limit can be set to this quantity.

Putting numbers into the formula, we note that in all practical systems of units the value of β is unity, and for m_G we can insert simply the mass of bodies, about 0·03 kg in the actual experiment; the torsion suspension, of fused silica, had a torsion constant μ of $2{\cdot}4 \times 10^{-8}$ Nm; l was about 40 mm; R is $1{\cdot}5 \times 10^8$ km and Ω is $2{\cdot}0 \times 10^{-7}$ s^{-1}. Then we find

$$\beta_1 - \beta_2 = 0{\cdot}006\,\phi.$$

It is of interest to see what would be obtained if the two bodies differed by enough to give a noticeable effect with a simple pendulum, e.g., $\beta_1 - \beta_2 = 10^{-4}$. The corresponding oscillation here would have $\phi = \frac{1}{60}$, or 1 degree of arc, which is very obvious with crude lamp and scale methods. For instance, with a mirror on the torsion suspension and a lamp and scale at 1 m the deflection would be 33 mm. In these experiments, Roll, Krotkov and Dicke paid extraordinary attention to the detection of minute angular displacements, and their final result shows that the sensitivity of their detector was something like 10^{-9} radians, equivalent to 20 Å (2×10^{-9} m) movement in a conventional lamp and scale arrangement. Their final conclusion was that the two materials used, gold and aluminium, had β differing from one another by less than 3×10^{-11}. It is an additional cause for admiration of this beautiful and delicate experiment that the gravitational effect used to reveal any difference in β was not the everyday attraction of the Earth but the 1600 times smaller attraction of the Sun.

Naturally, to achieve this high precision, great care had to be taken, and the original paper is well worth studying to see what is involved in a really sensitive measurement. One particularly interesting point is that the effect sought was an oscillation with a period of 24 hours, and this is the period of a great many possible disturbances. For example, any magnetic moment due to minute traces of iron as impurities in the suspended system experiences a torque from the Earth's magnetic field, which itself is subject to a 24-hour oscillation caused by currents in the ionosphere, which are dependent on the degree of ionization produced by solar radiation. The choice of gold and aluminium for the bodies was partly conditioned by the possibility of freeing them of magnetic impurities. Again, any gradient in the gravitational field may result in a different horizontal force on the two ends of the suspension and thus in a torque. Such gravitational gradients are caused by massive objects in the vicinity, which themselves have a tendency to vary with a 24-hour period if they are the inhabitants of nearby laboratories or rain and snow lying on and in the ground. As the diagram of the suspension shows, the whole arrangement was given triangular symmetry, with two masses of aluminium and one of gold, because a triangular body is not affected by gravitational gradients. These are outstanding examples of troublesome

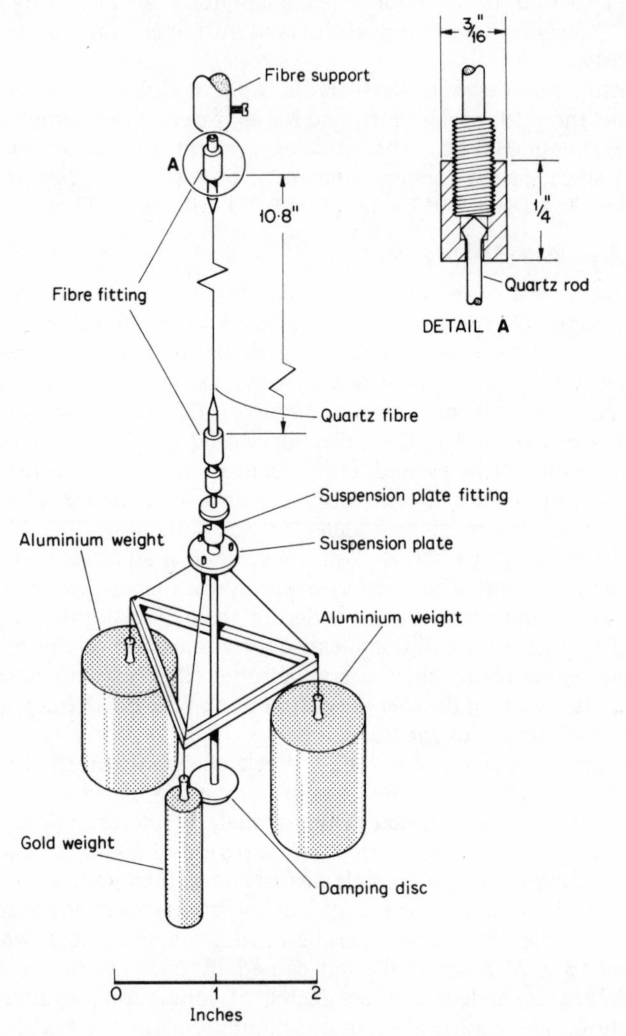

interference that must be eliminated, but there were many finer points to bear in mind.

Finally, a word about the choice of gold and aluminium is in order, though it is far outside the scope of this book. It was obviously desirable to choose materials that were as different as possible in constitution, so as to have the best chance of revealing variations in β. With what is now known about the nuclear constitution, a light and a heavy element were to be preferred. Aluminium has a nucleus consisting of 13 protons and 14 neutrons, and there are 13 electrons in the extra-nuclear structure. Gold has 79 protons and 118 neutrons in its nucleus, and 79 extra-nuclear electrons. Thus the proportion of neutrons to protons is different in the two (1·08 in Al, 1·50 in Au). This is not the whole difference, however; for if we add up the masses of all the particles in an aluminium atom and compare it with the actual mass of an atom (the masses have been measured with extraordinary accuracy by means of mass-spectrometers and by other experiments), we find that the atom is 0·895% underweight. The reason is that if we were to bring the particles together to build up an atom, they would attract one another and come together with very great kinetic energy, which would have to be removed to make the atom stable. By Einstein's famous mass-energy relationship the removal of energy E reduces the mass by E/c^2. Therefore there is a negative contribution to the mass of the atom and this contribution has its origin in loss of energy. In gold, the difference is slightly less: 0·850%. The difference between 0·895 and 0·850% may not seem much, but it is vast in the context of an experiment that is sensitive to 2×10^{-11}. The conclusion to be drawn is that neutrons, protons, electrons and energy all have the same ratio of gravitational and inertial mass. One can go even further in the light of modern theories of the part played by ephemeral particles (positrons, mesons) in the structure of the nucleus, and conclude that all forms of mass are to a high degree of probability precisely equivalent in their gravitational and inertial roles. One may therefore with confidence adopt for all normal purposes the convention that goes back to Newton of representing both roles by a single mass.

Cavendish Problems: 104–106.

READING LIST

Cavendish Experiment: S. J. PLIMPTON and W. E. LAWTON, *Physical Review*, **50**, 1066 (1936).

Eötvös Experiment: P. G. ROLL, R. KROTKOV and R. H. DICKE, *Annals of Physics*, **26**, 442 (1964).

5

Dynamics of particles

Central orbits

The use of Newton's laws to study the motion of a particle in a given field of force is essentially a mathematical rather than a physical question. If the field of force is defined as a function of position by the specification of $\mathbf{F}(\mathbf{r})$, the problem is to solve the equation $m\ddot{\mathbf{r}} = \mathbf{F}(\mathbf{r})$. With modern computers, it is easy to trace out the trajectory of a particle with any desired precision, and if the need is simply to know the trajectory, given the velocity of the particle at the starting point, there is little point usually in doing anything except feed the information to a computer and await the answer. However, every initial velocity results in a different trajectory, and computing many trajectories may be an uneconomical way of finding out whether the trajectories are governed by any general rules. This is not to suggest that computers should not be used in some circumstances; indeed, a few experimental trajectories traced out with the help of a computer may be the way of guessing what general rules may exist. But ultimately a certain amount of mathematical analysis is the surest way of revealing the regularities and necessary properties of a family of trajectories. We shall study here the analysis of a particular type of tra-
236 jectory, that executed by a particle in a central field of force $\mathbf{F}(\mathbf{r}) = \mathbf{r}f(r)$,

since this has certain simple rules to obey, and moreover there are particular laws of force, such as the harmonic force, $f(r)$ = a negative constant, and the inverse-square law, $f(r) \propto r^{-3}$ (note the presence of r in the definition of the force law; in these two cases F varies as r and as r^{-2} respectively) which occur in important contexts. Newton's universal law of gravitation is an inverse-square law, and it is the solution of the orbits of the planets and other celestial bodies, and observational verification of these solutions, that provides the best critical test of classical mechanics.

A particle subject to a central force executes a trajectory that lies in a plane. For if we construct at any instant the plane containing **r** and **v**, the particle is moving in this plane and the acceleration is also in the plane. Thus **v** changes but remains in the plane, and so does the particle. We need consider only the equation of motion in the plane, and if we describe the position of the particle by polar coordinates ρ and θ, the kinematic results of pp. 14 and 15 apply. The tangential component of acceleration is zero, since the force is purely radial, and therefore the radius vector sweeps out area at a constant rate and the angular momentum $L(=m\rho^2\dot{\theta})$ is constant.

As for the radial component of acceleration, we equate this to F/m where $F(\rho)$ is the central force, i.e.,

$$\ddot{\rho} - \rho\dot{\theta}^2 = F/m,$$

or

$$m\ddot{\rho} = F + L^2/(m\rho^3). \tag{5.1}$$

To determine the form of the orbit, we manipulate this equation with the help of the constancy of L, and it is convenient to define u as $1/\rho$, so that $\mathrm{d}\rho/\mathrm{d}u = -\rho^2$. Then, by the rule for evaluating products of differential coefficients,

$$\begin{aligned} \dot{\rho} &= \frac{\mathrm{d}\rho}{\mathrm{d}u}\frac{\mathrm{d}u}{\mathrm{d}\theta}\frac{\mathrm{d}\theta}{\mathrm{d}t} \\ &= -\rho^2\frac{\mathrm{d}u}{\mathrm{d}\theta}\frac{\mathrm{d}\theta}{\mathrm{d}t} \\ &= -\frac{L}{m}\frac{\mathrm{d}u}{\mathrm{d}\theta}. \end{aligned} \tag{5.2}$$

Therefore,

$$\begin{aligned} \ddot{\rho} &= -\frac{L}{m}\frac{\mathrm{d}^2u}{\mathrm{d}\theta^2}\frac{\mathrm{d}\theta}{\mathrm{d}t} \\ &= -\frac{L^2u^2}{m^2}\frac{\mathrm{d}^2u}{\mathrm{d}\theta^2}. \end{aligned} \tag{5.3}$$

Substituting this in (5.1), we see that

$$\frac{d^2u}{d\theta^2} = g(u), \tag{5.4}$$

where $g(u) = -u - (mF/L^2u^2)$.

This equation tells us that if we project a particle of known mass at such a velocity as to give it a certain angular momentum L, then given the law of force $F(\rho)$, the second derivative $d^2u/d\theta^2$ is everywhere determined. As we shall see presently, this is enough, together with the initial conditions of position and velocity, to define the variation of u with θ uniquely, and thus the orbit is determined in principle.

An alternative derivation of the equation of motion starts from the conservation laws. We know that central forces are derivable from a potential $\phi(\rho)$ such that $-d\phi/d\rho = F$, and that the sum of ϕ and the kinetic energy is invariant; moreover for central forces the angular momentum $L = m\rho^2\dot{\theta}$ is also invariant. If we write the kinetic energy in terms of the instantaneous tangential and radial components of the velocity,

$$\begin{aligned} T &= \tfrac{1}{2}m[(\rho\dot{\theta})^2 + (\dot{\rho})^2] \\ &= \tfrac{1}{2}m[L^2/(m\rho)^2 + (\dot{\rho})^2] \\ &= L^2[u^2 + (du/d\theta)^2]/(2m). \end{aligned} \tag{5.5}$$

Then the total energy E, which is a constant, is $\phi + T$, so that

$$\left(\frac{du}{d\theta}\right)^2 = \frac{2m}{L^2}(E - \phi) - u^2. \tag{5.6}$$

By differentiation (5.6) will be found to be the first integral of the second-order equation (5.4).

The actual solution of (5.4) or (5.6) for any form of $F(u)$ is a purely routine matter for a computer, if an analytical solution does not spring to mind. It is of interest to demonstrate the routine of solving a differential equation of the form (5.4), which is about the simplest type of equation to solve numerically. We consider the value of u at equal intervals of θ, which can be as close together as we like; the closer they are the more work to run off a solution but the more accurate the solution. If the spacing of successive values of θ is a, and the values of u at three successive points are u_{n-1}, u_n and u_{n+1}, we know that for small enough a the quotient $(u_n - u_{n-1})/a$ is a good approximation to the gradient $du/d\theta$ halfway between these points, at $(n - \frac{1}{2})a$; similarly $(u_{n+1} - u_n)/a$ is approximately the

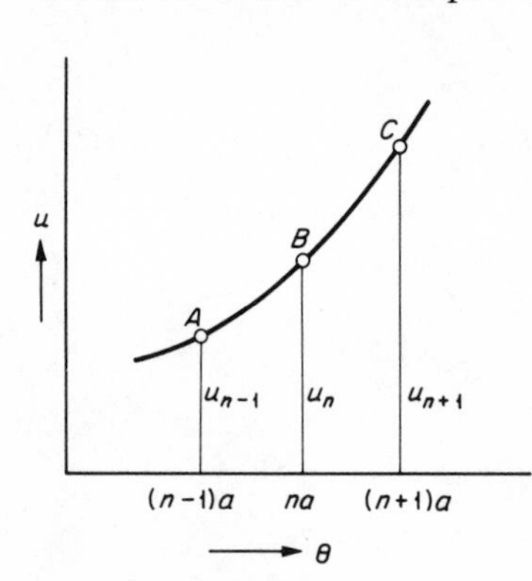

gradient at $(n + \frac{1}{2})a$. Then the second derivative, $\mathrm{d}^2u/\mathrm{d}\theta^2$, at na is approximately the mean rate of change of $\mathrm{d}u/\mathrm{d}\theta$ between $(n - \frac{1}{2})a$ and $(n + \frac{1}{2})a$, i.e.,

$$(\mathrm{d}^2u/\mathrm{d}\theta^2)_{na} \approx (u_{n+1} - 2u_n + u_{n-1})/a^2. \tag{5.7}$$

If this explanation seems unconvincing, the reader may care to fit a parabola to the three points A, B and C and verify by differentiation that the same result is obtained. The smaller a is taken to be, the more exact is the fit of the parabola to the real curve.

We now take (5.7) as exactly true, and substitute in (5.4) to give

$$u_{n+1} = a^2g(u_n) + 2u_n - u_{n-1}. \tag{5.8}$$

This is just the rule for finding u_{n+1} when we know the previous two ordinates u_n and u_{n-1}, and the determination of all the u's is a routine process. In order to start the process, however, we must choose the first two ordinates u_1 and u_2; this is equivalent to defining the direction of motion of the particle at one point in its orbit, and, given L, it fixes the velocity vector at one point and uniquely defines the particular orbit being computed.

EXAMPLE

We compute the solution for a case where the analytical solution is known, the harmonic force $F \propto -\rho \propto -u^{-1}$. For simplicity we choose the constants so that $g(u) = -u + 1/u^3$, and compute the orbit that starts with two successive values of u equal to 2; this is equivalent to injecting the particle tangentially into its orbit. We shall choose for a the rather large value

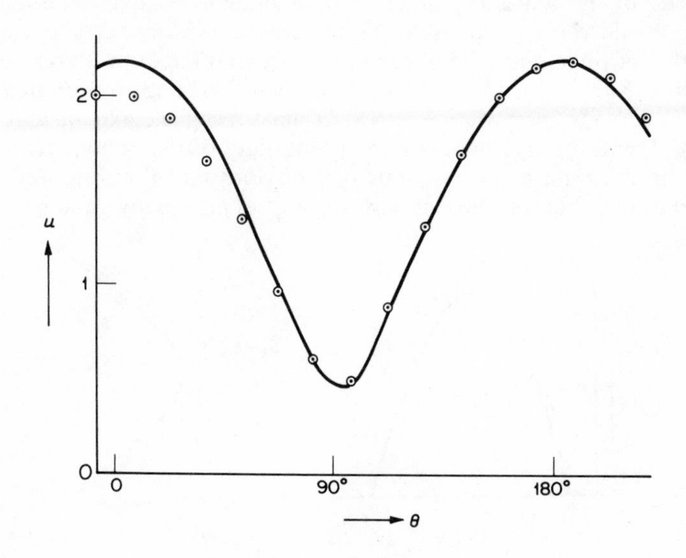

of $\frac{1}{4}$, i.e., nearly 15° intervals of θ. Then the tabulated solution starts as follows:

u	$a^2g(u)$
2	
2	−0·1172
1·8828	−0·1083
1·6573	−0·0899
1·3419	
etc.	

At each stage, the value of $a^2g(u)$ is computed from u, and is added to twice the value of u minus the preceding value of u. Thus the next step in the calculation is to compute $1/16[-1{\cdot}3419 + 1/(1{\cdot}3419)^3]$, which is $-0{\cdot}0580$, and to form $-0{\cdot}0580 + 2 \times 1{\cdot}3419 - 1{\cdot}6573$, i.e., $0{\cdot}9685$, which is the next value of u. The values obtained at 15° intervals are shown in the graph, together with the exact solution which, as may be verified by substitution in the differential equation, is the elliptical orbit $u^2 = A^2 \cos^2 \theta + 1/A^2 \sin^2 \theta$, with A chosen as 2.2 to fit the computed curve reasonably well.

The errors produced by choosing a rather large value of a show clearly as a lack of exact periodicity, the second maximum of u being distinctly higher than the first. We need not, however, take the computation further, since the purpose of the example was to illustrate the simple nature of the process, which is not made any more difficult when $g(u)$ has a form that allows no analytical solution in terms of known functions.

Recognizing then, that (5.4) provides a complete solution of the central orbit problem, and requires no more analysis, but only routine computation, to translate it into numbers, we proceed to examine the general character of the solutions, which may be bounded or unbounded. The reader should verify for himself that the same conclusions can be reached from the starting point of (5.6) rather than (5.4). If the force is everywhere repulsive (F positive), $g(u)$ and therefore $\mathrm{d}^2u/\mathrm{d}\theta^2$ are negative for all u and the $u(\theta)$ curve is everywhere convex. It must reach the axis of u at two angles, θ_1 and θ_2, say; these are the asymptotes of the unbounded orbit shown in the right-hand diagram. It is obvious that for this, as for all central orbits, rotation through any angle gives an equally good orbit.

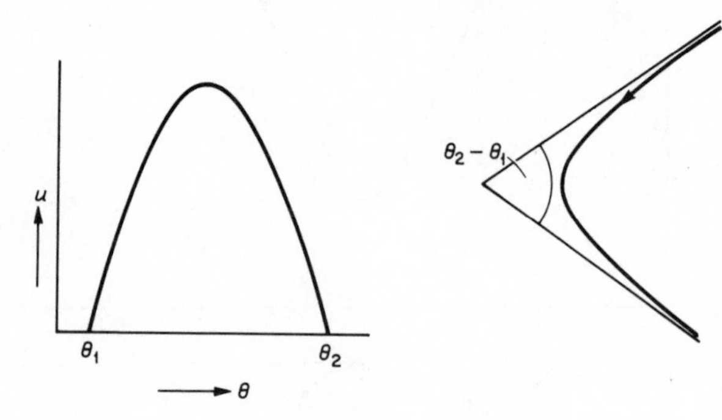

If the force is everywhere attractive (F negative), $g(u)$ may be either positive or negative at various u, and bounded solutions are possible for a wide range of attractive force laws.

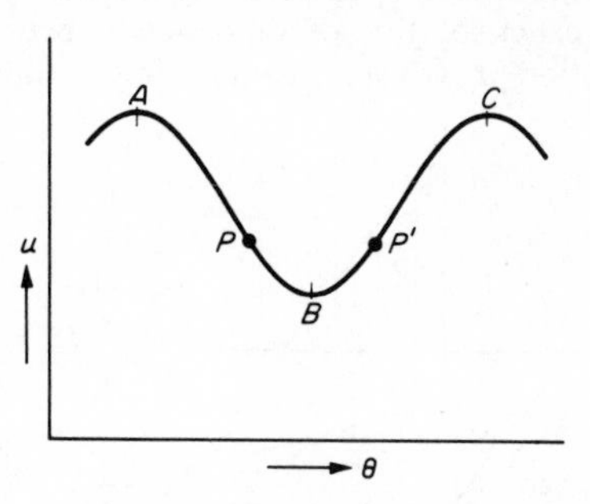

The schematic solution is what is obtained when $g(u)$ is sufficiently negative for large u, and sufficiently positive for small u, to reverse the slope before the curve for u hits the axis ($\rho \to \infty$) or rises without limit ($\rho \to 0$). It is clear from the form of $g(u)$ as well as for physical reasons that an attractive force that dies away too rapidly at great distances is unable to ensure a minimum such as B, while a force that attracts too strongly at small distances will pull the particle into the centre of attraction, i.e., will not ensure a maximum such as A.

PROBLEM

If the law of force is attractive, $F \propto -1/\rho^n$, show that a circular orbit is always possible, but that if $n \geqslant 3$ the orbit is unstable with respect to a small perturbation, and the particle either spirals into the origin or out to infinity. Show further that if $1 < n < 3$ both bounded and unbounded orbits are possible, but that if $n \leqslant 1$ all orbits are bounded.

The form of (5.4) shows that, however complicated the law of force, u can have at most one maximum and one minimum. For the solution is symmetrical about an extremum. If we draw the branch BC as a mirror image of the branch BA the two points P and P' have the same u and therefore the same $g(u)$; they also have the same second derivative $\mathrm{d}^2u/\mathrm{d}\theta^2$, and thus if AB satisfies the equation so does BC. The same argument shows that the successive branches repeat the pattern. The general bounded orbit is thus a rosette type of curve lying between two circles. There is no reason in general why the successive maxima should be separated by an angular distance that is simply related to 2π, and one must therefore expect unclosed rosette orbits as the general rule. It happens that the harmonic force and the inverse-square force give closed orbits, but very few other force laws do.

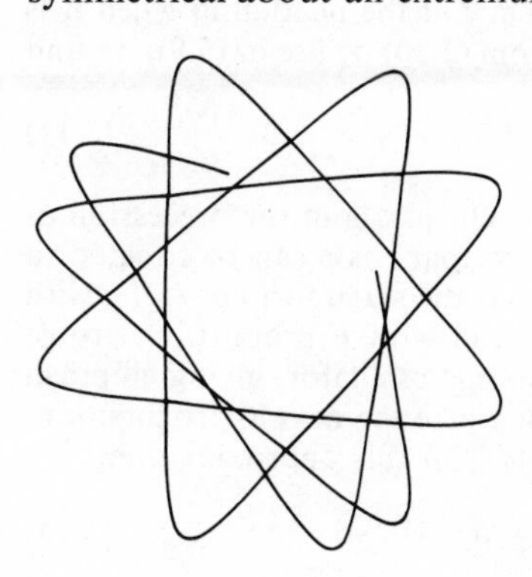

EXAMPLE

Precession of the orbit of a simple pendulum swung with large amplitude. The exact solution of almost every problem involving large amplitudes of oscillation proves very troublesome, and here we shall go no further than to demonstrate the precession of the orbit, without attempting numerical estimates of its rate. The first task is to write down the equation of motion of the bob, whose position is defined by the angles θ and ϕ. The angular momentum is constant, and this gives one equation,

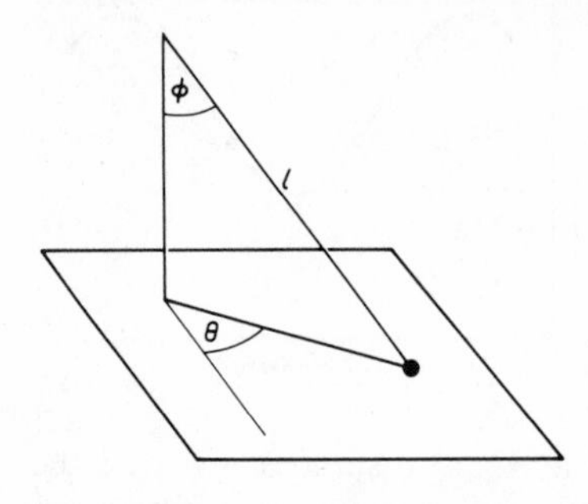

$$L = ml^2 \sin^2 \phi . \dot{\theta}. \tag{5.9}$$

Next we consider the component of acceleration, a_n, normal to the string in the vertical plane containing the pendulum. This may be evaluated easily by regarding the horizontal plane as a complex plane in which the position of the bob is defined by $\rho = l \sin \phi . \mathrm{e}^{i\theta}$. Differentiating twice to form $\ddot{\rho}$, and taking only the real part of the coefficient of $\mathrm{e}^{i\theta}$ in the resulting expression, we evaluate the horizontal component of a in the plane of the pendulum:

$$a\,(\text{horiz.}) = l[\cos \phi . \ddot{\phi} - \sin \phi(\dot{\phi}^2 + \dot{\theta}^2)],$$

while the vertical component of a_n is found by differentiating $z = l \cos \phi$:

$$a\,(\text{vert.}) = l[-\sin \phi . \ddot{\phi} - \cos \phi(\dot{\phi})^2].$$

Hence

$$a_n = a\,(\text{horiz.}) \cos \phi - a\,(\text{vert.}) \sin \phi = l[\ddot{\phi} - (\dot{\theta})^2 \cos \phi \sin \phi].$$

Since the corresponding component of force on the bob is $-mg \sin \phi$, the second equation of motion takes the form

$$\ddot{\phi} - \dot{\theta}^2 \cos \phi \sin \phi + \omega^2 \sin \phi = 0, \tag{5.10}$$

in which ω is written for $(g/l)^{1/2}$, the frequency of the pendulum when it is executing small oscillations. Eliminating $\dot{\theta}$ from (5.10) by use of (5.9), we find

$$\ddot{\phi} = -\omega^2 \sin \phi + \frac{L^2 \cos \phi}{m^2 l^4 \sin^3 \phi}. \tag{5.11}$$

It is the examination of this equation that tells us about the precession of the orbit. It will be noted that if ϕ is small, so that $\cos \phi$ can be equated to unity and $\sin \phi$ to ϕ, the resulting equation has the same form as (5.1), with ϕ serving as a measure of displacement, ρ, and with F proportional to ϕ. This is, as expected, the equation for a harmonic oscillator, giving elliptical orbits with a periodicity in θ of π. Let us then take the next approximation, writing $\sin \phi = \phi - \frac{1}{6}\phi^3$ and $\cos \phi = 1 - \frac{1}{2}\phi^2$. In this approximation,

$$\ddot{\phi} = -\omega^2 \phi(1 - \tfrac{1}{6}\phi^2) + \frac{L^2}{m^2 l^4 \phi^3}\,[1 + \mathcal{O}(\phi^4)]. \tag{5.12}$$

The meaning of the square bracket is that the quadratic term in the correction to the second expression on the right of (5.11) vanishes identically, and to obtain the right coefficient for the correction in ϕ^4 we should need to expand cos ϕ and sin ϕ to the next order. For not too large amplitudes, the dominant correction is that to the first expression on the right, and its negative sign implies that the pendulum behaves as a particle in a central field which is not strictly proportional to displacement, but which increases rather more slowly at larger amplitudes. If, then, we start with a schematic solution of (5.4) for a strictly harmonic force, we must modify it to take account of the large-

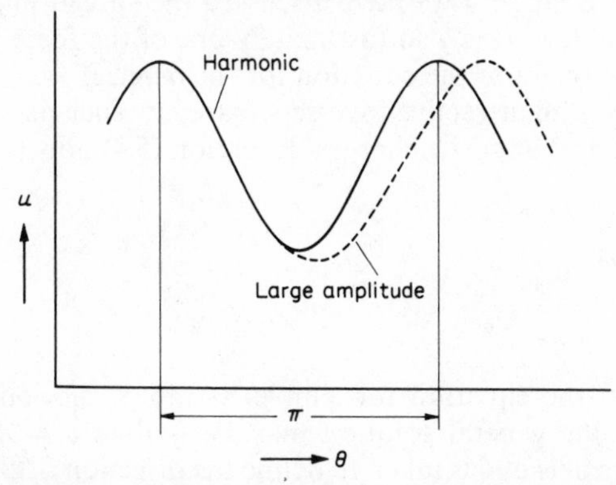

amplitude correction by noting that when ϕ is large, $u(=1/\phi)$ is small, and that around the minima, where $g(u)$ is positive, a weakening of F diminishes $g(u)$ and therefore $\mathrm{d}^2u/\mathrm{d}\theta^2$. Hence the curve is spread out and the period increased from its low-amplitude value π. This means that the elliptical orbit precesses in the same sense as that in which the particle describes the orbit. The reader may hang a simple pendulum from a point on the ceiling and verify that this really does happen.

Another way of looking at the orbit problem is to study the radial motion from the point of view of an observer sitting on the radius vector. From the kinetic expression (5.5) we may write

$$\begin{aligned} E &= \phi + L^2/(2m\rho^2) + \tfrac{1}{2}m\dot{\rho}^2, \\ &= \phi' + \tfrac{1}{2}m\dot{\rho}^2, \end{aligned}$$

in which the fictitious potential ϕ' contains the real potential ϕ and a 'centrifugal potential' $L^2/(2m\rho^2)$. The centrifugal force derived from this potential will be seen on the right-hand side of (5.1). The general meaning is clear: if the particle comes close to the origin, its angular momentum will ensure that it travels fast, and a considerable radial force will be needed to keep it close. The radial acceleration is to be thought of as caused by what is left of the real force after taking that part which is needed to keep ρ constant. We may

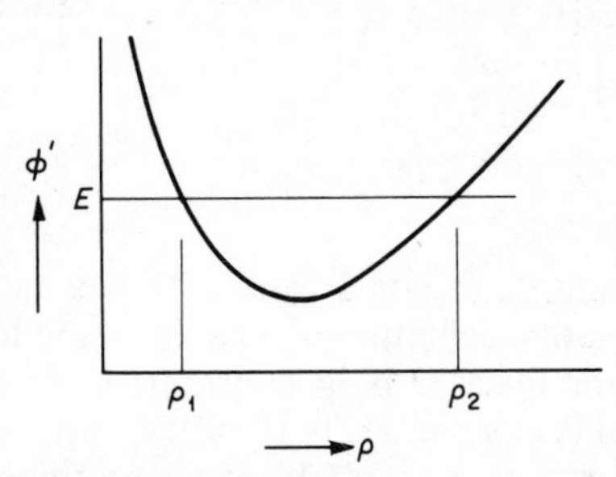

now regard the radial motion of the particle as the motion under Newton's laws of a particle in a potential ϕ'. If the force is attractive, ϕ rises as ρ increases, while the centrifugal potential is repulsive, rising to infinity as ρ goes to zero. A horizontal line at the level E intersects the curve of ϕ' at two points, ρ_1 and ρ_2, marking the bounds of the orbit. It is instructive to consider the problem on p. 65 in the light of this diagram.

Already in Chapter 4 we have discussed the special importance of the inverse-square law. It is also fortunately one of the force laws that allow analytical treatment of the equation for the orbit. If we write F as $-\mu u^2$, positive μ gives an attractive inverse-square law such as applies to gravitating bodies and opposite charges. Equation (5.4) now takes the form

$$\mathrm{d}^2u/\mathrm{d}\theta^2 = m\mu/L^2 - u,$$

or

$$\mathrm{d}^2v/\mathrm{d}\theta^2 = -v,$$

where $v = u - m\mu/L^2$.

This is just the equation for simple harmonic motion with unit frequency, and the general solution may be written $v = A\cos\theta$, with A positive, if the axis of θ is taken to define the direction of closest approach of the particle to the origin. Then the polar equation of the orbit follows immediately,

$$1/\rho = A\cos\theta + B, \quad \text{where } B = m\mu/L^2. \tag{5.13}$$

This is the equation of a conic, i.e., in Cartesian coordinates a curve quadratic in x and y, as may be seen by substituting $\rho^2 = x^2 + y^2$ and $\rho\cos\theta = x$. There are two types of conic, the closed curve which is the ellipse, and the open curve which is the hyperbola. Various degenerate forms (circle, parabola, straight lines) may appear as special cases but need not be considered separately; the prime distinction is between closed and open orbits, and the condition for orbits to be closed is that a solution of (5.13) with ρ positive must exist for all θ. This clearly implies that B must be positive (attractive force) and larger than A. The values to be assigned to A and B depend on the precise conditions under which the particle is projected into its orbit; there is no obvious physical interpretation of A. It should be noted that the origin of coordinates is a *focus* of the conic, as is most clearly seen from an alternative definition of a conic as the locus of a point P whose distance from the focus O is in constant proportion to its distance PN from a fixed line (*directrix*). It is easily seen, by multiplying both sides by ρ and rearranging, that (5.13) expresses this property. From now on we shall assume the reader to be familiar with the elementary properties of conics.

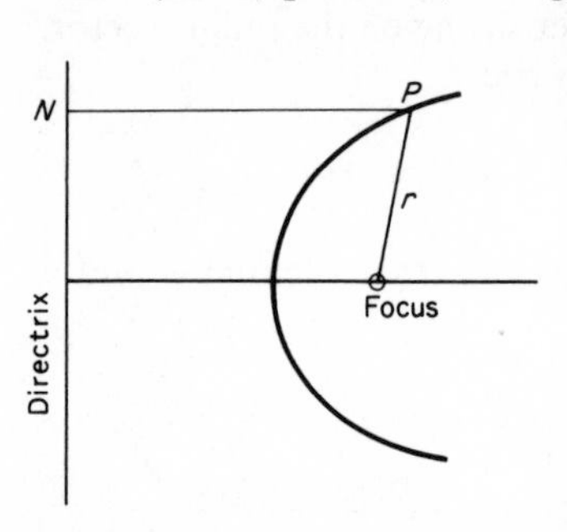

Elliptical inverse-square orbits and Kepler's laws

If $B > A$, ρ runs between $(B + A)^{-1}$ and $(B - A)^{-1}$ which are therefore the values of OQ and OP. The major axis $2a$ is the sum of these two, i.e.,

$$a = B/(B^2 - A^2). \qquad (5.14)$$

To find the semi-minor axis $b = NC$, we recollect a standard result, that OC equals the semi-major axis, so that application of Pythagoras' theorem to ONC gives

$$b = (B^2 - A^2)^{-1/2} = (a/B)^{1/2}. \qquad (5.15)$$

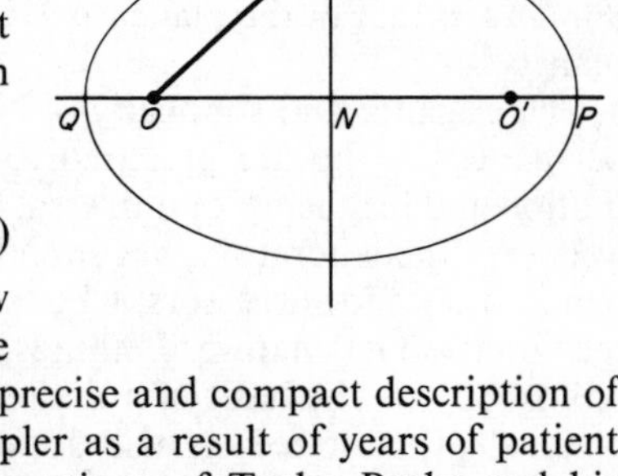

These results are enough to allow discussion of the movement of the planets round the sun. A remarkably precise and compact description of the planetary orbits was given by Kepler as a result of years of patient and imaginative analysis of the observations of Tycho Brahe and his predecessors. This very important and beautiful work is well described in Rogers' *Physics for the Inquiring Mind*, and we shall not go into it here, but simply quote the relevant results. It is desirable to stress the word relevant, since for a considerable time Kepler believed in the significance of a wholly irrelevant discovery he had made, that the sizes of the planetary orbits were governed by the geometry of nesting regular polyhedra.* Leaving this aside, however, we may enunciate Kepler's laws in the following form:

1. The planets move in ellipses with the Sun as one focus.
2. Equal areas are swept out in equal times by any one planet.
3. The periodic time of a planet is proportional to the three-halves power of its semi-major axis.

These three laws may now be interpreted in the manner first achieved by Newton. The second law is an indication that the force is central, and the first law shows that it is an inverse-square force. The third law depends on the proportionality of gravitational and inertial mass which we have already discussed at some length. To show this last point we note that the area of an ellipse is πab, so that from (5.15) the orbit area is $\pi a^{3/2}/B^{1/2}$ or,

61

55

* Perhaps because it has been the subject of so prolonged and minute an attention, the solar system has been found to exhibit several properties that now seem to be the result of chance rather than necessity. One may point to the close agreement between the angular diameters of Sun and Moon, and to Bode's remarkable law relating the distances of the planets (the nth planet out from the Sun is at a distance proportional to $4+3\times 2^{n-2}$, the first, Mercury, being at a distance of 4 units). These observations must be classed with Kepler's polyhedra as probably irrelevant, though it is always possible that the real explanation of the origin of the Solar system (a matter still in doubt) will provide a reason for Bode's law.

using (5.13), $\pi a^{3/2}L/(m\mu)^{1/2}$. Since the rate at which area is swept out is $L/(2m)$, the periodic time required to sweep out the whole orbit area

$$\tau = 2\pi a^{3/2}(m/\mu)^{1/2}.$$

For τ to vary as $a^{3/2}$ it is obviously necessary for the attractive force constant μ to be proportional to the inertial mass; if we write Newton's law of gravitation in the form $F = -Mm/(4\pi\gamma r^2)$, where M is the mass of the Sun and m that of the planet, $\mu = Mm/(4\pi\gamma)$ and m/μ is the same for all planets.

The elegance and simplicity of Newton's explanation of Kepler's laws persuade us, who are prepared to be persuaded, that his equations of motion and his theory of universal gravitation, applying in the cosmos as well as in the laboratory, are sound. It was not so with many of his contemporaries and successors, who, adhering to alternative philosophies of the fundamental nature of things, were not prepared to accept a new philosophy on the basis of a limited amount of experimental agreement. Naturally they seized, as also do modern scientists whose basic patterns of thought are challenged, on apparent discrepancies between theory and experiment, and used them as justification for maintaining their traditional beliefs. It is easy at this distance in time to criticize the adherents of the views of Descartes as wilfully obtuse, but one should remember that at the time Descartes was a towering authority, and his scheme one of the newest fashions in cosmology, against which the arguments of Newton were far too mathematical to appeal to any but the most skilled. It was necessary for a long time to elapse before Newtonianism was taken for granted, and probably the last stumbling-blocks to be removed were the apparently anomalous variations of orbit caused by the gravitational interaction of Jupiter and Saturn, and certain small anomalies in the Moon's motion. When Laplace explained these in detail within the framework of Newtonian dynamics, in 1784–7, 100 years after Newton's *Principia*, there could be no doubt left of the power of Newton's laws and their very high degree of precision, which could be relied on to account precisely for quite minute apparent discrepancies. Nowadays we have great faith in the trustworthiness of calculations of eclipses far in the future, and in the comparison of past eclipses with ancient historical records to fix precise times of certain events. The discovery of Neptune by Le Verrier in 1846 was possible because he believed in Newtonian mechanics sufficiently to look for an explanation of small fluctuations in the orbit of Uranus.

This is not to say that everything is completely accounted for in the Solar system. Mercury, for example, does not perform an exactly elliptical orbit, but one whose axes precess very slowly, at a rate of about 9 minutes of arc per century. Most of this can be accounted for by the influence of the other planets, but there remains 38 seconds of arc per century unexplained. This famous Precession of the Perihelion of Mercury was one of the few observational facts available to be cited by Einstein in support of his *General Theory of Relativity*, and although we cannot discuss this very difficult subject, it is worth pausing to note that we trust Newton's theory so fully as to take seriously a tiny anomaly such as this. Let us see how much modification would be needed to the inverse-square law to cause this precession to appear as a natural consequence of orbital theory.

EXAMPLE

Precession of the orbit in a not quite inverse-square law of force. Let the law be written in the form $F = -\mu/\rho^{2+h}$, where $h \ll 1$, and let the orbit be very nearly circular, as are most planetary orbits. Then we write u as $u_0 + x(\theta)$, where $x \ll u_0$, and (5.4) takes the form

$$\begin{aligned} d^2x/d\theta^2 &= -u_0 - x + m\mu(u_0 + x)^h/L^2, \\ &\approx -u_0 - x + m\mu u_0^h(1 + hx/u_0)/L^2, \\ &\approx \text{const.} - x[1 - m\mu h/(L^2 u_0^{1-h})]. \end{aligned}$$

In this approximation the solution is still sinusoidal, but the period is not 2π but (to first order in h) $2\pi[1 + m\mu h/(2L^2u_0^{1-h})]$, or $2\pi(1 + \frac{1}{2}h)$ since for a circular orbit $m\mu/(L^2u_0^{1-h})$ is easily found to equal unity. Thus the orbit precesses through an angle πh in every revolution. Applying this to Mercury, which goes round the Sun 415 times in a century, we see that to account for a precession of 38 seconds in a century, h need be only $1{\cdot}4 \times 10^{-7}$. We conclude that the exponent in the gravitational power law is likely to be 2 within 1 part in 10 million.

Hyperbolic orbits and Rutherford scattering

As an example of hyperbolic orbits we shall consider the passage of a positively-charged α-particle (nucleus of helium, made up of 2 protons and 2 neutrons) past an atomic nucleus containing Z protons, which repels the α-particle and causes its trajectory to be deflected into a hyperbolic orbit. The problem arises from the observation by Geiger and Marsden in 1909 that α-particles, which usually pass through thin foils with only very small deflections, on rare occasions suffer large angle deflections which are unaccountable without the supposition that a very strong force can be exerted by an atom on an α-particle. It was this that led Rutherford to suggest his nuclear model of the atom, in which the electron gas, occupying a sphere of about 10^{-10} m in radius, is held together by the attraction of a tiny, concentrated positive nucleus with a radius perhaps 10^4 times smaller. The rare deflections of α-particles through large angles are then to be interpreted as due to the passage of a particle very close to the nucleus. In the extreme case of a head-on collision, the α-particle would come to rest momentarily at such a distance that its Coulomb potential energy equalled the kinetic energy with which it was shot at the nucleus; it would then bounce back, having suffered 180° deflection.

It is instructive to estimate this closest distance of approach. Customarily the energy of a particle is measured in electron volts; a particle is said to have an energy of V electron volts when its kinetic energy is eV, e

being the electronic charge. Thus a particle carrying a charge n times that of an electron, and accelerated through a potential difference V, acquires kinetic energy neV, or nV electron volts. The α-particles used in the early scattering experiments were those that are spontaneously emitted from radioactive atoms with energies of a few million volts. The closest distance of approach of a one-million volt α-particle, with positive charge $2e$, to a nucleus with positive charge Ze (i.e., containing Z protons), is obtained by equating the potential energy $2Ze^2/(4\pi\varepsilon_0 r)$ to the kinetic energy $10^6 e$, so that

$$r = \frac{2Ze}{4\pi\varepsilon_0} \times 10^{-6} \approx 3Z \times 10^{-15}\ \text{m}.$$

Only if the nucleus is smaller than this can it exert a Coulomb repulsion strong enough to deflect an α-particle through a large angle; a more extended distribution of positive charge, if not supplemented by other than Coulomb forces, would be penetrated by the α-particle and probably destroyed or modified. This is, in fact, what happens when light elements (low Z) are bombarded by energetic charged particles. However, at the time of the scattering experiments this was still in the future, and Rutherford's appreciation of the very powerful force needed was enough to lead him to his model. Since the α-particle must approach so near the nucleus to experience a strong enough force to deflect it appreciably, it is a good assumption that the extra-nuclear electrons play no role, and that the scattering process is simply the result of an inverse-square repulsion, leading to a hyperbolic orbit. We may now do as Rutherford did, calculating how many particles might be expected to be scattered through a given angle, to compare with observations and check the basic assumption of the model that there is effectively a point charge exerting a Coulomb repulsive force.

The typical hyperbolic orbit shown illustrates how the α-particle moves in relation to the nucleus at O. At great distances, the orbit merges into the asymptotes CZ and CY, and ϕ is therefore the angle of deflection. It

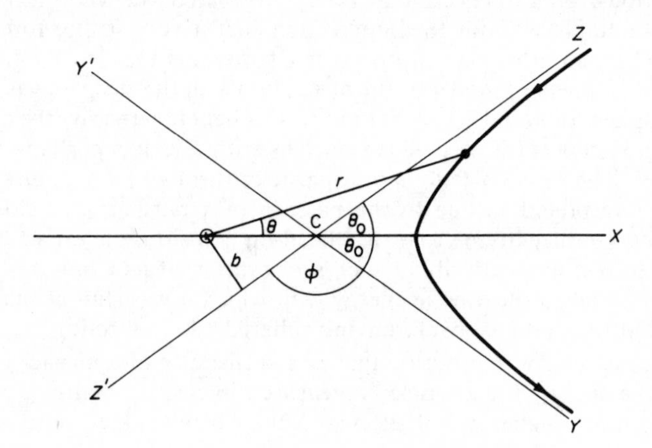

can be seen that if there were no repulsive force the particle would carry on along ZCZ' and pass within a distance b of the nucleus; b is called the impact parameter, and for a given nucleus and initial energy of α-particle, ϕ is determined by b. The smaller b is, the greater is the deflection ϕ, with the limiting deflection of π when $b = 0$ and the particle approaches the nucleus head-on. If therefore we wish to know how many α-particles will be deflected through an angle greater than a certain ϕ, we need to know how many would, if undeflected, pass closer to a nucleus than the appropriate distance b to cause deflection through ϕ. Let us therefore look at the scattering foil as an array of targets, each of radius b and drawn around a nucleus, and let us suppose the foil to contain few enough nuclei for the targets not to overlap (this is not difficult in practice with a thin foil). Then if unit area of foil contains N nuclei, the targets in unit area present an area $N\pi b^2$, and this is the fraction of α-particles which would hit a target if they were not deflected. That is, the probability of deflection through an angle greater than ϕ may be written

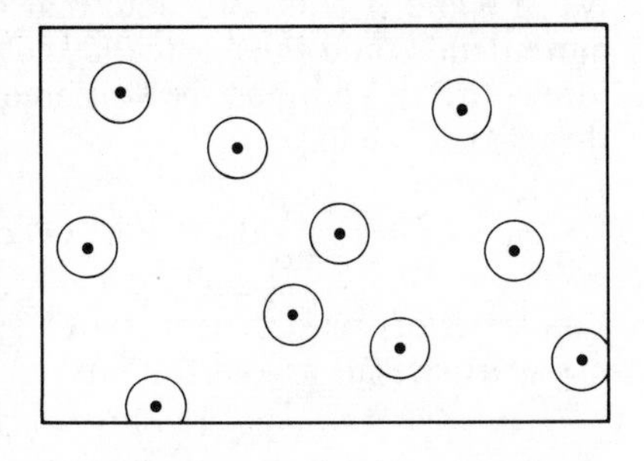

$$P(\phi) = N\pi[b(\phi)]^2. \tag{5.16}$$

We have then to find how b and ϕ are related in order to complete the solution. This is a mathematical problem that can easily get bogged down in the geometry of the hyperbola; there is, however, a neat solution that avoids this difficulty, but which is too elegant to be fairly recommended to the reader as a model to emulate. We consider the component of momentum parallel to the symmetry axis OX, which is reversed in the scattering process from $-mv \cos \theta_0$ to $+mv \cos \theta_0$, if v is the velocity of the particle at a great distance. The time-integral of the component of force parallel to OX must thus equal $2mv \cos \theta_0$, or $2mv \sin (\frac{1}{2}\phi)$, i.e.,

$$2mv \sin (\tfrac{1}{2}\phi) = \frac{2Ze^2}{4\pi\varepsilon_0} \int \frac{\cos \theta}{r^2}\, \mathrm{d}t = \frac{2Ze^2}{4\pi\varepsilon_0} \int \frac{\cos \theta}{r^2\, \mathrm{d}\theta/\mathrm{d}t}\, \mathrm{d}\theta.$$

Now $r^2\, \mathrm{d}\theta/\mathrm{d}t$ is L/m, a constant of the motion, whose value is obviously bv. Therefore,

$$2\, mv \sin (\tfrac{1}{2}\phi) = \frac{2Ze^2}{4\pi\varepsilon_0 bv} \int \cos \theta\, \mathrm{d}\theta = \frac{2Ze^2}{4\pi\varepsilon_0 bv} . 2 \sin \theta_0.$$

Writing $\cos (\frac{1}{2}\phi)$ for $\sin \theta_0$, and E for $\frac{1}{2}mv^2$, the α-particle energy, we have the required relationship

$$b = \frac{Ze^2}{4\pi\varepsilon_0 E} \cot (\tfrac{1}{2}\phi).$$

Then, substituting in (5.16), we have

$$P(\phi) = A \cot^2 (\tfrac{1}{2}\phi),$$

where $A = NZ^2e^4/16\pi\varepsilon_0^2E^2$, and is constant for a homogeneous group of α-particles (constant E) in a given scattering experiment with a single nuclear species.

To find the fraction of particles scattered in a given range of angles between ϕ and $\phi + \mathrm{d}\phi$, we note that a fraction $P(\phi)$ is scattered through more than ϕ, and $P(\phi) + (\mathrm{d}P/\mathrm{d}\phi)\,\mathrm{d}\phi$ through more than $\phi + \mathrm{d}\phi$. Therefore $-(\mathrm{d}P/\mathrm{d}\phi)\,\mathrm{d}\phi$ must be scattered in this particular range $\mathrm{d}\phi$. Calling this $n(\phi)\,\mathrm{d}\phi$ we have

$$n(\phi)\,\mathrm{d}\phi = -\frac{\mathrm{d}P}{\mathrm{d}\phi}\,\mathrm{d}\phi \propto \cot (\tfrac{1}{2}\phi) \operatorname{cosec}^2 (\tfrac{1}{2}\phi)\,\mathrm{d}\phi. \tag{5.17}$$

In a series of measurements by Geiger and Marsden, a collimated beam of α-particles hit a gold foil, and those that were deflected to strike the

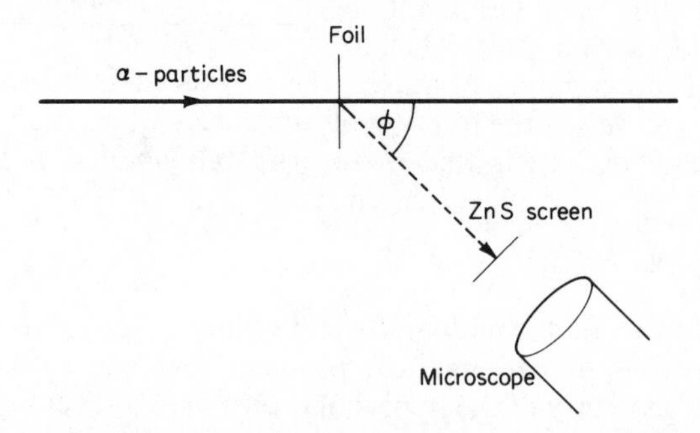

scintillating zinc sulphide screen were counted. The screen and microscope were moved to various values of ϕ at a constant distance from the screen, and the relative scattering at different angles determined. It should be noted that (5.17) gives the number scattered into the annulus between ϕ and $\phi + \mathrm{d}\phi$ (i.e., into a solid angle $2\pi \sin \phi\,\mathrm{d}\phi$), while what is recorded is the number scattered into a given solid angle. Thus the relative counting rates expected at different ϕ are obtained by dividing (5.17) by $\sin \phi\,\mathrm{d}\phi$, to give a counting rate proportional to $\operatorname{cosec}^4 (\tfrac{1}{2}\phi)$.

This is a very rapid function of ϕ, varying for example from 3445 when $\phi = 15°$ to 1·15 when $\phi = 150°$. Nevertheless, the scintillations counted at 11 angles in this range came at rates proportional to $\operatorname{cosec}^4 (\tfrac{1}{2}\phi)$ with a mean deviation of only 11%. Such a result may not seem particularly impressive to a reader who has grown up to believe that better than 1% is good agreement and worse than 1% bad agreement. This view of what constitutes confirmation of a theoretical formula is altogether too naive, since it does not take account of what alternative formulae might be expected to predict. To take a ridiculous example, if one had calculated the refractive index, μ, of helium gas at n.t.p. and obtained a theoretical estimate of 1·00006, and then measured it to be 1·00003, it would carry very little weight to claim that the theory was correct within three thousandths of one percent. In this case the theory is concerned to calculate $\mu - 1$ and the result obtained is wrong by a factor of two (which is a very bad answer for this comparatively straightforward calculation). In the case of the scattering experiments the situation is reversed; we have a function that ranges over a very wide factor, and different functions covering the same sort of range of variation might easily differ from $\operatorname{cosec}^4(\frac{1}{2}\phi)$, not by 11% but by factors of 2 or 10 or more. Perhaps a better perspective is placed on the comparison by using logarithms to get all the data on a single graph. Here the measurements of log C, where C is

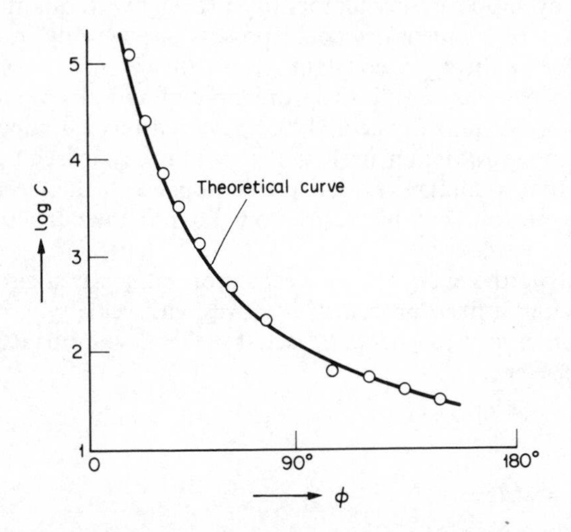

the counting-rate, are plotted against ϕ, and $\log [A \operatorname{cosec}^4(\frac{1}{2}\phi)]$, with A suitably chosen, is shown for comparison. It is clear that the functional form of the observed behaviour accords very closely indeed with the theory. At all events, this agreement served to establish the nuclear model of the atom, whose subsequent development by theory and experiment has left no doubt at all of its correctness.

The Bohr atom*

Once the nuclear atom was established, however, it became a critical problem how the extra-nuclear electrons could be maintained in a stable

* Many teachers of quantum physics hold that Bohr's theory should be left as a historical curiosity, on the grounds that, once studied, it is difficult to forget and becomes a positive hindrance to understanding the more powerful theories that succeeded it. Perhaps it is the very beauty of Bohr's ideas that constitute their menace, and the reader deserves therefore to be warned not to devote too much attention to this section. Later, when he has understood Schrödinger's treatment of the hydrogen atom, he may return and see for himself that, with all its faults, Bohr's theory deserves to be remembered as the work of a real master; moreover, it still has the power, when seen in the light cast by Schrödinger, to reveal truths that are all too hard to grasp by the use of more sophisticated approaches.

state for an indefinite time, since it was to be expected that as they were accelerated in their motion round the nucleus they would, like all accelerated charged particles, emit electromagnetic energy and soon fall into the nucleus. Bohr refused to let this problem worry him unduly; he proposed to assume that at the atomic level the laws of physics were such as to prevent the catastrophe, and he developed in 1913 a set of rules which would give the correct answer, when applied to the simplest of atoms, hydrogen, consisting of a single electron circling round a single proton. Bohr's theory sprang from the earlier ideas of Planck who had shown in 1900 how certain difficulties in the theory of radiation from hot bodies could be overcome by supposing oscillators to exist only with quantized energies: an oscillator of frequency ω could possess energy only in integral multiples of $\hbar\omega$, $\hbar$ being a constant of nature (Planck's constant h was originally defined as $2\pi\hbar$); it therefore emitted and absorbed energy only in quanta of $\hbar\omega$, and Einstein (1905) gave reasons for supposing that it was inherent in radiation itself that it could only exist in quanta. Bohr proposed that a similar restrictive rule applied to the electron orbiting round the proton, but his restriction did not take the same form as Planck's.

Let us write the energy of an electron of mass m in a circular orbit of radius r round a fixed proton. If it moves with velocity v, its centripetal acceleration v^2/r must be produced by the Coulomb attractive force $e^2/(4\pi\varepsilon_0 r^2)$. Hence

$$mv^2/r = e^2/(4\pi\varepsilon_0 r^2),$$

or

$$\tfrac{1}{2}mv^2 = e^2/(8\pi\varepsilon_0 r). \tag{5.18}$$

This is the kinetic energy; the potential energy is $-e^2/(4\pi\varepsilon_0 r)$ and has magnitude twice the kinetic energy, whatever r may be. The total energy E is thus $-e^2/(8\pi\varepsilon_0 r)$, and this (taken as positive) is the energy needed to pull the electron away to infinity and bring it to rest there. Because E is negative the electron is bound to the proton and cannot spontaneously escape, and $-E$ is the *binding energy*. Now Bohr proposed that of all the classically possible circular orbits, of all radii, only those occurred whose angular momentum was an integral multiple of $\hbar$. This is his version of the restrictive rule that applies to circular orbits, analogous to the quantization of Planck's oscillator, and it is specifically designed to give the required answer. Since the angular momentum is mrv, we write the condition for an orbit to exist as $mrv = n\hbar$, with n an integer. But from (5.18)

$$mrv^2 = e^2/(4\pi\varepsilon_0).$$

Dividing the latter equation into the square of the former we eliminate v and obtain for the permitted radii,

$$r = \frac{4\pi\varepsilon_0\hbar^2}{me^2}n^2,$$

and for the permitted binding energies

$$-E = A/n^2, \tag{5.19}$$

where A is a constant, $me^4/(32\pi^2\varepsilon_0^2\hbar^2)$, whose magnitude is 13·60 electron volts when calculated from the best known values for m, e, ε_0 and $\hbar$.

The lowest energy state (*ground state*) of the atom is obtained by putting $n = 1$, and it is supposed that although the electron is constantly accelerated in its orbit it does not radiate energy because there is no lower energy state to which it could go. If, however, it starts in the state where $n = 2$ or higher, it can radiate and drop to a lower state. But here again Bohr makes a slashing assumption—the frequency of the radiated wave is not to be related, as one might expect, to the frequency of the electron in its orbit, but is to be determined by the quantized character of radiation itself; the energy in a transition appears as one quantum, and this is enough to fix its frequency. Thus if the electron drops from the mth to the nth state, we write

$$\hbar\omega = E_m - E_n = A\left(\frac{1}{n^2} - \frac{1}{m^2}\right). \tag{5.20}$$

According to this theory, the spectral lines emitted by incandescent atomic hydrogen should have frequencies related to integers by (5.20). But this had been known for some time, Balmer having discovered this sort of regularity in the spectral lines in 1885*. Moreover, the value of A found from the spectral lines is in marvellous agreement with the value deduced from atomic constants, especially when correction is made for the fact that the electron does not revolve about a stationary proton, but both revolve about their common centre of mass. Since the proton is 1836 times more massive than the electron, the correction is small.

PROBLEM

Show that if the angular momentum of (electron + proton) is $n\hbar$, the orbit radii remain unchanged when allowance is made for motion of the proton, but that A is changed by an amount that is equivalent to replacing m by $m_e m_p/(m_e + m_p)$, where m_e and m_p are the masses of electron and proton.

The success of Planck's and Bohr's arbitrary modifications of classical mechanics naturally led to attempts to generalize the rules for

* The practice of looking at regular sets of numbers discovered empirically and attempting to find by trial and error a mathematical rule connecting them is sometimes very productive, as with Balmer's formula, sometimes amusing but unproductive, as with Bode's law for the planetary radii, and sometimes a vicious practice that only serves to degrade the intellectual powers of the victim. Any student tempted in this direction would do well to treat numerology, as it is derisively called, as a dangerous and habit-forming drug.

quantization so as to set up a coherent system. It would carry us too far afield to go deeply into this, but an important idea due to Ehrenfest (1914) is worth following up, if only because it furnishes examples of a slightly different sort of calculation in classical mechanics. Ehrenfest's view may be explained briefly by reference to a harmonic oscillator. Suppose we alter the force constant very slowly, so that the frequency and amplitude change slowly, it will be very awkward if the system started quantized (i.e., with energy a multiple of $\hbar\omega$) but changed its energy and frequency at different rates so that the quantum condition was no longer obeyed. If this happened, would the oscillator suddenly emit energy and switch to a properly quantized condition, and if so, how would it know when to do so? Ehrenfest considered this to be an unresolvable problem and hence suggested that it never arose in practice; the only physical variables that were capable of being quantized were, in his view, those that remained constant during this sort of slow variation of the parameters. Such variables are called *adiabatic invariants*. We shall now show how Planck's rule for oscillators and Bohr's rule for orbiting particles conform to this principle.

234

The Bohr orbit is particularly simple. If we imagine the strength of the attractive force to be slowly increased, the orbit will shrink, but because the force is still radial at all times, angular momentum L will remain unchanged. Thus angular momentum is an adiabatic invariant for a central orbit and can be quantized. It has the same dimensions as $\hbar$, and we may write a possible quantization scheme in the form

$$L/\hbar = \text{any one of a discrete set of numbers.}$$

The theory of adiabatic invariants as stated here does not allow one to predict what these numbers should be, though we might hope they would be simple, and in fact in Bohr's theory they are the integers.*

As a second example we take a harmonic oscillator, which also allows a demonstration of a useful mathematical device, the W.K.B. method (Wentzel, Kramers and Brillouin were some of those who developed its application in quantum mechanics). The equation of motion for a harmonic oscillator of frequency ω takes the form:

$$\ddot{x} + \omega^2 x = 0, \tag{5.21}$$

and we are now interested in the solution when the force constant is changed so that ω varies very slowly with time. How does the ampli-

* It might be objected that this is a highly unrealistic argument since we are in no position to change the force constant gradually; but in semiconductors one finds electrons in something very like a Bohr orbit round a charged impurity, and here the force constant is modified by the dielectric constant of the semiconductor, which itself can be slowly altered by heating or by applying pressure. So, slow (adiabatic) changes of the parameters are possible, and the theory is seen to have some contact with reality.

tude alter with frequency? It is convenient to remember that if ω is constant a possible solution is $A\,e^{i\omega t}$, representing a vector of constant length A rotating in a complex plane with angular frequency ω; the real part of this solution can be taken as the actual variation of x. Now if ω is not quite constant we may expect that the solution will look like a rotating vector whose length A changes slowly, while the instantaneous angular velocity will be very close to the instantaneous value of ω. If this is so, after time t the vector will have turned through an angle $\int \omega(t)\,dt$, and we may guess that it will be useful to write the solution in the form

$$x = A(t)\,e^{i\int\omega(t)\,dt}. \tag{5.22}$$

It should be recognized that no approximation is involved here—any solution can be written like this. We only approximate when we come to solving the differential equation for $A(t)$. From (5.22) we have that

$$\dot{x} = (\dot{A} + i\omega A)\,e^{i\int\omega(t)\,dt}$$

and

$$\ddot{x} = (\ddot{A} + i\dot{\omega}A + 2i\omega\dot{A} - \omega^2 A)\,e^{i\int\omega(t)\,dt},$$

so that, from (5.21), we see that

$$\ddot{A} + i\dot{\omega}A + 2i\omega\dot{A} = 0. \tag{5.23}$$

Now we approximate, tentatively, by supposing that if ω changes very slowly indeed, A will also change so slowly that $\ddot{A}$ may be neglected in (5.23). If this is so, the remaining terms give a readily soluble equation for A,

$$\dot{\omega}/\omega + 2\dot{A}/A = 0,$$

or

$$A^2\omega = C, \quad \text{a constant.} \tag{5.24}$$

Before proceeding, we must verify that the solution (5.24) justifies our neglect of $\ddot{A}$. This is fairly clear, since if ω changes by an amount comparable with itself in a long time T, this is also the time in which A and $\dot{A}$ change by amounts comparable with themselves, so that $\dot{A}$ is of the order A/T and $\ddot{A}$ of the order A/T^2. Thus the three terms in (5.23) are of order A/T^2, $A\omega/T$ and $A\omega/T$ respectively, and we can make the first as small as we wish in comparison to the other two by making T big enough. Thus (5.24) is the correct solution in the limit of very slow changes of ω. This is the W.K.B. solution of the problem, and it illustrates a very useful approximate treatment of an otherwise awkward differential equation.

From (5.24) we can calculate how the energy of the oscillator varies. Since the amplitude is A, the velocity of the particle as it passes through the centre of oscillation, where its energy is wholly kinetic, is ωA, and the energy is thus $\frac{1}{2}m\omega^2A^2$. So from (5.24) we have that

$$E = \tfrac{1}{2}mC\omega.$$

Hence E/ω is an adiabatic invariant, with the dimensions of $\hbar$, and we may suggest the quantization rule

$$E/(\hbar\omega) = \text{any one of a discrete set of numbers.}$$

Again, the discrete set was thought by Planck to be simply the integers, though later work has improved this to the half-integers, $n + \frac{1}{2}$.

Here, then, we have two examples of how a general argument may be used to determine possible quantization schemes. It was developed considerably further up to about 1926, and still has a number of useful applications even though as a quantum theory it has been completely superseded by the new quantum mechanics that came in at that time. It suffered from two basic weaknesses, that it could only be applied to systems whose motion was in some degree periodic, and that it only gave the rule for choosing which were to be permitted out of an infinity of classical solutions to a dynamical problem. The second weakness is fatal, for it is known that on the atomic scale particles may exhibit behaviour which is precluded by classical dynamics. Thus a particle of total energy E may be confined on one side of a potential barrier whose height is greater than E, so that to get through it involves traversing a region where its kinetic energy would be negative. This cannot happen classically, but is permitted in quantum systems (*tunnel effect*); it is, for example, the mechanism by which an α-particle leaves the nucleus of a radioactive atom. No theory that simply picks out certain classical solutions can cope with this phenomenon.

The great years 1926–7 saw the invention and development at explosive pace, by Heisenberg, Schrödinger, Dirac and others, of a far more general quantum mechanics which takes in its stride all the special tricks of Planck, Bohr and their followers, and accounts with ease for the tunnel effect and many other atomic phenomena. It is a rarity among revolutionary theories in being so obviously successful right from the start that there has never been any significant opposition to its basic principles of calculation, though its philosophical implications have been vigorously debated ever since. As a result of its success, it is possible to look back on classical mechanics and state rather precisely how correct Newton's ideas
143 are when applied to small things. Heisenberg's *Uncertainty Principle* sums the matter up conveniently—if you know the position of a particle within a short distance Δx, you cannot know its momentum at the same time to better than $\hbar/\Delta x$. And if you find yourself applying classical mechanics to a problem that demands dividing space and momentum into increments whose product is less than $\hbar$, you are unlikely to get the right answer. But on the everyday scale, where a micron (10^{-6} m) is a small distance, and a

body weighing a microgram, moving at a rate of 1 mm a year, may be considered to have rather a small momentum ($\frac{1}{3} \times 10^{-19}$ kg m s^{-1}), the product of momentum and distance, $\frac{1}{3} \times 10^{-25}$ J s, is still enormous compared with $\hbar$ (about 10^{-34} J s) and Newtonian mechanics may be taken to apply to the motion for almost all practical and theoretical purposes.

In the light of these remarks, it is pertinent to ask whether the argument for the nuclear atom, based on a classical theory of α-particle scattering, can be regarded as valid. The criticism may be phrased in terms of the uncertainty principle. If the kinetic energy, $\frac{1}{2}mv^2$, of the α-particle is T, its momentum, mv, is $(2mT)^{1/2}$ and for a particle with energy of 10^6 electron volts, or $1{\cdot}6 \times 10^{-13}$ J, and mass 7×10^{-27} kg the momentum is $4{\cdot}7 \times 10^{-20}$ kg m s^{-1}. Since the momentum is changed significantly in a scattering process, we may take this figure as a rough measure of the precision with which we must determine momentum, and correspondingly the precision needed for distance measurements is the distance of nearest approach, say 10^{-13} m. The product of these two is $4{\cdot}7 \times 10^{-33}$ J s which is rather larger than $\hbar$, but not perhaps so much larger as to give us any confidence in the classical theory. Nevertheless, the experiments showed that the theory is well-obeyed, and in this case it happens that the theory gives the right answer; for an inverse-square law of force, classical and quantum mechanics both lead to the same scattering formula. It is occasionally possible, then, to use classical mechanics even when it is quite invalid, but one needs considerable experience or good luck to get away with it in such conditions.

Tides

So far, our analysis of orbital motions has assumed the bodies to be either particles or at any rate rigid spheres. One effect of imperfect rigidity is

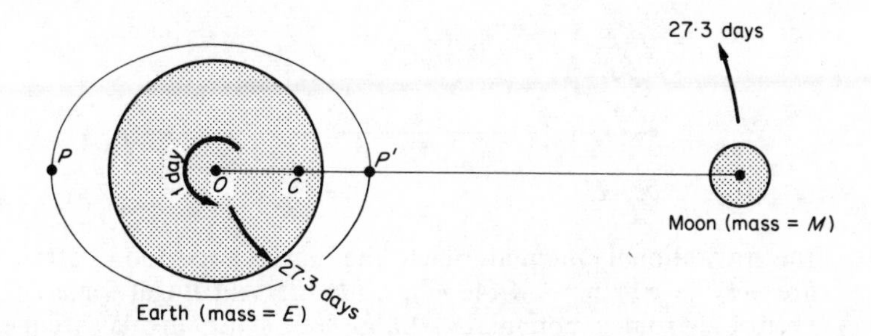

the tidal motion due to the Moon and, less marked, the Sun. Let us consider a very much simplified model of what in reality is an extremely complex problem; we suppose the Earth and Moon to perform circular orbits about their common centre of mass, C, while the Earth rotates about an axis normal to the plane of the linear orbit, and of course carries the Sea with it. For convenience we observe the system in a frame of reference rotating with the 27·3-day lunar period, and in this frame the equilibrium of the two bodies is seen to result from the balance of their gravitational attraction and their centrifugal repulsion from C. If the Earth were a point mass, this would be all; but clearly a body situated at P suffers greater centrifugal repulsion and less gravitational attraction than if it were at O, and is therefore not in balance but subject to an outward-pulling force. Similarly a point at P' is pulled toward the Moon, i.e., also away from O. It is this imbalance of forces that causes the sea to pile up into two opposing bulges, and with the rotation of the Earth any point on the surface of the sea travels through the bulges twice a day and experiences two high tides.

To take the analysis further, we note that, in the absence of the Moon, the Earth's daily rotation would cause no tides. There is, to be sure, a considerable centrifugal effect—enough to flatten the Earth into an oblate spheroid whose polar radius is 21 km (1 part in 297) shorter than the equatorial radius—but this is a steady effect from which the lunar tides are the perturbations that concern us here. Therefore we may begin by treating an Earth that presents the same face to the Moon at all times, just as the Moon does to the Earth; in the chosen reference frame, the whole system is now stationary. Any point at the surface of the sea is at rest, and rest is possible in a liquid only when there are no resultant forces acting on any element. In particular, there are no forces parallel to the surface, which is thus an equipotential surface, whose shape it is our task to find. Now unit mass situated at the point Q (r, θ) is subject to three forces, as shown in the diagram, each of which can be defined by a potential. Thus,

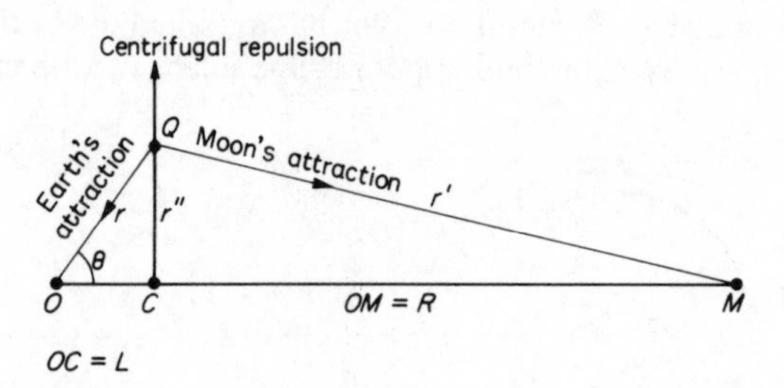

the gravitational potentials due to the Earth's and Moon's attraction are $-E/(4\pi\gamma r)$ and $-M/(4\pi\gamma r')$, while the centrifugal force $\omega^2 r''$ is derivable from a potential $-\frac{1}{2}\omega^2 r''^2$ (ω is the angular frequency $2\pi/27{\cdot}3\ \text{day}^{-1}$). Hence, expressing all potentials in terms of r and θ,

$$\phi(r, \theta) = -\frac{E}{4\pi\gamma r} - \frac{M}{4\pi\gamma}(R^2 + r^2 - 2\,Rr\cos\theta)^{-1/2} - \tfrac{1}{2}\omega^2(L^2 + r^2 - 2Lr\cos\theta).$$

The second term can be expanded by the binomial theorem, after taking $(R^2 + r^2)$ outside the brackets, to give an expansion of ϕ in powers of $\cos\theta$:

$$\phi(r, \theta) = -\left[\frac{E}{4\pi\gamma r} + \frac{M}{4\pi\gamma}(R^2 + r^2)^{-1/2} + \tfrac{1}{2}\omega^2(L^2 + r^2)\right]$$

$$-\left[\frac{M}{4\pi\gamma}\frac{Rr}{(R^2 + r^2)^{3/2}} - \omega^2 Lr\right]\cos\theta$$

$$-\frac{3M}{8\pi\gamma}\frac{R^2r^2}{(R^2 + r^2)^{5/2}}\cos^2\theta + \cdots.$$

The first square bracket is dominated by terms that express the Earth's attraction, $E/(4\pi\gamma r)$, and the other terms may be dropped without great error. The second term would vanish identically if we were to replace $(R^2 + r^2)^{3/2}$ by R^3, since R is determined by the balance of gravitational and centrifugal forces at the centre of the Earth, where $r = 0$; i.e., $M/(4\pi\gamma R^2) = \omega^2 L$. In fact at sea level $r/R \sim 1/60$, so that we may approximate to $(R^2 + r^2)^{-3/2}$ by $1/R^3 - 3/2(r^2/R^5)$, and in addition drop r^2 in the denominator of the term in $\cos^2\theta$. After all this, a simpler but still very adequate form emerges:

$$\phi(r, \theta) = -\frac{E}{4\pi\gamma r}\left[1 - \frac{3}{2}\frac{M}{E}\left(\frac{r}{R}\right)^4\cos\theta + \frac{3}{2}\frac{M}{E}\left(\frac{r}{R}\right)^3\cos^2\theta + \cdots\right],$$

from which it is clear that the term in $\cos^2\theta$ is some 60 times greater than that in $\cos\theta$; the principal tide-generating force is described by a potential of the form $\cos^2\theta$, having maxima at $\theta = 0$ and π, and this confirms our qualitative explanation of the existence of two tides a day. We now drop the term in $\cos\theta$.

In order to estimate the variations in sea level, we note that as θ is changed, ϕ must stay constant, and therefore r must change very slightly from its mean value of 6360 km so that

$$\frac{1}{r}\left[1 + \frac{3}{2}\frac{M}{E}\left(\frac{r}{R}\right)^3\cos^2\theta\right]$$

is constant. As $\cos^2\theta$ varies between 0 and 1, the quantity in brackets suffers a tiny fractional variation amounting to

$$\frac{3}{2}\frac{M}{E}\left(\frac{r}{R}\right)^3,$$

about 1 part in 12 million, which must be compensated by a corresponding fractional variation in r outside the brackets. Thus the difference between high and low tides, by this reckoning, should be about one 12-millionth of the radius of the Earth, or $\frac{1}{2}$ m.

Although this is a reasonably good estimate, the behaviour of real tides is complicated and highly variable from place to place, the tidal range being almost zero throughout much of the Mediterranean, but rising to 15 m at Chepstow, at the head of the Severn Estuary, and even higher in the Bay of Fundy. It is easy to appreciate the limitations of the calculation, even when applied to a world totally covered in water. For the movement of the bulges, as they follow the movement of the Moon, does not involve much bodily translation of the sea relative to the Earth; rather, the sea is carried round by the Earth while the bulges remain pointing in the direction of the Moon. In other words, the tides are the manifestation of a wave, with a wavelength half the circumference of the Earth, which is excited by the Moon and forced to sweep round the Earth daily; at the equator the speed must therefore be 460 m s^{-1}. Now the speed of a wave in a trough of depth d, much shallower than the wavelength, is $\sqrt{(gd)}$, and d must be about 20 km for the natural wave-speed to match what is imposed by the driving force. If the sea were uniformly of this depth, or even better if the depth at different latitudes were adjusted to give a natural synchronism everywhere, resonance would occur and a tidal wave of enormous amplitude would develop. A much deeper ocean would allow waves to travel so fast that the tide would be able to follow the equilibrium configuration without difficulty, while in a much shallower ocean only a very small tide could be propagated. In reality, of course, there is no continuous path of water round the Earth, and the tidal movement of the oceans and of more or less enclosed seas depends on the extent to which they are capable of natural oscillations with a period of half a day, such as would be excited to resonance by the tide-generating force. The Bay of Fundy provides a good example of such a resonance.

The large tide at Chepstow and at the head of many other estuaries has a different origin in the amplification of the tidal wave as it sweeps up the narrowing channel, and has its energy concentrated into an ever smaller volume of water. In special circumstances, in the case of the Severn, Amazon and Yangtze rivers for example, the speed of the current as the tide rushes into the shallower upper reaches may equal the natural speed of wave propagation, and then a vertical wall of water tends to develop and move rapidly, often destructively, upstream. This is the *bore* or *eagre*, whose theory is unfortunately beyond the scope of this account. Enough should have been said, however, to make clear that a full understanding of real tides involves much more than the elementary analysis of simplified models, but depends on the painstaking accumulation and study of much detailed information from all over the world and from scale models which are often more susceptible to experimental investigation.

Cavendish Problems: 17, 18, 22, 103

READING LIST

Numerical Solution of Equations: D. R. HARTREE, *Numerical Analysis*, Clarendon Press, Oxford.

Conic Sections: E. M. HARTLEY, *Cartesian Geometry of the Plane*, Cambridge U.P.
Tycho Brahe and Kepler: E. M. ROGERS, *Physics for the Inquiring Mind*, Princeton U.P.
Rutherford Scattering and the Nuclear Atom: M. BORN, *Atomic Physics*, Blackie.
Tides: Article 'Tide', in *Encyclopedia Britannica.*
General: K. R. SYMON, *Mechanics*, Addison-Wesley.

6

Analysis of complex dynamical systems

In the last chapter we were concerned solely with the motion of a particle in a fixed field of force. If, however, the system consists of more than one particle, with mutual interactions, the ultimate solution involves the more complicated problem of treating the motion of each particle separately, in the changing field of force produced by all the others. It may be said at once that this is usually intractable; even the three-body problem, e.g., two planets moving round the sun and attracting each other so as to perturb each other's orbits, is a very heavy matter. When an assembly of many particles is involved, it is often possible to make statistical statements about the solution. For example, Maxwell and Boltzmann showed by different arguments that in a gas at temperature T, the fraction of molecules with speed between v and $v + \mathrm{d}v$ is proportional to

$$v^2 \exp\left[-mv^2/(2kT)\right] \mathrm{d}v.$$

They did not attempt to follow in detail the motion and collision of individual molecules, and the power of *statistical mechanics*, of which this is an early example, lies in the results it can derive without considering the details. We shall not enter into this matter at all, but consider only such systems as are simple enough, or can be approximated by simple enough models, to allow something like a detailed analysis.

A number of general principles are useful to bear in mind as applicable to a wide range of problems:

(1) *Conservation of linear momentum.* An isolated system has invariant linear momentum, under all circumstances. If the system is not isolated but is acted upon by external forces whose resultant is a known function of time, the rate of change of momentum $\mathbf{P}(=M\dot{\mathbf{R}}$, if M is the total mass of the system and $\dot{\mathbf{R}}$ the coordinate of its centroid) is equal to the force; thus in a uniform gravitational field, $\dot{P} = Mg$. If one chooses to view the system from a frame of reference having acceleration g, the system behaves as though completely isolated. A shell fired from a gun and subsequently exploding in mid-air would appear to anyone in free flight alongside it to be at rest, and to explode in all directions so that the centre of the mass of the fragments remained at rest. 27

(2) *Conservation of angular momentum.* An isolated system has invariant angular momentum about any fixed point, under all circumstances, or if not isolated changes its angular momentum about a given fixed point at a rate determined by the total torque of external forces about that point. 28

(3) *Conservation of energy.* An isolated system of particles interacting by means of conservative forces has invariant total energy. If the particles are moving in a fixed conservative force field (e.g., a gravitational field) their potential energy in that field must be taken into account. If one is considering a system of bodies having internal structure, it is a matter for careful discussion whether in any interaction some of the energy of motion of the centres of mass will be transferred to random motion (heat) within the bodies so that conservation of energy occurs on an atomic but not obviously on a macroscopic scale. This conservation law is one that demands caution, unlike the momentum conservation laws which call for no delicate provisos. 45

EXAMPLES

(1) *The collision of two spheres.* When two otherwise isolated bodies collide, momentum is always conserved; but only if the collision is elastic is energy conserved. Indeed we define an elastic collision as one in which energy is conserved, and it is up to us to determine, usually by experiment, whether a given collision is elastic or not. Soft, plastic materials collide inelastically; two lumps of putty, hurled together, stick and lose much kinetic energy. But two billiard balls, or two steel ball bearings, projected at equal and opposite speeds so as to collide with each other, bounce away with very much the same speed as before—energy is very nearly conserved. If we assume perfectly elastic collisions, conservation of linear momentum and of energy enable a complete

solution to be found very readily. As an example we take a ball moving on a plane and colliding with an identical ball at rest. In a velocity diagram the initial velocities are represented by O and P, O being the origin (i.e., the vector OP represents the velocity of the moving ball). An observer moving with the centroid of the two balls would have his velocity represented by O', the mid-point of OP and would see the balls approaching each other at equal and opposite velocities. The total momentum would be zero, and would remain zero after collision. Therefore the points representing the velocities after collision may be any Q, Q', provided that O' is the mid-point of Q, Q'. Now we have seen in Chapter 3 that if energy is conserved in one inertial frame it is conserved in all, and an elastic collision must therefore have $O'Q$ and $O'Q'$ of the same magnitude as $O'P$ and $O'O$; that is, Q and Q' lie at opposite ends of a diameter of a circle of which OP is another diameter. Then, returning to the laboratory frame of reference, the velocities after collision are OQ and OQ', and are always perpendicular. The actual angle through which the moving ball is deflected is determined by the particular point of impact. B, which is initially at rest, receives a blow along the radius through the point of impact if the balls are smooth, and

45

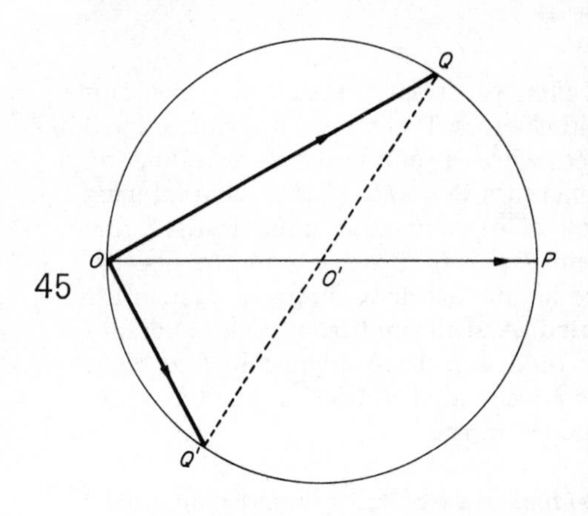

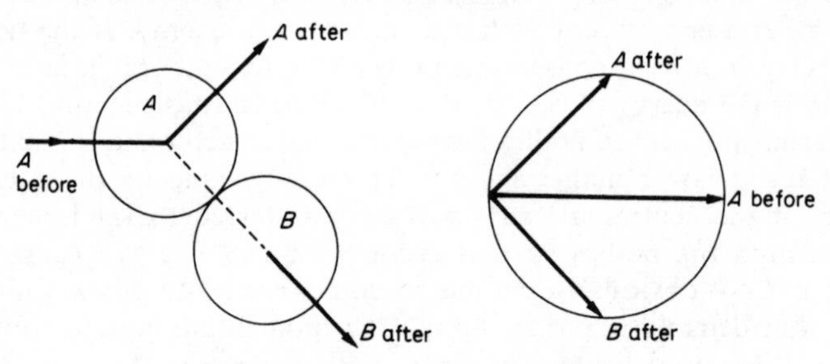

therefore moves off in this direction, as shown, while A is deflected into the normal to this direction. The magnitudes of the velocities are obviously determined by the circle diagram on the right.

PROBLEMS

(1) Show that if the collision is inelastic, and kinetic energy is converted to heat on collision, the angle between the velocity vectors after collision is less than $\pi/2$.

(2) Show that a heavy ball of mass M striking a stationary light ball of mass m cannot be deflected through an angle greater than $\sin^{-1}(m/M)$.

If a collision is inelastic, conservation of momentum still holds but not conservation of energy. We need therefore to know something more about the system to determine what actually happens. Newton suggested after experiments that if we resolve the velocities along the line of impact, the parallel component of relative velocity is reversed in sign and multiplied by a factor $e(<1)$ characteristic of the colliding bodies, while the component normal to the line of impact remains unchanged. Perfectly elastic bodies have their *coefficient of restitution*, e, equal to unity. We can illustrate this by viewing the

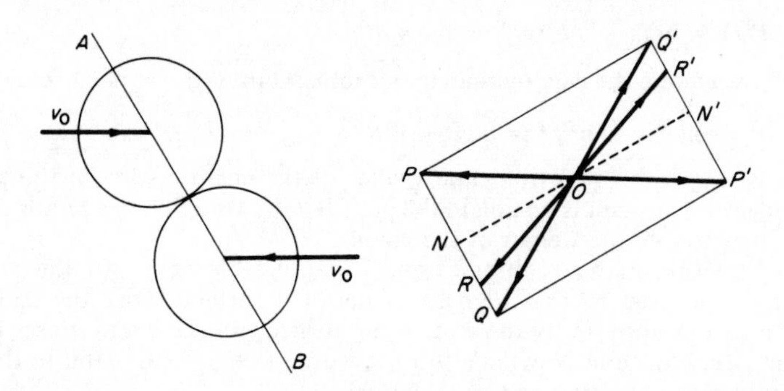

oblique collision of two identical bodies in the centroid frame of reference. AB is the line of impact, and the state of affairs before impact is represented by the two velocity vectors OP, OP'. Complete the rectangle $PQP'Q'$ with sides parallel and perpendicular to AB. Then for an elastic collision the final velocities are OQ and OQ', but for an inelastic collision OR and OR', where $NR = e.NQ$ and $N'R' = e.N'Q'$.

> This rule of Newton's can only apply to smooth bodies. If they are rough or sticky enough to adhere on oblique impact there is no conservation of the tangential component of relative velocity, since conservation is dependent on there being no tangential force at impact. The value of e is not strictly independent of the relative velocity. Hard rubber balls are sold which have remarkable bouncing powers, especially at very low impact speeds—when dropped on the floor from a low height they take an extraordinarily long time to stop murmuring, as they execute tiny bounces at a great speed. If dropped from a height of 20 m they bounce fairly well, but by no means exceptionally (air resistance may account for some of the apparent failure to bounce as well as expected). At the other extreme is 'crazy putty' which has no bounce in it at low speeds but a very high elasticity (e very nearly unity) when dropped from a moderate height. In between these substances there are the majority of ordinary materials for which Newton's rule is a very fair approximation.

(2) *A flexible chain is held by one end so that its lower end just touches a scale pan, and is then released; how does the reading of the scale vary with time?* Here

we have a straightforward example of a problem that is more readily solved by taking an overall view than by trying to worry about the details. If the chain has mass m per unit length and length l, it is acted upon by external forces mgl downwards due to gravity and W upwards, W being the force exerted by the scale pan on the chain, which by N3 is the same as the weight recorded by the scale. At a time t after release the top of the chain has fallen $\frac{1}{2}gt^2$, and this is the length resting in a heap in the scale pan. The rest, of length $l - \frac{1}{2}gt^2$, is falling at a speed of gt, so that the total momentum is given by

$$P(t) = m(l - \tfrac{1}{2}gt^2)gt.$$

We now equate the rate of change of momentum to the applied force:

$$\dot{P} = mgl - \tfrac{3}{2}mg^2t^2 = mgl - W;$$

i.e., $W = \frac{3}{2}mg^2t^2 = 3$ times the weight of the chain resting in the pan. As soon as all the chain is resting in the pan, W falls abruptly by a factor of 3 and becomes simply the weight of the chain.

Notice that energy conservation is inapplicable here. All the potential energy possessed by the chain by virtue of its height above the pan at the beginning disappears by the end, being turned into heat and noise. But the momentum law and Newton's third law are of course still valid in this non-equilibrium, non-conservative situation.

(3) *As a result of frictional forces between an artificial satellite and air molecules in the upper atmosphere, the satellite moves more rapidly.* This apparent paradox is readily analysed in terms of angular momentum. The Earth's gravitational pull exerts a central force on the satellite, which, if this were all, would maintain constant angular momentum. But the air friction acts in the opposite direction to the satellite's velocity and therefore exerts a torque which tends to reduce its angular momentum about the centre of the Earth. This is enough to determine how the orbit radius and the velocity of the satellite change. For if the attractive force is written as μ/r^2 and the satellite has velocity v, the relation between orbit radius and velocity is obtained from the centripetal acceleration, $mv^2/r = \mu/r^2$, or

$$v = (mr/\mu)^{-1/2}.$$

But

$$L = mvr = (m\mu r)^{1/2},$$

and therefore, as L is reduced, so is r in proportion to L^2; at the same time v, being proportional to $r^{-1/2}$, increases as $1/L$.

Alternatively, one may note that the total energy of the satellite decreases by reason of air friction, and that for a circular inverse-square orbit the total energy is the negative of the kinetic energy and is $-\mu/(2r)$. Thus the kinetic energy increases and r decreases.

76

The instantaneous effect of each collision of the satellite with a gas molecule is to slow down the satellite; but the longer-term effect is to bring it closer to Earth and increase its speed. An extreme example illustrates this point. If we consider collisions to be so rare that the satellite executes more than one orbit between them, we may imagine it in a circular orbit and then slightly slowed by a collision at P. The result is to throw it into an elliptical orbit which runs nearer the earth at the opposite side and goes faster there. It

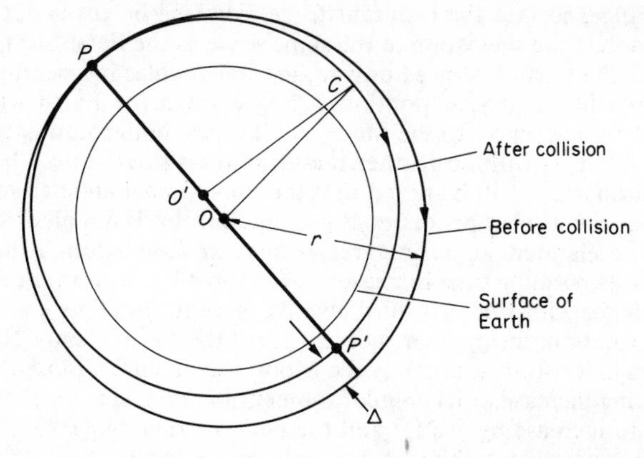

is this secondary effect which is greater, and the mean speed is increased by the collision. As the speed is reduced by δv at the moment of collision the attractive force is unchanged, and ρ, the instantaneous radius of curvature of the orbit, must decrease from its initial value r to keep v^2/ρ constant; thus $\delta\rho = 2r\delta v/v$. Suppose then that at P', the opposite end of the ellipse, the orbit has fallen through Δ; the ellipse has semi-major axis $a = r - \frac{1}{2}\Delta$, and one of the foci is still at O, of course, while the other is at O' such that $O'P' = OP = r$. The two foci are therefore $\pm\frac{1}{2}\Delta$ from the centre of the ellipse, and since OC is equal to the semi-major axis we see that the semi-minor axis b is given by $b^2 = (r - \frac{1}{2}\Delta)^2 - (\frac{1}{2}\Delta)^2 = r^2 - r\Delta$. Now the radius of curvature at the end of the major axis is b^2/a, i.e., $r(1 - \frac{1}{2}\Delta/r)$ to first order in Δ. This is $\frac{1}{2}\Delta$ less than the radius before the collision, and the decrease in radius must be equated to $2r\delta v/v$ to give $\Delta/r = 4\delta v/v$. The collision reduced the speed and the angular momentum by a fraction $\delta v/v$; L is thereafter conserved so that at P' the fractional reduction in radius must be accompanied by a corresponding increase in speed by a fraction $4\delta v/v$, i.e., $3\delta v/v$ above the initial speed before collision. On average, then, there is a gain in speed.

(4) *The effect of tidal friction on the distance of the Moon.* This problem is closely related to the last. If the tidal bulges could exactly follow the Moon they would not influence its motion appreciably. But the inevitable dissipation of the energy of tidal motion into whirlpools and smaller eddies leads to a frictional drag between the sea and the solid Earth, so that the bulges are dragged ahead of the Moon's position, as shown very schematically in the diagram. The nearer bulge attracts the Moon slightly more strongly than the

81

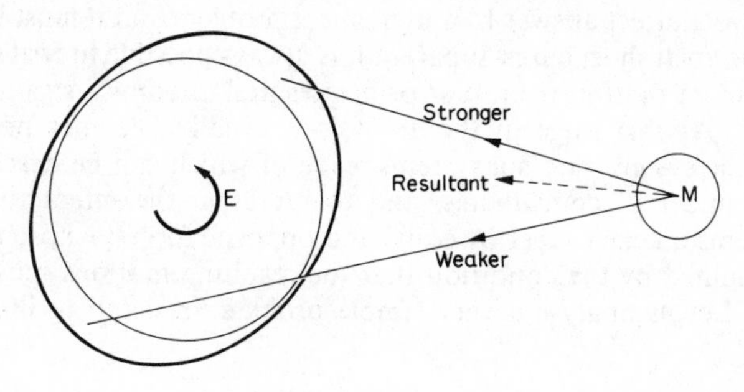

further bulge, so that the resultant force from the bulges is not central, but tends to accelerate the Moon in the same sense as the Earth's rotation. At the same time the Earth is slowed down, the total angular momentum being conserved, and if the process continues long enough the Earth will ultimately present the same aspect to the Moon at all times, both rotating together once in about 47 days, with no further tidal friction, since the tidal bulges will remain stationary. It will be noted that the quoted final angular velocity is less than either constituent processes at present, but this is not an error; for as the Moon is accelerated, acquiring greater angular momentum, it moves further away and its periodic time increases as the three-halves power of its distance, in accordance with Kepler's third law. At present, for all its greater mass and speed of rotation, the angular momentum of the Earth is only 21% of that of the Moon in its orbit; ultimately the Moon will acquire almost all of this, and the resulting increase of its angular momentum by a factor 1·21 will cause the distance to increase by $(1{\cdot}21)^2$ and the periodic time by $(1{\cdot}21)^3$.

69

The process is very slow; it has been estimated that as a result of tidal friction the length of the day has increased by 1 s in the last 120,000 years. This result must, however, be seen in the context of the age of the Moon, perhaps 4×10^9 years, in which time tidal friction could well have been responsible for the recession of the Moon from a point very close to the Earth (if indeed it was formed by being torn from the molten Earth—a hypothesis not universally accepted) to its present position. At the same time the tides set up by the Earth in the Moon's molten, or at least deformable, rocks could have accounted for its slowing down so as to present the same face to the Earth at all times.

PROBLEM

Show that the tides set up by the Sun will have the reverse effect of bringing the Moon back towards the Earth, until in the end, if no other catastrophe has intervened, the Moon will crash into the Earth. [Energy arguments yield this result very readily, but it is instructive to follow the process in detail to see how the Sun causes Earth and Moon to spin ever faster.]

These examples have been given to show how general principles may often give a direct answer to a dynamical problem, but it must be stressed that if no such short cut is apparent it is always possible to write down the equations of motion from first principles and attempt a straightforward solution. At this stage in the analysis it usually becomes necessary to dissect the system into subsystems, each of which can be described by a small number of coordinates, and to represent the interactions of the subsystems on each other by equal and opposite forces, whose magnitude is determined by the condition that the resulting motions must be compatible. Let us analyse a very simple problem in detail to illustrate the idea.

EXAMPLE

22

The trolley experiment discussed in Chapter 3 is a complex dynamical system which can be broken down into the trolley, bits of string, a pulley, and a weight (if we have any doubts we can take the trolley to pieces, too). We imagine the string cut, and forces T_1, T_2 etc. applied to the cut ends of such magnitude that they cause identical motions to those executed by the real

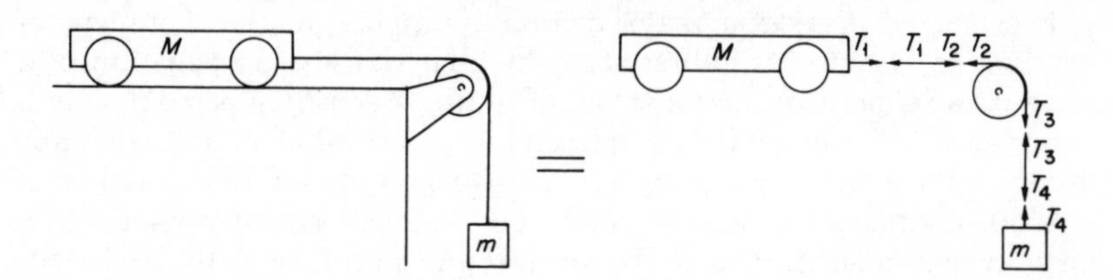

system; in this case, if we assume the string inextensible, we must adjust the tensions T so that its various parts remain at their correct distances. If the string is very light it needs no significant resultant force to give it the same acceleration as the trolley or the weight, and we approximate by setting $T_1 = T_2$, $T_3 = T_4$. The difference $T_3 - T_2$ provides what torque is required to overcome any friction in the pivot of the wheel and to give the wheel the angular acceleration which is compatible with the string not slipping. For convenience here we assume no friction and a light wheel, and put $T_3 = T_2$. If this is a false assumption, we have to replace it by a better description of the wheel and pivot and find how T_2 and T_3 are really related. For the moment, however, we have $T_1 = T_2 = T_3 = T_4$ (= T, say) and we have to adjust T so that the trolley and the weight have the same acceleration, a, i.e.,

$$T = Ma \quad \text{and} \quad mg - T = ma;$$

or

$$a = \frac{m}{M + m} g \quad \text{and} \quad T = \frac{M}{M + m} mg.$$

Note that if the trolley is loaded first to a total mass M_1 and then to M_2, and subjected in both cases to the same force due to a given m, the ratio of accelerations is $(M_2 + m)/(M_1 + m)$, and will differ for different choices of m. It was for this reason that in Chapter 3 we assumed accelerations much less than g, equivalent to choosing $m \ll M$. Having reached this point in our analysis of Newton's laws, we might well go back to the experiment to see if our corrected formula fits the facts even better, and if we satisfy ourselves on this point we are confirmed in our belief in the laws.

PROBLEM

A double pendulum consists of a string of length l carrying a bob, with an identical bob and string hung from it. The whole executes small oscillations in a vertical plane. Show that there are two modes of pure simple harmonic oscillation in one of which the bobs swing in phase and in the other in antiphase, and that the frequencies of these two modes are those of a simple pendulum of length $l/(2 \pm \sqrt{2})$. Show also how these lengths are related to the point where the line of the lower string cuts the vertical axis. Set up the experiment and verify the result.

The idea of dissection may be usefully applied to the treatment of continuous systems, as is illustrated by analysing the propagation of a wave on a string. Consider a string of uniform density ρ per unit length under tension T, and let it be displaced in the vertical plane so that at any instant t its shape is given by $y(x, t)$. We now derive the equation of motion, assuming y to be very small; there is then virtually no horizontal displacement associated with the vertical movement, since the difference

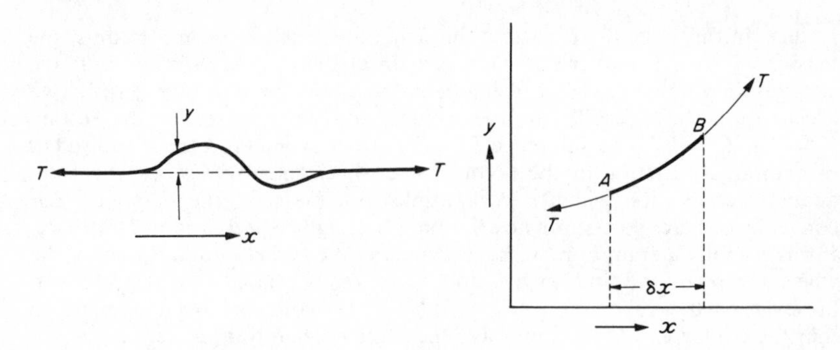

between the true length of the string and its projected length along the x-axis depends on the square of the angle of tilt. To the same degree of approximation we may take the instantaneous motion at any point to be normal to the direction of the string, with no acceleration of the string along its own length. If therefore we isolate a short section δx the same force is needed at each end to prevent longitudinal acceleration, and clearly this force must be the tension T applied at the ends. If the section of string is curved, there is a vertical component of force acting on it. For the vertical component of T at A is

$$-T(\partial y/\partial x)_A$$

while at B it is

$$+T(\partial y/\partial x)_B,$$

i.e.,

$$T(\partial y/\partial x)_A + T(\partial^2 y/\partial x^2)\,\delta x.$$

The resultant is

$$T(\partial^2 y/\partial x^2)\,\delta x$$

upwards, and this is responsible for vertical acceleration of the element, whose mass is $\rho\ \delta x$. Hence

$$\rho\ \delta x(\partial^2 y/\partial t^2) = T(\partial^2 y/\partial x^2)\ \delta x,$$

or

$$\partial^2 y/\partial x^2 = (\partial^2 y/\partial t^2)/V^2, \qquad (6.1)$$

where $V = (T/\rho)^{1/2}$.

This is the so-called *wave equation* in one dimension, the most fully studied of all the equations that can describe various types of wave-motion. This particular equation describes a non-dispersive wave, that is, one that propagates without change of shape whatever that shape may be. Any small disturbance, like the kink shown earlier, runs along at the characteristic velocity V, as may be inferred from the general solution of (6.1) which can be written $y = f_1(z) + f_2(z')$, where f_1 and f_2 are any continuous functions whatever, and $z = x - Vt$, $z' = x + Vt$. This can be checked directly by differentiation; thus to take the first term only, if

$$y = f_1(z),$$
$$\partial y/\partial x = (\mathrm{d}f_1/\mathrm{d}z)(\partial z/\partial x) = \mathrm{d}f_1/\mathrm{d}z,$$

and

$$\partial^2 y/\partial x^2 = \mathrm{d}^2 f_1/\mathrm{d}z^2.$$
$$\partial y/\partial t = (\mathrm{d}f_1/\mathrm{d}z)(\partial z/\partial t) = -V\,\mathrm{d}f_1/\mathrm{d}z,$$

and

$$\partial^2 y/\partial t^2 = V^2\,\mathrm{d}^2 f_1/\mathrm{d}z^2.$$

Therefore (6.1) is satisfied by $f_1(z)$, and the same goes for $f_2(z')$. The meaning of the solution $f_1(z)$ is that at time $t = 0$, $z = x$ and $f_1(x)$ represents the instantaneous shape of the disturbance; at a subsequent moment t, $z = x - x_0$ where $x_0 = Vt$, and the disturbance is $f_1(x - x_0)$ which is the same as before but shifted to the right through x_0 i.e., Vt. The function therefore represents a disturbance of constant shape moving to the right with velocity V. Similarly $f_2(z')$ represents another disturbance moving to the left with velocity V. The complete solution, being the sum of both, shows that if two such disturbances meet they pass through one another and emerge on the other side unchanged. The theory, it should be remembered, involves an approximation which is valid only for small disturbances; if the disturbance is larger, it changes its form as it propagates. It is very common in wave-motions of various types (sound waves in air, for example) for the change of form of large-amplitude waves to result in a sharpening of the wave-form at the leading edge until it is very abrupt; this is a *shock-wave*, whose theoretical treatment is difficult and mathematically interesting, and which is very important in discussing explosions, sonic booms and other noisy disturbances.

When analysing the trolley problem, we imagined we had cut the string and determined subsequently what forces would be needed to produce compatible motions of the parts of the system, so that the string might just as well not have been cut. This procedure gives operational meaning to what may be regarded as too vague a definition of the *tension*, T, in the string if we think of it only in terms of the force by which the atoms on one side of a section attract those on the other. The value of such an operational definition becomes apparent in more general situations where it will not always happen that if we imagine a system dissected, the various parts can be persuaded to undergo compatible motions simply by exerting a tensile force. Consider a uniform beam of weight W_0

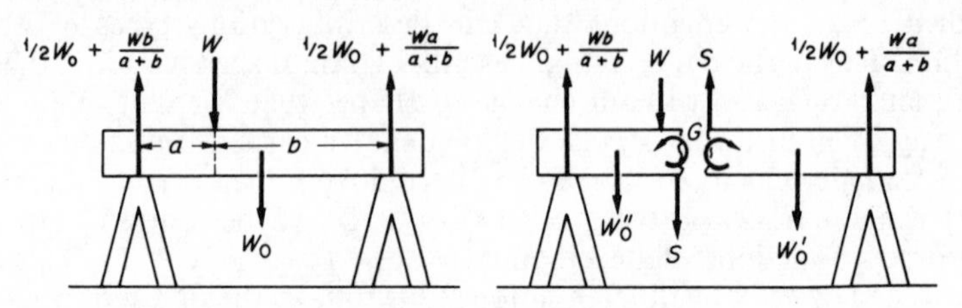

resting on two trestles and loaded with a weight W. Without any dissection we can determine what forces are exerted by the trestles on the beam, which, being at rest, must be acted upon by such forces as produce no resultant force or torque. The solution is shown in the diagram, W being divided in proportion to the lever arms a and b. If now we cut the beam in two and seek to find what forces are needed at the cut ends to keep the separate pieces (of weights W_0' and W_0'') in equilibrium, it is clear that there is in general a resultant vertical force on the right-hand piece and that a vertical force (*shearing force*) S equal to $W_0' - \frac{1}{2}W_0 - Wa/(a + b)$ must be applied to neutralize it. But this is not enough, for the three forces shown do not in general exert a vanishing torque. The moment of all forces about, say, the cut end must be neutralized by a torque G (*bending moment*) applied to it. The mechanism by which G is produced is the compression of the upper layers of material and stretching of the lower layers, so that the left piece of the beam presses on the right piece at the top and pulls at the bottom, setting up a pure torque with no resultant force. Since different layers of the beam must be differently strained to produce this couple, the beam is bent; the curvature shown is that required to produce a clockwise moment. The relation of the curvature to the bending moment is a basic part of the theory of engineering structures, into which we shall go no further.

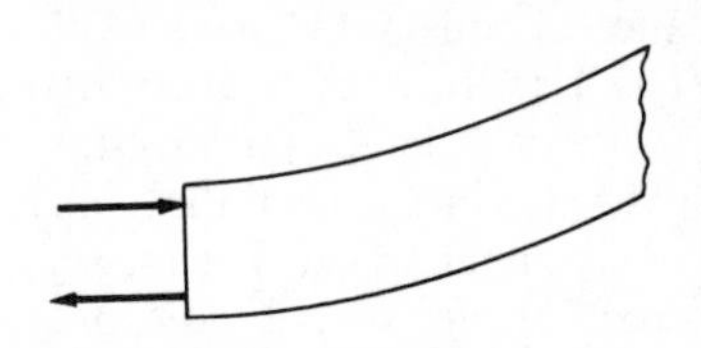

In the most general situation when the beam is simultaneously bent, pulled and twisted, the stresses at any cross-section can be resolved into a force and a torque whose vector directions may be anywhere, so long as they serve to maintain a cut portion in equilibrium or, if it is moving, cause it to move in a way that matches the movement of the other cut portion on which equal and opposite force and torque are acting.

To leave the beam and generalize still further, we may define the state of internal stress at a point in a solid or liquid by the same device of imagining it cut locally, and asking what must be done to prevent the cut from showing. The answer is to apply a force along one cut edge and an equal and opposite force along the other, adjusting the strength and direction appropriately. A rubber sheet, for example, lying unstretched can be cut without the slit opening up, but if it is pulled out of its original shape

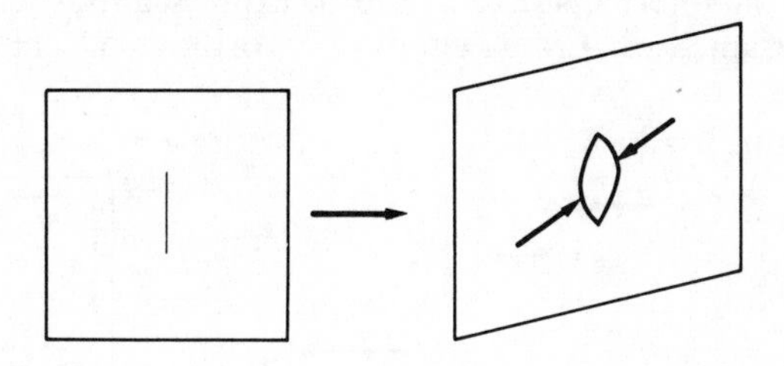

the slit will open and need forces applied to its edges to close it again. From what was said about the beam, it might be supposed that a torque would also be needed, but this is not so; for the bending moment in the beam is the result of having an extended cut along which the local internal force varies; it is a tension at one end and a compression at the other, in addition to the shearing force. By contrast, the cut in the rubber sheet is short enough for the state of strain to be constant along it. So, too, if we consider the local state of affairs on the cut surface of the beam, the local force that must be applied to an elementary area dS is a simple force $\mathbf{P}\,dS$ of magnitude proportional to dS; $\mathbf{P}$ is then a force per unit area (*stress*).

In the rubber sheet illustration of the effect of a cut, and how to undo it, we made the cut in a particular direction. What would happen if we

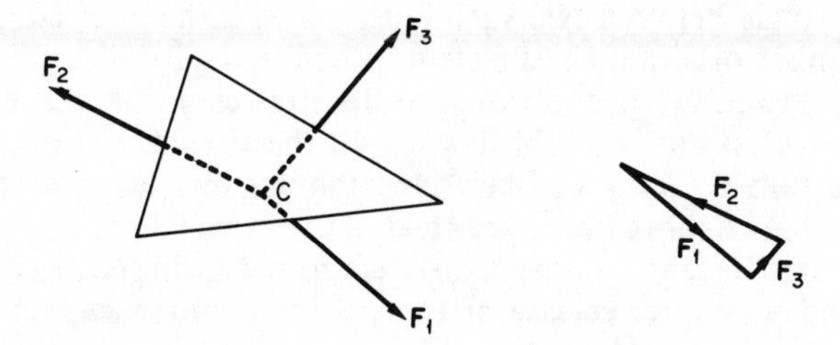

cut in a different direction? In general the force needed to undo the effect would be different. However, once we have determined the required force for two different cuts, the answer is known for all others. Imagine a small triangle cut from the sheet; then the forces distributed along the cut edges can be compounded, so far as their effects on the triangle as a whole are concerned, into three forces acting one at the centre of each edge. If the triangle is not to move, the lines of action of the three forces must be concurrent and the forces must have zero resultant. Thus if $\mathbf{F}_1$ and $\mathbf{F}_2$ are known, $\mathbf{F}_3$ must pass through C and be of such a magnitude that the three vectors form a closed triangle. This is enough to determine $\mathbf{F}_3$*; it is, in fact, more than enough. For the directions of $\mathbf{F}_1$ and $\mathbf{F}_2$ determine the position of C and hence the direction of $\mathbf{F}_3$, so that the shape of the triangle of forces is determined and the relative magnitudes of $\mathbf{F}_1$ and $\mathbf{F}_2$ are not adjustable. There are clearly quite firm limits to the permissible patterns of the internal stress, as can be expressed most conveniently by using Cartesian axes. Let us choose two orthogonal cuts to define the

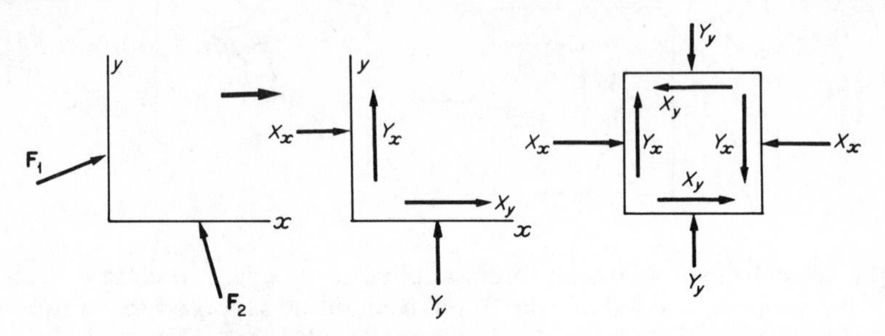

axes, and resolve the forces acting along unit length of cut into two normal stresses X_x and Y_y, and two shear stresses Y_x and X_y (X_x is a force in the x-direction acting on a line cut normal to x; Y_x is a force in the y-direction acting on a line cut normal to x). By considering the equilibrium of a square of the material we see that these four stress components are not independent, but that $X_y = Y_x$.

These concepts are readily extended to three dimensions, and the relation of the internal stresses to the resulting distortion of the material (*strain*) forms the basis of the theory of elastic and other deformable bodies. It is enough to mention here that the degree of distortion can be related to the changes in length of lines drawn in the material and the changes of angle between two such lines. As with the internal stresses, a limited number of strains is sufficient to define the local state of distortion. If the strains are proportional to the stresses, in the sense that any one element of strain is doubled when all the stresses are doubled, the material is said to obey Hooke's law, though this law was originally stated in the much more limited context of the extension of a spring being proportional to its tension. Just as an overstressed spring ceases to respond linearly, and may after release of tension be found to have acquired a

* The implication that this is true only in equilibrium is in fact incorrect. As the triangle is made smaller, while keeping its shape, the force per unit length stays constant, so that $\mathbf{F}_1$ etc. diminish in proportion to the linear dimension. But the mass of the triangle diminishes as the square of its dimension, and thus for an infinitesimal triangle the forces must balance if the triangle is not to have infinite acceleration. When the sheet is not in equilibrium, for instance when parts of it are vibrating in the plane, the accelerations are produced, not by violations of these relations between the stresses, but by variations of the stresses from point to point, so that the resultant force on a small element is non-vanishing.

permanent set, or if the stress is maintained it may continue to flow plastically and may eventually break, so a solid may be strained beyond its elastic limit and become set, or if the stress is maintained it may continue to flow plastically and may eventually break. Here we begin to approach the problems that engineers describe under the heading *strength of materials* and which are part of a rapidly growing field of *materials science*.

A fluid is a special case of a substance that can maintain no shear stress in equilibrium. If we take a square vessel filled with fluid and apply tangential forces to shear it, as shown, the fact that it is filled with fluid

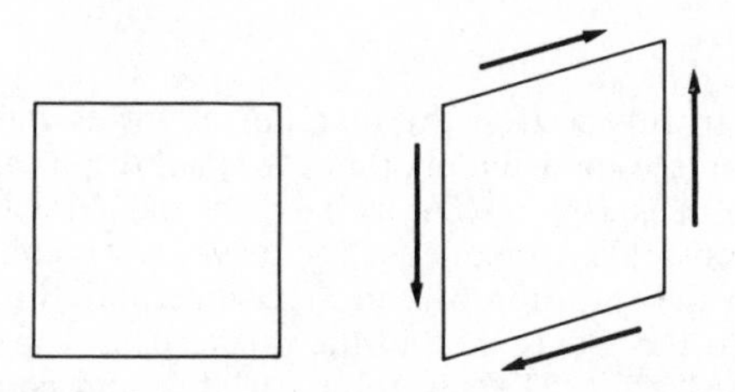

involves no extra force once the fluid has come to rest. While it is still flowing, its viscosity may make extra shearing force necessary, and we may characterize a fluid as a substance that can only support shearing stresses when it is in motion.

PROBLEMS

(1) Show that a small triangular prism in a fluid is only in equilibrium if the normal force per unit area (*pressure*) acting on each face is the same, and that therefore the pressure in a stationary fluid is the same in all directions.

(2) Show that if the pressure in a fluid varies with position, the force acting on a volume element $\mathrm{d}V$ is $(-\operatorname{grad} P)\,\mathrm{d}V$, so that P can be considered as a potential defining a conservative field of force.

41
104

EXAMPLES

(1) *A horizontal rubber membrane is stretched uniformly and loaded with weights. What shape does it take up?* If the membrane is stretched without shearing, an initially circular hole remains circular after stretching, and can be restored to its original form by pulling uniformly all round its edge, without any tangential forces. Hence the state of stress is one of uniform tension, T per unit length, say. Consider the equilibrium of a small rectangular element of sides δx and δy. Just as with the problem of a wave on a string, if the membrane is only slightly displaced we can take T to be the same everywhere, even after displacement. Equilibrium is maintained by the curvature of the membrane producing a vertical component of the tension to balance the load, which we write as $W\,\delta x\,\delta y$, W being the vertical force per unit area, a function of position. Take a section of the element in a vertical plane containing

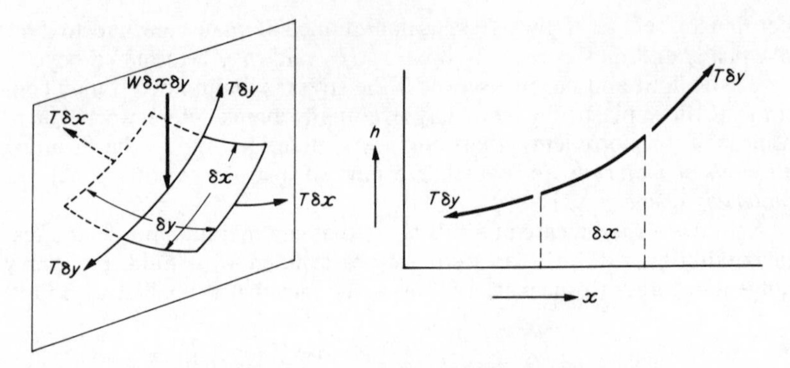

the x-direction. In this section the diagram of forces resembles that for a string, and we can see immediately that for small displacements the upward component of force is $T\,\delta y \times (\partial^2 h/\partial x^2)\,\delta x$, the partial derivative expressing the fact that we have taken a plane section in which y is constant and are concerned only with the variation of h in the x-direction. Similarly by taking a section parallel to the y-axis we find the other edges to produce an upward force $T\,\delta x \times (\partial^2 h/\partial y^2)\,\delta y$. Taking these together and equating to the load $W\,\delta x\,\delta y$, we find the differential equation for the displacement of the membrane,

$$T\left(\frac{\partial^2 h}{\partial x^2} + \frac{\partial^2 h}{\partial y^2}\right) = W(x, y). \tag{6.2}$$

Notes. (a) Since the value of W at a given point is not influenced by the choice of x and y axes, we expect the left-hand side of (6.2) to take the same value for any choice of axes. This may readily be verified by expanding h in a power series about some point (x_0, y_0):

$$h = h_0 + a(x - x_0) + b(y - y_0) + c(x - x_0)^2 + d(x - x_0)(y - y_0) + e(y - y_0)^2 + \cdots.$$

Then at (x_0, y_0),

$$\nabla^2 h = \partial^2 h/\partial x^2 + \partial^2 h/\partial y^2 = 2(c + e).$$

If we now rotate the axes through θ, we must replace x by $x\cos\theta + y\sin\theta$ and y by $y\cos\theta - x\sin\theta$ to obtain the power series expansion for h in the rotated coordinate system. Direct evaluation shows that the coefficient of x^2 is now $c\cos^2\theta + e\sin^2\theta - d\cos\theta\sin\theta$, and the coefficient of y^2 is $e\cos^2\theta + c\sin^2\theta + d\cos\theta\sin\theta$. The sum of the two is still $(c + e)$ and therefore $\nabla^2 h$ takes the same value. We can choose θ so that the coefficient of xy vanishes, and the axes are then said to lie along the principal directions of curvature.

(b) For small displacements h (more correctly, for such h that the gradient of the surface is small), the geometrical meaning of $\partial^2 h/\partial x^2$ is the reciprocal of the radius of curvature in the plane containing the x-axis, and $\partial^2 h/\partial y^2$ is the reciprocal of the radius in the plane containing the y-axis. Writing these radii as R_1 and R_2, we see that $1/R_1 + 1/R_2$ is invariant to any rotation of the axes, and (6.2) may be written

$$T\left(\frac{1}{R_1} + \frac{1}{R_2}\right) = W(x, y). \tag{6.3}$$

This is a well-known result for the curvature of a soap film, which is also ideally a membrane having uniform isotropic tension T. If the load is caused by a pressure difference ΔP between the two sides, $W = \Delta P$.

(c) If the membrane is unloaded except at those points where it is held displaced, (6.2) takes the form of Laplace's equation in two dimensions,

$$\nabla^2 h = \partial^2 h/\partial x^2 + \partial^2 h/\partial y^2 = 0, \tag{6.4}$$

which must be solved subject to boundary conditions fixed by the specified displacements. Since Laplace's equation commonly occurs in many branches of physics, it is of value to be able to generate solutions by displacing a stretched membrane or blowing a soap film on a suitably bent wire frame. This is the principle underlying analogue methods of solving Laplace's equation, which we shall meet again in electrostatics. 156

(d) The vanishing of $\nabla^2 h$ implies that at all points $R_1 = -R_2$ (*anticlastic curvature*). Let us imagine how the curvature at a point varies as we take the plane in which the curvature is determined to be first parallel to the x-direction and then to be rotated until it is parallel to the y-direction. The curvature $1/R$ in the section must change sign in the process, and there must be a section in which R is infinite, i.e., the membrane is uncurved;

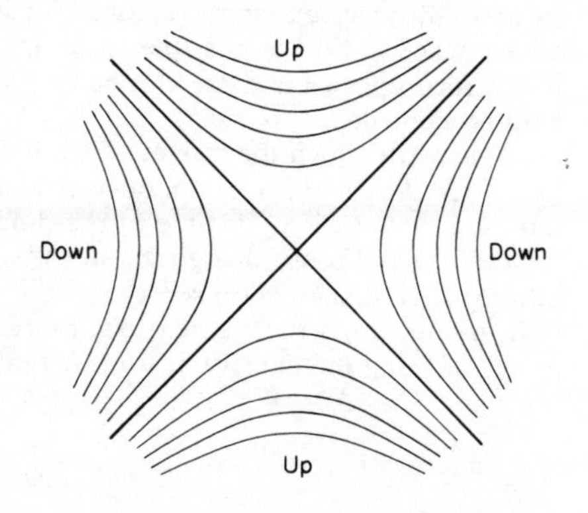

clearly at right angles to this direction there will be another section without curvature. If we draw the tangent plane at this point, the contours representing the distance of the surface from the plane will be rectangular hyperbolas (at least close to the point—obviously they will depart from this simple pattern at great distances) with asymptotes in the directions where the curvature is zero.

(e) In general, mathematical determination of the shape of a membrane or soap film involves the solution of (6.2) or (6.4), but there are special cases where simpler procedures suffice. Consider, for example, a light membrane supported horizontally by a circular hoop of radius b and loaded at the centre with a circular weight w of radius a. If we consider the force exerted by the membrane outside a circle of radius r on the membrane within, it has a vertical component $T\,dh/dr$ per unit length, giving a total vertical force $2\pi rT\ dh/dr$, which must support the weight w. Hence

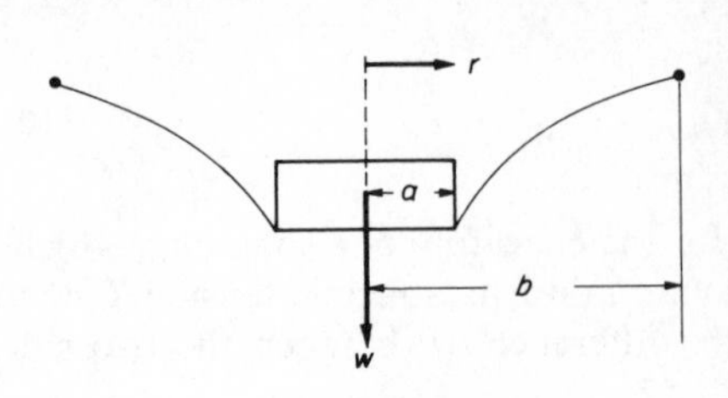

$$r\,dh/dr = w/(2\pi T),$$

or

$$h = w \ln r/(2\pi T) + h_0.$$

Therefore the depression of the centre, $h(b) - h(a)$, is

$$w \ln (b/a)/(2\pi T).$$

(2) *Dynamics of a non-viscous fluid.* We start to develop the consequences of the result posed as a problem above, that when shearing stresses are absent (as in a non-viscous fluid) or negligible, the force causing motion is due to the gradient of the pressure, being $-\text{grad}\, P$ per unit volume of fluid. Now when the fluid moves under the influence of this force-density its velocity $\mathbf{v}(\mathbf{r})$ cannot be entirely arbitrary; for instance, the fluid cannot continue indefinitely to flow into a given small region. If it is compressible, a certain degree of convergence of flow is possible, accompanied by a local rise of pressure, but if it is incompressible there can be no excess of inflow over outflow for any volume element.

Let us write down the rate at which fluid enters or leaves a region bounded by a fixed closed surface S. Consider a surface element $d\mathbf{S}$ at a point where the fluid velocity is $\mathbf{v}$. In time δt the fluid contained in a pill-box defined by $d\mathbf{S}$ and $\mathbf{v}\ \delta t$ will pass through the surface element, and since the volume of the pill-box is $(\mathbf{v}.d\mathbf{S})\ \delta t$, $\mathbf{v}.d\mathbf{S}$ represents the volume of fluid leaving the region in unit time through the element $d\mathbf{S}$. Since the mass of this amount of fluid is $\rho\mathbf{v}.d\mathbf{S}$, ρ being the density, it is clear that the total rate at which the mass of

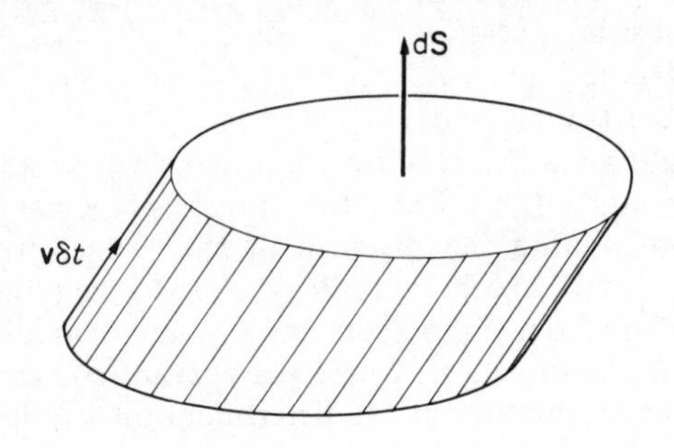

fluid inside S is being depleted is just $\int_S \rho\mathbf{v}.d\mathbf{S}$. For an incompressible fluid, ρ is constant, and there can be no depletion, so that $\int_S \mathbf{v}.d\mathbf{S}$ must vanish over all closed surfaces drawn within the fluid.

Let us apply this argument to an infinitesimal volume element. The value of $\rho\mathbf{v}.d\mathbf{S}$ for one of the faces bounded by δy and δz is $-\rho v_x\, \delta y\, \delta z$ (flow into the element) and for the other $(\rho v_x + \partial(\rho v_x)/\partial x.\delta x)\, \delta y\, \delta z$, the difference being due to the change in ρv_x with position. The net outflow over these two faces is $(\partial(\rho v_x)/\partial x)\, \delta V$, where δV is written for the volume element $\delta x\, \delta y\, \delta z$. Adding the net outflow for the other faces, we see that the rate of depletion of material from the volume element is div $(\rho\mathbf{v})\, \delta V$, div $(\rho\mathbf{v})$ being written for the function $\partial(\rho v_x)/\partial x + \partial(\rho v_y)/\partial y + \partial(\rho v_z)\, \partial z$. The *divergence* of a vector field $\mathbf{A}(\mathbf{r})$ is a scalar function of position defined by $\partial A_x/\partial x + \partial A_y/\partial y + \partial A_z/\partial z$, and describes the net outward flux of the vector $\mathbf{A}$ through the surface of an element of volume δV, divided by the magnitude of δV. It is in fact a measure of flux per unit volume, and the example of fluid flow by which we have chosen to introduce the concept is the clearest illustration of the physical meaning of this mathematical construction. It is easy to satisfy oneself, by dividing up a larger volume into infinitesimal elements, that the net flux of $\mathbf{A}$ through the surface S, $\int_S \mathbf{A}.d\mathbf{S}$, is the volume integral of the divergence of $\mathbf{A}$, $\int_V$ div $\mathbf{A}\, dV$, taken over the volume bounded by S. This is Gauss' theorem for vector fields, referred to henceforth as IX.

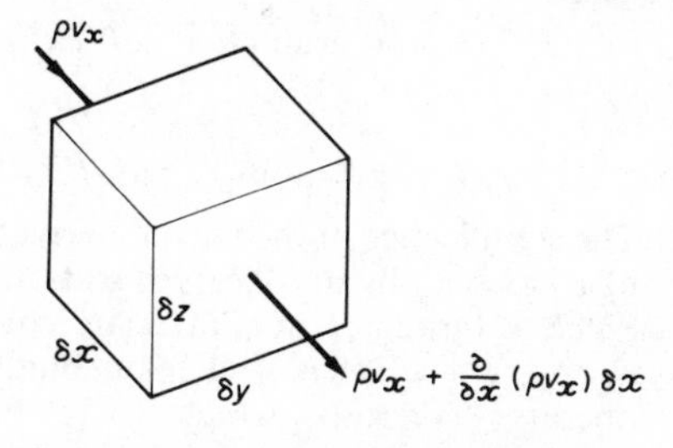

For an incompressible fluid, div $\mathbf{v} = 0$ everywhere, and the flow is said to be non-divergent. Let us, however, consider first a slightly compressible fluid for which ρ may be taken as sensibly constant when we write the rate of depletion of material as ρ div $\mathbf{v}$, rather than the strictly correct div $(\rho\mathbf{v})$. In time δt the amount of fluid leaving a volume element δV is $(\rho$ div $\mathbf{v})\, \delta V\, \delta t$, which represents a proportionate decrease in density of (div $\mathbf{v})\, \delta t$, since the total mass in the element is $\rho\, \delta V$. The *bulk modulus*, K, of a fluid is defined in terms of the pressure increase needed to bring about a given proportionate increase in density,

$$K\, \delta\rho/\rho = \delta P,$$

and we may therefore express the pressure change accompanying the divergent fluid flow as $-K$ (div $\mathbf{v})\, \delta t$. The equation governing the rate of change of pressure at this point therefore takes the form

$$\partial P/\partial t = -K \text{ div } \mathbf{v}. \tag{6.5}$$

Another equation relating P and $\mathbf{v}$ may be derived by considering the acceleration of a volume element under the influence of the force density, $-$grad P. Some careful thought is needed here; we wish to describe how the velocity at a given point $\mathbf{r}_0$ changes with time, and this is not the same as the rate of change for a given element of the fluid which moves as it accelerates. If the velocity at $\mathbf{r}_0$ is $\mathbf{v}$ at a certain instant, the velocity at the same point after

99

an interval δt will be the velocity of a fluid element that was initially located at $\mathbf{r}_0 - \mathbf{v}\,\delta t$ and has moved to $\mathbf{r}_0$ in the interval. But while it moved it was subjected to a force density $-\text{grad}\,P$, which caused its velocity to change by an amount $(-\text{grad}\,P)\,\delta t/\rho$. There are thus two contributions to the change of $\mathbf{v}$ at $\mathbf{r}_0$, and it is convenient to start by considering just one component, v_x say, of $\mathbf{v}$. First, the variation of the scalar v_x with position around $\mathbf{r}_0$ at any instant can be characterized by grad v_x, and the value of v_x at $\mathbf{r}_0 - \mathbf{v}\,\delta t$ differs from that at $\mathbf{r}_0$, at the same instant, by $(\text{grad}\,v_x)(-\mathbf{v}\,\delta t)$; secondly, the element acquires an extra x-component of velocity, of magnitude $-(\delta t/\rho)\,\partial P/\partial x$.

Thus

$$\delta v_x = -[\mathbf{v}\,.\,\text{grad}\,v_x + (\partial P/\partial x)/\rho]\,\delta t;$$

i.e.,

$$\partial v_x/\partial t = -\mathbf{v}\,.\,\text{grad}\,v_x - (\partial P/\partial x)/\rho. \tag{6.6}$$

The significance of the partial derivative is that we evaluate the rate of change of v_x as seen by an observer stationed at $\mathbf{r}_0$, not one moving with the fluid. Since $\mathbf{v}\,.(\text{grad}\,v_x)$ in Cartesian coordinates takes the form $(v_x\,\partial v_x/\partial x + v_y\,\partial v_x/\partial y + v_z\,\partial v_x/\partial z)$, it may equally well be written in terms of the scalar operator $(\mathbf{v}\,.\,\text{grad})$, which is to be interpreted as $(v_x\,\partial/\partial x + v_y\,\partial/\partial y + v_z\,\partial/\partial z)$, and in this notation the meaning of (6.6), when written in complete vector form, is unambiguous,

$$\partial \mathbf{v}/\partial t = -(\mathbf{v}\,.\,\text{grad})\mathbf{v} - \text{grad}\,P/\rho. \tag{6.7}$$

As illustrations of hydrodynamical problems we consider briefly two applications of (6.5) and (6.7). First, the propagation through a fluid of sound waves, with such small amplitude that the first term in (6.7), which has magnitude proportional to v^2, is negligible. Then by taking the time derivative of (6.5) and substituting (6.7) we derive an equation for P,

$$\partial^2 P/\partial t^2 = -K\,\text{div}\,(\partial \mathbf{v}/\partial t) = K\,\text{div grad}\,P/\rho = K\nabla^2 P/\rho, \quad \text{by X;}$$

i.e.,

$$\nabla^2 P - \frac{1}{V^2}\frac{\partial^2 P}{\partial t^2} = 0, \tag{6.8}$$

where

$$V = (K/\rho)^{1/2}.$$

Equation (6.8) is the generalization to three dimensions of the non-dispersive wave equation (6.1), and describes the propagation of pressure waves (*sound*) with a velocity V. The simplest solution describes a plane wave travelling in the x-direction with its wavefronts lying in planes parallel to the y–z plane. If the amplitude across a wavefront is constant, all derivatives with respect to y and z vanish, so that (6.8) takes the one-dimensional form (6.1) and has the same general solution. But (6.8) also describes spherical waves emanating from a point source, and an infinite variety of other types of wave in which the pressure varies with x, y, z and t all at once. For further discussion of the mathematical theory of wave propagation the reader should consult specialized texts.

A typical value for the bulk modulus of a liquid is 10^9 N m^{-2} (for water at room temperature it is about 2×10^9, for pentane $\frac{1}{3} \times 10^9$), and for the density 10^3 kg m^{-3}, so that we may expect sound velocities in the neighbourhood of 10^3 m s^{-1} ($V = 1500$ m s^{-1} for water). Gases under normal conditions have bulk moduli about 10^4 and densities 10^3 times less, and the sound velocities are consequently not very different, about three times less ($V = 331$ m s^{-1} for air at 0°C).

A second application of (6.5) and (6.7) is to a special type of flow which has been extensively studied because of its mathematical simplicity, but which is rather a rarity in practice. This is steady irrotational flow of an incompressible fluid, in which the velocity vector $\mathbf{v}$ at any point does not change with time and has, moreover, the property, analogous to that of a conservative force, that $\oint \mathbf{v} . d\mathbf{r} = 0$ round any closed loop. It is therefore possible to introduce a scalar *velocity potential*, ϕ, such that $\mathbf{v} = \text{grad}\, \phi$. In steady flow $\partial P/\partial t$ must vanish and therefore also div $\mathbf{v}$, from (6.5), so that ϕ must obey the equation

$$\text{div grad}\, \phi = 0$$

or, by X,

$$\nabla^2\phi \equiv \partial^2\phi/\partial x^2 + \partial^2\phi/\partial y^2 + \partial^2\phi/\partial z^2 = 0,$$

which is Laplace's equation in three dimensions.

If $\mathbf{v}$ is irrotational (6.7) can be cast in a neat form by transforming the term $(\mathbf{v}.\text{grad})\mathbf{v}$. Consider for example the x-component:

$$\begin{aligned}(\mathbf{v}.\text{grad})v_x &= v_x\, \partial v_x/\partial x + v_y\, \partial v_x/\partial y + v_z\, \partial v_x/\partial z,\\ &= v_x\, \partial v_x/\partial x + v_y\, \partial v_y/\partial x + v_z\, \partial v_z/\partial x,\end{aligned}$$

since

$$\partial v_x/\partial y = \partial^2\phi/\partial x \partial y = \partial v_y/\partial x,$$

and similarly

$$\partial v_x/\partial z = \partial v_z/\partial x.$$

Therefore,

$$(\mathbf{v}.\text{grad})v_x = \frac{1}{2}\frac{\partial}{\partial x}(v_x^2 + v_y^2 + v_z^2) = \frac{1}{2}\frac{\partial}{\partial x}(v^2),$$

and

$$(\mathbf{v}.\text{grad})\mathbf{v} = \tfrac{1}{2}\,\text{grad}\,(v^2).$$

In steady irrotational flow, then, the left-hand side of (6.7) is zero and the right-hand side takes the form $-\text{grad}\,(\tfrac{1}{2}v^2 + P/\rho)$. It follows immediately that $\tfrac{1}{2}v^2 + P/\rho$ takes the same value at all points in the fluid, a result known as *Bernoulli's theorem*. Physically the significance is clear—unit volume of fluid in its motion is subjected to a conservative field of force derived from a potential P, and therefore the sum of kinetic energy $\tfrac{1}{2}\rho v^2$ and potential energy P is invariant. It is not so obvious why this result should hold not only along the streamlines but normal to them as well, but the analysis shows that this is indeed so.

The solution of Laplace's equation for irrotational non-viscous flow round a sphere gives a pattern of streamlines which is quite symmetrical about the equatorial plane R. In consequence the velocity at points such as A and A', equidistant from R, is the same, and by Bernoulli's theorem so is the pressure. The symmetry of pressure has the inescapable consequence that there is no resultant force exerted by the fluid on the sphere. Indeed, Euler proved the

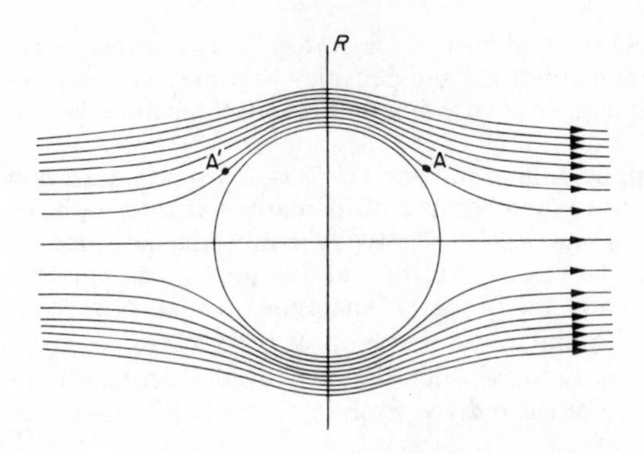

paradoxical result that no obstacle, of any shape, should experience a force in irrotational non-viscous flow. This result does not extend to the torque which an unsymmetrically placed obstacle may experience. A disc, for example, in irrotational flow distorts the pattern so as to produce rather still regions at B and B', but not on the opposite sides C and C'. The pressure being higher at B than at C, there is a resultant torque turning the disc athwart the stream. The Rayleigh Disc is useful as a means of detecting and measuring certain types of flow, especially motion due to sound waves.

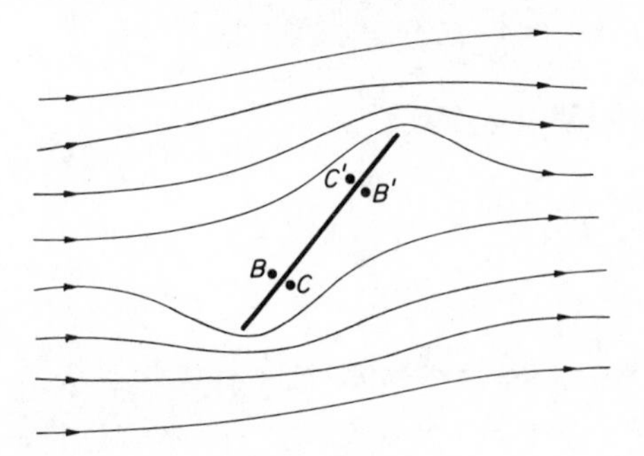

In reality, fluids are viscous and do not flow freely round the surface of an obstacle, but exhibit a quite different pattern and exert a tractive force proportional to the speed of flow if this is slow enough. When the speed is increased turbulence sets in, with a wake of eddies breaking away from the trailing end of the obstacle. All our foregoing analysis is now totally inadequate and the experimental and theoretical investigation of turbulence forms a large, difficult and important branch of classical physics and of modern hydraulic and aeronautical engineering, but these matters are still far from being fully investigated.

The smaller the viscosity, the lower the speed at which turbulence appears —in a very viscous fluid the turbulent eddies are discouraged by the friction between different elements of fluid as they slide against each other. If we could eliminate viscosity altogether, the state of irrotational non-viscous flow over which so much care has been lavished would prove to be unrealizable, since at no speed, however slow, would the flow be stable against the onset of turbulence. There is, however, one exception; the flow of liquid helium at temperatures below 2·18 K (where it suffers a phase change into a liquid modification. He11, having uniquely extraordinary properties) is controlled to a marked degree by quantum effects. Just as in the hydrogen atom only certain angular momenta are permitted, so in He11 the vortex motions (eddies) are restricted to certain rotation speeds. If the flow is so gentle as not to be able to establish the slowest permitted vortex, it genuinely is irrotational and non-viscous, and it has been shown that suspended obstacles do indeed experience no resultant force, but only a torque if they are asymmetrical.

Dynamics of a rigid body

We have seen that a swarm of interacting particles is subject to certain overriding rules—the acceleration of the centroid is determined by the total applied force, and the change of angular momentum about a fixed point by the total moment of the applied forces about that point:

$$(3.4) \quad M\ddot{\mathbf{R}} = \mathbf{F}, \qquad (3.6) \quad \dot{\mathbf{L}} = \mathbf{G}.$$

27

Before specializing the results for a swarm of particles to apply to a rigid body, in which the particles exert such mutual forces as to maintain their relative positions, it is convenient to extend the argument and to show that the angular momentum theorem (3.6) applies not only to a fixed point but also to the centroid of the particles, even though it is accelerating under the influence of **F**.

We define the position of each particle in the swarm by its distance $\boldsymbol{\rho}_i$ from the centroid,

$$\mathbf{r}_i = \mathbf{R} + \boldsymbol{\rho}_i.$$

From the definition of the centroid, $\sum_i m_i\boldsymbol{\rho}_i = 0$ at all times, therefore $\sum_i m_i\ddot{\boldsymbol{\rho}}_i$ also vanishes. Then the equation of motion of a particle, expressed in the form (3.5), becomes

$$m_i(\mathbf{R} + \boldsymbol{\rho}_i)\wedge(\ddot{\mathbf{R}} + \ddot{\boldsymbol{\rho}}_i) = (\mathbf{R} + \boldsymbol{\rho}_i)\wedge\mathbf{F}_i$$

in which $\mathbf{F}_i$ includes mutual as well as external forces.

We now multiply out the brackets and sum over all particles, noting that

$$\sum_i m_i\mathbf{R}\wedge\ddot{\mathbf{R}} = M\mathbf{R}\wedge\ddot{\mathbf{R}} = \mathbf{R}\wedge\mathbf{F}, \qquad \text{from (3.4)},$$

$$\sum_i m_i\mathbf{R}\wedge\ddot{\boldsymbol{\rho}}_i = \mathbf{R}\wedge\sum_i m_i\ddot{\boldsymbol{\rho}}_i = 0,$$

$$\Big(\sum_i m_i\boldsymbol{\rho}_i\Big)\wedge\ddot{\mathbf{R}} = 0,$$

$$\sum_i m_i\boldsymbol{\rho}_i\wedge\ddot{\boldsymbol{\rho}}_i = \frac{\mathrm{d}}{\mathrm{d}t}\sum_i m_i\boldsymbol{\rho}_i\wedge\dot{\boldsymbol{\rho}}_i \equiv \dot{\mathbf{L}}_0,$$

$$\sum_i \mathbf{R}\wedge\mathbf{F}_i = \mathbf{R}\wedge\mathbf{F},$$

and

$$\sum_i \boldsymbol{\rho}_i\wedge\mathbf{F}_i \equiv \mathbf{G}_0.$$

In these equations, $\mathbf{L}_0$ is the angular momentum about the centroid as measured by an observer in a non-rotating frame moving with the centroid, and $\mathbf{G}_0$ is the moment of the external forces about the centroid, if we assume as before that mutual forces between the particles make no contribution to $\mathbf{G}_0$. Putting everything together, we find that

$$\dot{\mathbf{L}}_0 = \mathbf{G}_0.$$

The centroid of a swarm of particles accelerates under the influence of external forces as though it were a particle of the same mass as the swarm acted upon by the vector sum of the forces, and the angular momentum about the centroid changes at a rate given by the moment of the applied forces about the centroid. We shall see immediately that this result is enough to determine the motion of a rigid body under the influence of specified forces.

A rigid body may be considered to be a swarm of particles constrained by their mutual interactions to maintain the same relative positions. The movement of the body over a short time-interval is uniquely described by the distance moved by any point fixed in the body, plus the rotation of the body about an axis passing through this point, the direction of the axis and the angle of rotation being needed to specify the precise movement. Thus the dynamical behaviour is specified instantaneously by two vectors, a velocity $\mathbf{V}$ and an angular velocity $\boldsymbol{\Omega}$. In view of the special importance of the centroid we shall always take this to be the fixed point in the body to which the vectors refer.

A body in equilibrium under the action of a number of forces, $\mathbf{F}_i$, applied to points $\boldsymbol{\rho}_i$, has no linear or angular acceleration so that both $\sum_i \mathbf{F}_i$ and $\sum_i \boldsymbol{\rho}_i \wedge \mathbf{F}_i$ must vanish. It is easy to show that if the resultant moment of the forces about the centroid vanishes, it vanishes about any other point in the body.

The analysis of the forces in structures that are in equilibrium is a special branch of mechanics called *statics*. The vanishing of the resultant force and torque on any part of the structure may provide enough information to determine all the forces in, for example, a girder bridge, so that the dimensions of the girders can easily be chosen to give them sufficient strength. But more commonly, in *redundant structures*, this information is not in itself enough, and must be supplemented by a study of the deformations suffered by the members under load. It is this point that marks the beginning of the serious analysis of engineering structures, as distinct from routine application of the elementary laws of statics. We shall not treat the solution of problems in statics any further, but return to the more complicated problems arising from the accelerated motion of rigid bodies.

We know how $\mathbf{V}$ is related to the applied forces, $M\dot{\mathbf{V}} = \mathbf{F}$, and we shall take for granted that this part of the motion can be calculated readily; but we must examine the behaviour of $\boldsymbol{\Omega}$ in more detail. We therefore view the motion from the point of view of an observer moving with the centroid. Any particle in the body at a distance $\boldsymbol{\rho}_i$ has instantaneous velocity, as a result of $\boldsymbol{\Omega}$, given by

$$\dot{\boldsymbol{\rho}}_i = \boldsymbol{\Omega} \wedge \boldsymbol{\rho}_i,$$

and we may therefore write $\mathbf{L}_0$ as $\sum_i m_i \boldsymbol{\rho}_i \wedge (\boldsymbol{\Omega} \wedge \boldsymbol{\rho}_i)$. This can be transformed by VII into

$$\mathbf{L}_0 = \sum_i m_i [\rho_i^2 \boldsymbol{\Omega} - \boldsymbol{\rho}_i(\boldsymbol{\rho}_i . \boldsymbol{\Omega})].$$

Let us write this out explicitly in coordinate form for just one particle (dropping the subscript), and writing the coordinate $\boldsymbol{\rho}$ as (ξ, η, ζ):

$$L_{0x} = m[(\xi^2 + \eta^2 + \zeta^2)\Omega_x - \xi(\xi\Omega_x + \eta\Omega_y + \zeta\Omega_z)]$$

i.e.,

$$L_{0x} = m[(\eta^2 + \zeta^2)\Omega_x - \xi\eta\Omega_y - \xi\zeta\Omega_z]$$

similarly

$$L_{0y} = m[-\eta\xi\Omega_x + (\xi^2 + \zeta^2)\Omega_y - \eta\xi\zeta\Omega_z] \qquad (6.9)$$

and

$$L_{0z} = m[-\zeta\xi\Omega_x - \zeta\eta\Omega_y + (\xi^2 + \eta^2)\Omega_z].$$

Each component of $\mathbf{L}_0$ is seen to be linearly related to each component of $\boldsymbol{\Omega}$, a result which can be written in compressed notation by introducing a tensor $\mathbf{I}$, the *moment of inertia*. By definition, a tensor like $\mathbf{I}$ is specified by nine components, which, written out in a matrix, have the formal structure

$$\mathbf{I} = \begin{pmatrix} I_{xx} & I_{xy} & I_{xz} \\ I_{yx} & I_{yy} & I_{yz} \\ I_{zx} & I_{zy} & I_{zz} \end{pmatrix},$$

and when we write $\mathbf{L}_0 = \mathbf{I} . \boldsymbol{\Omega}$ we mean that each component of $\mathbf{L}_0$ is to be formed from $\mathbf{I}$ and $\boldsymbol{\Omega}$ by the operation:

$$\left.\begin{aligned} L_{0x} &= I_{xx}\Omega_x + I_{xy}\Omega_y + I_{xz}\Omega_z, \\ L_{0y} &= I_{yx}\Omega_x + I_{yy}\Omega_y + I_{yz}\Omega_z, \\ L_{0z} &= I_{zx}\Omega_x + I_{zy}\Omega_y + I_{zz}\Omega_z. \end{aligned}\right\} \qquad (6.10)$$

Comparing (6.9) and (6.10) we see that the contribution of one mass point m at (ξ, η, ζ) to the inertia tensor $\mathbf{I}$ is given by

$$m\begin{pmatrix} \eta^2 + \zeta^2 & -\xi\eta & -\xi\zeta \\ -\eta\xi & \xi^2 + \zeta^2 & -\eta\zeta \\ -\zeta\xi & -\zeta\eta & \xi^2 + \eta^2 \end{pmatrix}, \qquad (6.11)$$

and $\mathbf{I}$ is formed by adding up the contributions of all mass points in the rigid body, for example, $I_{xy} = -\sum_i m_i \zeta_i \eta_i$.

Perhaps the reader needs reassurance at this point. We are not about to embark on calculations involving tensor calculus, which at first sight is apt to appear repellently complex. We have developed the idea of the inertia tensor to show how it arises, but will proceed immediately to simplify it, stating the essential mathematical property of a tensor without proof. The expression (6.11) depends for the precise value of its individual components on the choice of axes with reference to which ξ, η and ζ are measured, and given a sufficiently symmetrical body it is obvious that we can choose axes so that all the terms like $\xi\eta$ vanish when the summation is performed. The rectangular block with axes as shown is

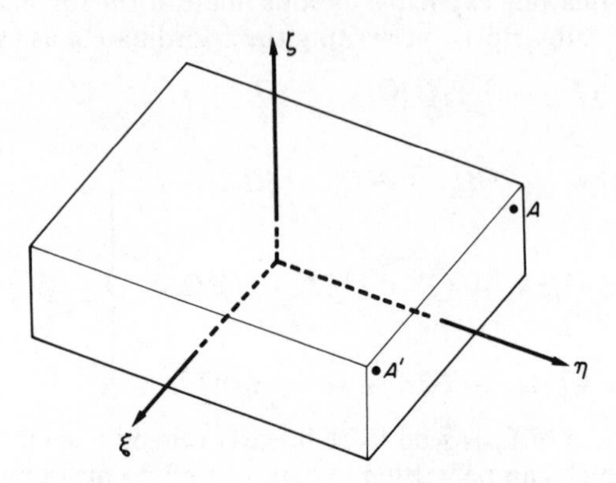

such a case, for every mass point at (ξ, η, ζ) corresponds to another at $(-\xi, \eta, \zeta)$ and these two cancel each other in $\sum_i \xi_i\eta_i$; A and A' are such points. It can be shown that what is obvious here is always true; axes can always be found for a symmetrical tensor ($I_{xy} = I_{yx}$ etc.) such that only I_{xx}, I_{yy} and I_{zz} are non-vanishing, and the tensor is then said to be *diagonal.* It may be somewhat tedious in the general case to determine how the *principal axes* of inertia lie in the body, but we shall assume henceforth that we have carried out this task, and that the inertia tensor has the form

$$\mathbf{I} = \begin{pmatrix} A & 0 & 0 \\ 0 & B & 0 \\ 0 & 0 & C \end{pmatrix}. \tag{6.12}$$

We shall further choose ξ, η and ζ axes so that A is the largest and C the smallest. The meaning of A, B and C is easily seen from (6.11). If a mass point m_i lies at a distance X_i from the ξ-axis through the centroid, A is $\sum_i m_i X_i^2$. We may define a mean distance k_ξ (*radius of gyration*) by the equation

$$Mk_\xi^2 = A, \qquad \text{or} \qquad k_\xi^2 = \sum_i m_i X_i^2 / \sum_i m_\zeta.$$

Similar definitions give us k_η and k_ζ.

PROBLEM

Show that the kinetic energy of a rigid body, as seen by an observer moving with the centroid, is $\sum_i \frac{1}{2} m_i \dot{\rho}_i \cdot \dot{\rho}_i$ or $\frac{1}{2}\boldsymbol{\Omega} \cdot \mathbf{L}_0$ or $\frac{1}{2}\boldsymbol{\Omega} \cdot (\mathbf{I} \cdot \boldsymbol{\Omega})$.

So long as a rigid body is so constrained or acted upon by applied forces as to rotate about a principal axis of inertia, no very tricky problems arise. The following examples start with a few simple cases of this sort before proceeding to discuss a typical problem in which this is not true.

EXAMPLES

(1) *A uniform cylinder rolling downhill.* When the cylinder has velocity V, its angular velocity is V/a. To calculate the radius of gyration about its axis, we observe that the amount of material between r and $r + \mathrm{d}r$ is proportional to

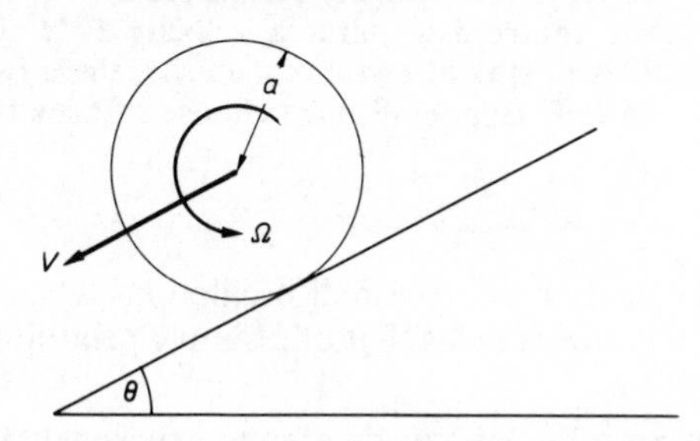

$r\,\mathrm{d}r$, and that therefore k^2, which is the mean value of r^2 for this uniform body, is $\int_0^a r^2 \times r\,\mathrm{d}r / \int_0^a r\,\mathrm{d}r$, i.e., $\frac{1}{2}a^2$. When the cylinder has velocity V its kinetic energy of rotation is $\frac{1}{2}\boldsymbol{\Omega} \cdot \mathbf{L}_0$, i.e., $\frac{1}{2}Mk^2\Omega^2$, i.e., $\frac{1}{4}MV^2$, while its kinetic energy of translation is $\frac{1}{2}MV^2$. When it has fallen through a height h from rest, its kinetic energy must be Mgh. Therefore,

$$Mgh = \tfrac{1}{2}MV^2 + \tfrac{1}{4}MV^2,$$

or

$$V = (\tfrac{4}{3}gh)^{1/2}.$$

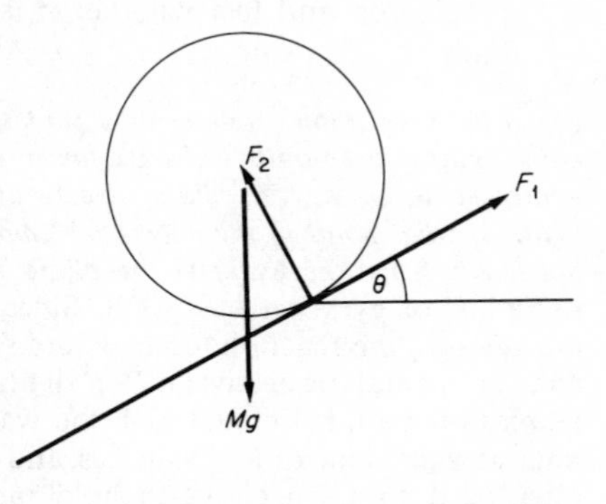

The acceleration is two-thirds as great as for a non-rotating body or one of negligible moment of inertia.

An alternative solution considers the forces acting on the cylinder, its weight Mg and the force exerted by the plane which we resolve into parallel and perpendicular components. The linear acceleration down the plane is caused by the resultant force $Mg \sin\theta - F_1$, and therefore $\dot{V} = g\sin\theta - F_1/M$. The angular acceleration is caused by the torque $F_1 a$, and therefore $\frac{1}{2} Ma^2\dot{\Omega} = F_1 a$, or $\dot{\Omega} = 2F_1/(Ma)$. If there is to be no slipping, $\dot{\Omega}a = \dot{V}$, so that $F_1 = \frac{1}{3} Mg \sin\theta$ and $\dot{V} = \frac{2}{3} g \sin\theta$. Once more, the acceleration is two-thirds as great as for a non-rotating body.

(2) *Centre of percussion.* A body supported from C is struck a blow at X. Where must X be for there to be no initial reaction at the support C? Clearly we must choose X so that in the absence of a reaction at C the initial velocity given to C by the blow vanishes.

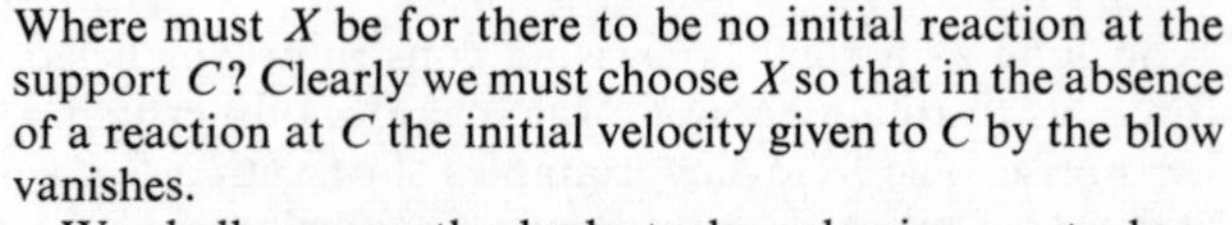

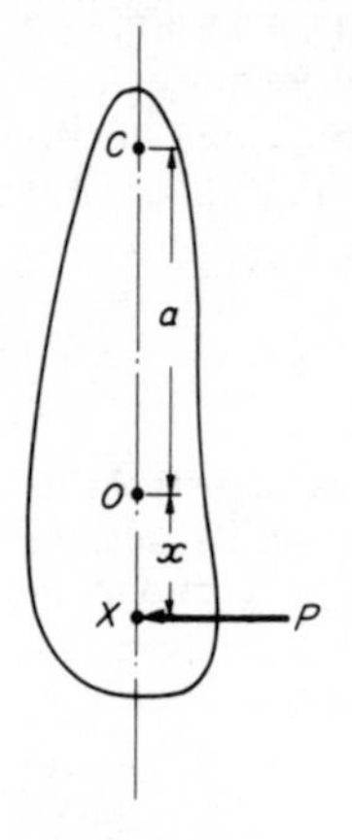

We shall assume the body to be a lamina, or to have C on its axis of symmetry, or otherwise to be so shaped that the initial rotation is about a principal axis of inertia. If k is the radius of gyration about an axis through the centroid O, normal to the paper, the angular velocity Ω about O immediately after the blow is such that $Mk^2\Omega = Px$, if P is the strength of the impulse, i.e., the time-integral of the force. As a result of this rotation C moves from left to right with velocity $a\Omega$, i.e., $aPx/(Mk^2)$. At the same time the centroid acquires a velocity P/M to the left, and if C is to stay at rest momentarily, these two velocities must cancel. Hence we must choose x so that $ax = k^2$.

PROBLEM

Show that if the body executes small oscillations when suspended from C, the frequency is the same as that of a simple pendulum of length CX.

This result may be used to determine experimentally the centre of percussion of a cricket or baseball bat, which is the ideal place for striking the ball so that the hands are not stung by the shock. Suspend the bat from a point on the handle where you judge your grip to be centred (tie a rod transverse to the bat and swing it on this between two chairbacks), and at the same time adjust a simple pendulum to swing at the same frequency. Hence find the centre of percussion and test whether it is indeed a good place to strike the ball.

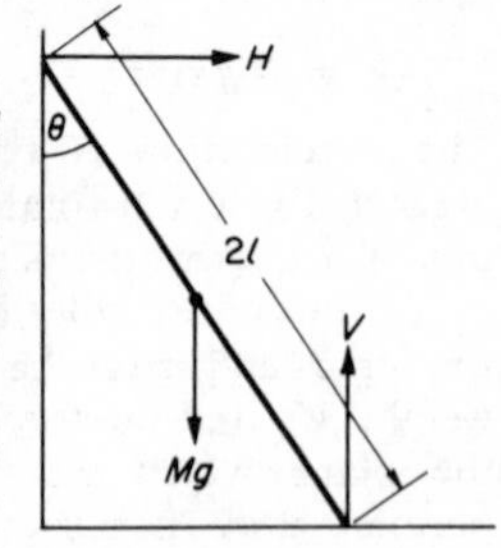

(3) *A uniform plank stands on a smooth floor vertically against a smooth wall; its lower end is pushed gently so as to start it sliding freely away from the wall. At what point in the subsequent motion does the upper end leave the wall?* If the plank has length $2l$ its radius of gyration $k = l/\sqrt{3}$. Since the surfaces are smooth, the reaction forces V and H are vertical and horizontal respectively. We shall suppose the plank to remain in contact with the wall and determine at what value of θ H vanishes, and would thereafter need to be negative to hold the plank in contact. This will fix the moment at which contact ceases.

At any instant the kinetic energy is made up of a part $\frac{1}{2}Ml^2\dot{\theta}^2$ contributed by the motion of the centroid (whose velocity has horizontal and vertical components $\mathrm{d}/\mathrm{d}t(l\sin\theta)$ and $\mathrm{d}/\mathrm{d}t(l\cos\theta)$, i.e., $l\cos\theta.\dot{\theta}$ and $-l\sin\theta.\dot{\theta}$) and a part $\frac{1}{2}Mk^2\dot{\theta}^2$ contributed by rotation; therefore,

$$T = \tfrac{1}{2}M(l^2 + k^2)\dot{\theta}^2 = \tfrac{2}{3}Ml^2\dot{\theta}^2.$$

This may be equated to the decrease in potential energy due to the centroid having dropped a distance $l(1 - \cos\theta)$; therefore,

$$\tfrac{2}{3}Ml^2\dot{\theta}^2 = Mgl(1 - \cos\theta).$$

Let us use this result to study the horizontal component of acceleration of the centroid, which is caused entirely by H. The horizontal component of velocity is $l\cos\theta.\dot{\theta}$, i.e., $(\frac{3}{2}gl)^{1/2}\cos\theta(1 - \cos\theta)^{1/2}$. This is zero when $\theta = 0$ and rises, as θ increases, to a maximum when $\cos\theta = \frac{2}{3}$ before falling again. At the maximum the acceleration vanishes, and at this point, i.e., when $\theta = 48°$, the plank leaves the wall.

Note. The expression for the kinetic energy, $T = \frac{1}{2}M(l^2 + k^2)\dot{\theta}^2$, may be derived in a different way by observing that a small movement $\delta\theta$ may be accomplished by rotating the plank about the point X. This is a special case of an easily proved general statement that a lamina moving in a plane and constrained to touch two lines has an instantaneous centre of rotation at the intersection of the normals to the lines at the points of contact. It may also be easily proved that the moment of inertia of a body about an axis parallel to a principal axis of inertia but a distance a from the centroid is $M(k^2 + a^2)$. In this case X is the instantaneous centre about which the angular velocity is $\dot{\theta}$, and the moment of inertia about an axis through X is $M(l^2 + k^2)$.

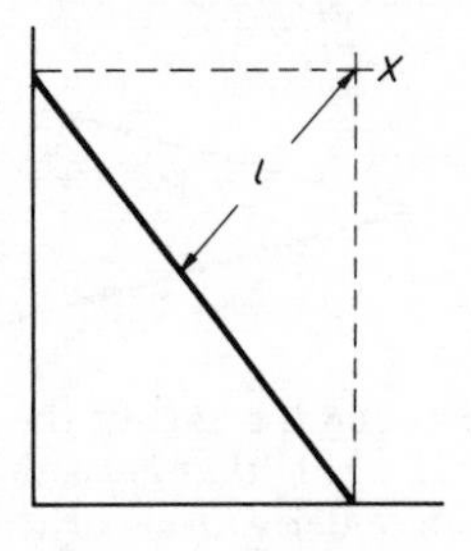

(4) *The gyroscope.* These three examples have involved elementary situations where the body rotated about a principal axis of inertia. The next example is also of this nature but is sufficiently important in its own right to deserve more extended discussion, especially as the characteristic precessional behaviour is not an immediately obvious consequence of the laws of dynamics. No description of the sensation of trying to tilt the axis of a gyroscope does justice to the peculiar way in which it writhes away from the direction in which it is being twisted; the reader is advised to find and play with a gyroscope before studying its theory further. Assuming he has done so, all that is needed is to remind him of the phenomenon of precession such as the apparatus in the diagram illustrates. The gyroscope is a heavy wheel, mounted on a free pivot and counterpoised. So long as the counterpoising is exact, the gyroscope continues to spin about a fixed axis. But if one adds an extra weight to the counterpoise, instead of the left-hand end falling the system settles down, after a few transient up-and-down oscillations, to a steady rotation about a vertical axis. As soon as the extra weight is removed the precession

stops, and if the counterpoise is further lightened precession starts in the opposite sense. As the gyroscope slows down through friction the precessional motion becomes faster.

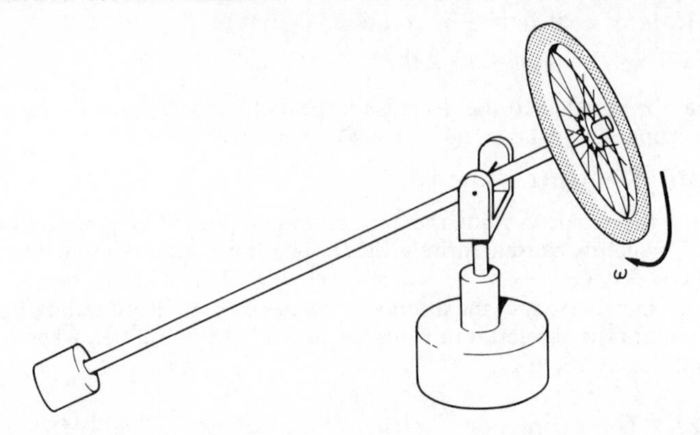

The essential feature of the gyroscope is the large angular momentum about its axis, which dominates its dynamical behaviour. Any attempt to turn the axis from **L** to **L**′ must involve adding angular momentum **ΔL**, and if the spin is rapid and the moment of inertia large, **ΔL** may be quite sizeable for a small twist of the axis. We note also that **ΔL** is produced by a torque whose axis is parallel to **ΔL** and lies in the paper, yet it results in **L** swinging to a new position **L**′ by rotation about an axis normal to the paper. It is this seeming discrepancy between the axis of the torque and the resulting axis of rotation of the body as a whole that gives the gyroscope its peculiar feel. We now give three alternative derivations of the basic formula for precession.

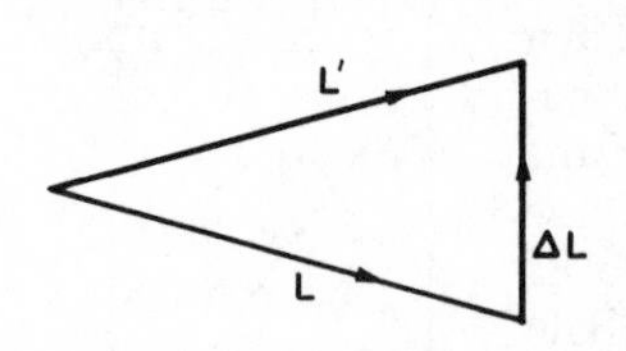

(a) We may discuss the precession very simply in terms of angular momentum vectors. Suppose the axis of the gyroscope makes an angle ϕ with the vertical, so that the vector $I\boldsymbol{\omega}$ represents its angular momentum. Adding an extra weight to the counterpoise exerts a torque **G** on the system, whose axis is horizontal and normal to the gyroscope axis. It is therefore tangential to the horizontal circle in the diagram. In time δt the extra angular momentum given to the system is **G** δt, and the vector diagram shows that if δt is small the next position of the angular momentum vector brings it to a neighbouring point on the circle. But now the direction of **G** has correspondingly turned, so that the process repeats itself, and the head of the vector simply runs steadily around the circle. This is the precession of the axis of the gyroscope, and the actual rate, $\boldsymbol{\Omega}$, may be written down without difficulty. For the radius of the circle is $I\omega \sin \phi$, and the angle

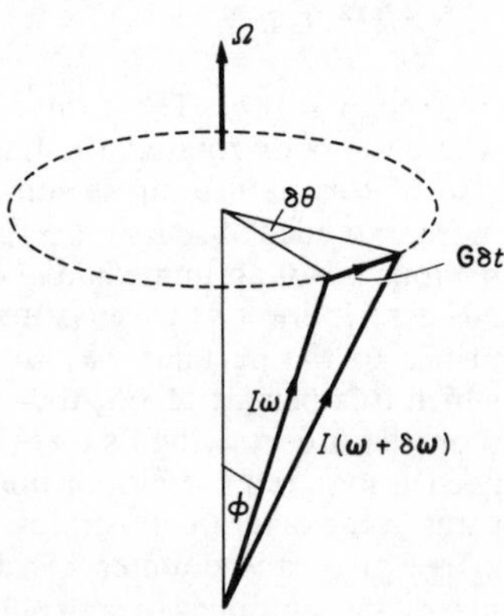

$\delta\theta$ through which the axis precesses in δt is $G\,\delta t/(I\omega \sin\phi)$. Hence the precessional angular velocity is determined:

$$\Omega = \delta\theta/\delta t = G/(I\omega \sin\phi).$$

This may be tidily expressed by the formula

$$\mathbf{G} = I\boldsymbol{\Omega} \wedge \boldsymbol{\omega}. \qquad (6.13)$$

(b) Let us break the problem down to its simplest constituents so as not to have to rely on the labour-saving device of an angular momentum vector. In this way we may appreciate more deeply why the precession occurs. It is in fact easier to work the other way round, imagining precession to be taking place, and determining by following the trajectory of a mass point what forces need to be applied so that it shall pursue the required course. Since our purpose is to illustrate the process rather than to derive a general formula we shall simplify the problem by taking the axis of the gyroscope as horizontal, i.e., putting $\phi = \pi/2$. Imagine a point mass m at a distance r from the gyroscope axis, and choose coordinates such that at time $t = 0$ the axis lies along x and the mass is on the y-axis. Then after time t the wheel has turned through ωt about its own axis and a precession of Ωt has occurred. The coordinates of the mass point are thus

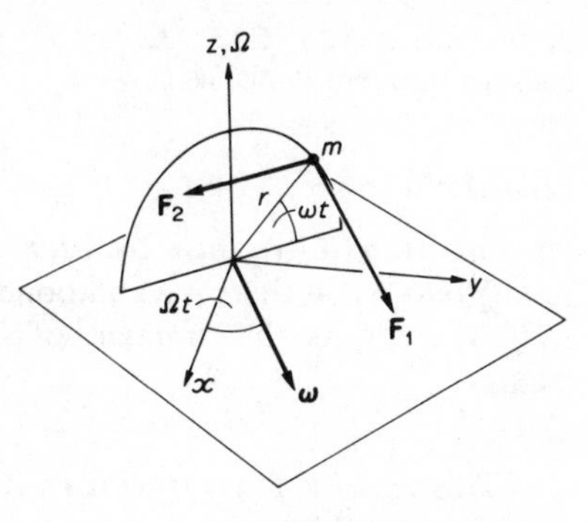

$$\left.\begin{aligned} x &= -r\cos\omega t \sin\Omega t,\\ y &= r\cos\omega t\cos\Omega t,\\ z &= r\sin\omega t. \end{aligned}\right\} \qquad (6.14)$$

Differentiating twice we have the components of acceleration,

$$\begin{aligned} \ddot{x} &= r(\omega^2 + \Omega^2)\cos\omega t\sin\Omega t + 2r\omega\Omega\sin\omega t\cos\Omega t,\\ \ddot{y} &= -r(\omega^2 + \Omega^2)\cos\omega t\cos\Omega t + 2r\omega\Omega\sin\omega t\sin\Omega t,\\ \ddot{z} &= -r\omega^2\sin\omega t. \end{aligned}$$

Now these accelerations are produced in any one mass point by the local stresses exerted on it by its neighbours, and we must suppose the wheel to be strong enough to exert the required forces. Some of the accelerations are due to the rotation alone; these are the centripetal accelerations present in any spinning wheel, and we ignore them. It is the extra forces due to $\boldsymbol{\Omega}$ that concern us, and these can be seen to combine into horizontal forces,

$\mathbf{F}_1$ of magnitude $2mr\omega\Omega \sin\omega t$ parallel to $\boldsymbol{\omega}$,

and

$\mathbf{F}_2$ of magnitude $mr\Omega^2\cos\omega t$ normal to $\boldsymbol{\omega}$.

Now all the forces between particles are mutually balancing, and when we sum the forces and their moments over all particles what is left is the external influence needed to cause the precessional motion. Since in the expressions for $\mathbf{F}_1$ and $\mathbf{F}_2$, ωt now means only the instantaneous angular coordinate of the mass point, we may sum over all mass points round a circle of radius r by taking the average of $\sin \omega t$ and $\cos \omega t$; as expected $\sum \mathbf{F}_1 = \sum \mathbf{F}_2 = 0$—no total force is needed because the centroid is at rest. But the forces $\mathbf{F}_1$ have a resultant moment about the horizontal diameter of the gyroscope, i.e., an axis normal to $\boldsymbol{\omega}$:

$$G = \sum F_1 r \sin \omega t = \sum 2mr^2\omega\Omega \sin^2 \omega t.$$

Since the average value of $\sin^2 \omega t$ is $\frac{1}{2}$, this can be written as a summation over all mass points,

$$G = \sum mr^2\omega\Omega = I\omega\Omega$$

in agreement with (6.13). As for F_2, it is a purely internal matter, since its average moment vanishes.

PROBLEM

Starting from (6.14), show by direct evaluation of the angular momentum $\mathbf{L} = \sum m\mathbf{r} \wedge \dot{\mathbf{r}}$ that there is a component L_z, due to precession, of magnitude $I_z\Omega$, where I_z is the moment of inertia of the gyroscope about the z-axis.

> This result is fairly obvious without such a detailed calculation. For the instantaneous velocity of a mass point m may be resolved into a component due to $\boldsymbol{\omega}$ which makes no contribution to L_z, and a component due to $\boldsymbol{\Omega}$ which would have the same magnitude if the gyroscope were not spinning ($\omega = 0$) but simple revolving at angular velocity $\boldsymbol{\Omega}$ about the z-axis: it is the latter that is wholly responsible for L_z.

(c) Instead of analysing the motion of a particle in detail, we may look at the system from the point of view of an observer rotating at the precessional angular velocity. He will see the gyroscope with its axis at rest, and will recognize that the Coriolis and centrifugal forces on any particle must be balanced by extra internal stresses set up by the precession. Since the Coriolis force on a particle whose speed is $r\omega$ is $-2mr\omega\Omega \sin \omega t$, the force $\mathbf{F}_1$ is derived immediately. $\mathbf{F}_2$ is what balances the centrifugal force $-mr\Omega^2 \cos \omega t$. The rest of the calculation follows as before.

(5) *Precession of the equinoxes.* The axis of the Earth is inclined at $23\frac{1}{2}°$ to the ecliptic, the plane of its orbit round the Sun, and points towards a position very close to the North Star, the hub of the heavens about which the stars are nightly seen to wheel. It has not always been so; some 13,000 years ago what we now call the North Star was 47° away from the hub since, as was known to the ancients, the Earth's axis does not remain parallel to a fixed direction relative to the stars, but precesses in a cone of semi-angle $23\frac{1}{2}°$ about a direction normal to the ecliptic, with a period of 25,800 years. The origin of this precession lies in the flattening of the Earth caused by the centrifugal forces of its own spin during the ages when it was a liquid mass. If we inscribe a sphere in the Earth (here assumed, quite falsely, to have uniform density) the material inside the sphere behaves gravitationally as a point mass; what is outside, however, does not since those portions of the annular bulge lying

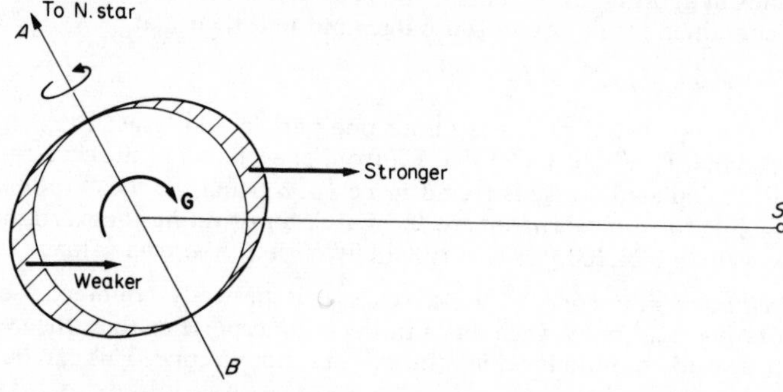

nearer the Sun are attracted more strongly than the more distant portions, and the resultant force does not pass through the centre of the Earth. In consequence a torque **G** acts on the Earth, and causes it to precess in accordance with (6.13). The diagram shows the situation producing a maximum torque, at the winter solstice. At the summer solstice, on the opposite side of the orbit, the torque has another maximum, and at both solstices it tries to pull the axis into a direction normal to the ecliptic; in between it varies sinusoidally between the maximum and zero (at the equinoxes), with a mean value equal to half the maximum.

To make a rough estimate of the precessional rate it is convenient to divide the excess material, shaded in the diagram, into two halves lying on either side of the plane represented by AB, and to lump each together at its own centroid.

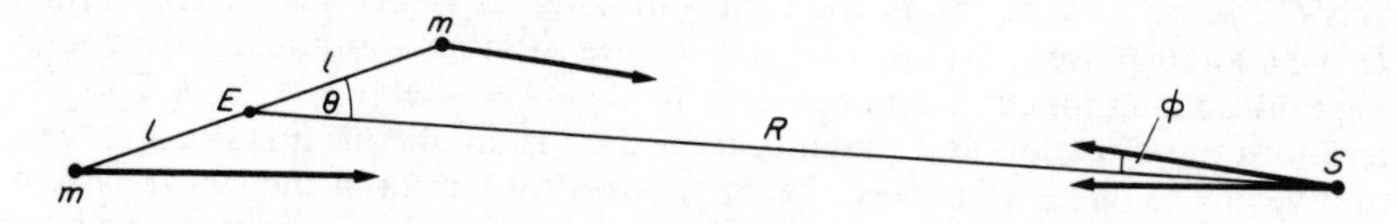

Now the resultant force exerted by the Sun on these two masses must, by N3, be equal and opposite to the force exerted by the two masses on the Sun. We may therefore calculate the torque as the resultant moment about the centre of the Earth of the two forces acting on the Sun. Thus the nearer mass exerts a force $mS/[4\pi\gamma(R - l\cos\theta)^2]$ and gives rise to a torque $(RmS\sin\phi)/[4\pi\gamma(R - l\cos\theta)^2]$. Solving the triangle EmS we see that $\sin\phi = (l\sin\theta)/[R - l\cos\theta]$, so that the torque is $(RmS\,l\sin\theta)/[4\pi\gamma(R - l\cos\theta)^3]$, or approximately $(mS\,l\sin\theta)[1 + (3l\cos\theta)/R]/(4\pi\gamma R^2)$. The corresponding result for the further mass is obtained by changing the sign of l, so that when the two are added the resultant torque takes the form

$$G = 6mS\,l^2\sin\theta\cos\theta/(4\pi\gamma R^3),$$

and the mean value over a year, $\overline{G}$, is half this. From (6.13) the precessional angular velocity, Ω, is $\overline{G}/(I\omega\sin\theta)$, where I is the Earth's moment of inertia and ω its angular velocity on its own axis. The period of precession, T, is $2\pi/\Omega$, i.e.,

$$T = 8\pi^2\gamma R^3 I\omega/(3mSl^2\cos\theta).$$

This can be put in tidier form by remembering that, from the dynamics of the Earth's orbit, 1 year $= (16\pi^3\gamma R^3/S)^{1/2}$. Further, we write I as Ek^2 where k is the radius of gyration of the Earth, and m as $\frac{1}{2}fE$, where f is the fraction of the Earth contained in the equatorial bulge, and find then that

$$T = 365\ k^2/(\tfrac{3}{2} fl^2 \cos\theta) \text{ years.}$$

Now the equatorial radius is about one part in 300 larger than the polar radius, so that f is about 1/150. The centroid of each half of the excess material is readily calculated, and l is found to be $\frac{3}{4}a$, a being the mean radius of the Earth, while for a uniform sphere $k^2 = \frac{2}{5}a^2$. Substituting these estimates, we reach a value of 28,200 years, surprisingly close to the true value.

(6) *Oscillating gyroscope*. If the gyroscope is perfectly counterpoised by a non-rotating rigid body, the inertia of the counterpoise enables the system to execute a steady motion in which the axis describes a cone. This can be readily

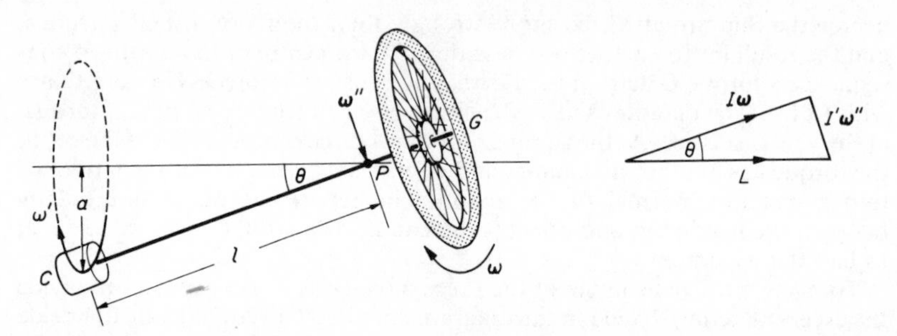

seen to be a consequence of conservation of angular momentum. Imagine the axis CG to be moving round the cone with angular velocity ω'. Then in time δt the counterpoise moves $\omega' a\ \delta t$ as if it were rotating about an axis through the point of support, P, with angular velocity $\omega'' = \omega' a/l$ or $\omega' \sin\theta$. This determines the direction and magnitude of the instantaneous angular momentum vector of the counterpoise. If the moment of inertia of the whole system about an axis through P, normal to CG, is I', the vector diagram shows how the two components of angular momentum are compounded into a resultant **L** along the axis of the cone. As CG swings round, the resultant is automatically constant and the dynamical problem is solved. The angular velocity of the motion depends on the angle of the cone according to the formula, derived from the vector diagram,

$$I'\omega'' = I\omega \tan\theta,$$

or

$$\omega' = I\omega/(I' \cos\theta).$$

The counterpoise describes a circle in the same sense as the gyroscope's spin. If a massive, but rather slowly spinning, gyroscope is counterpoised and at rest, and an extra weight is dropped on to the counterpoise, the resulting precession is at first accompanied by an up-and-down leaping motion (*nutation*) of the counterpoise which is this conical movement superposed on the steady precession; friction at the pivot soon damps the conical motion. With a small and rapidly spinning toy gyroscope the effect is hardly noticeable, since the oscillation frequency ω' is comparable to the angular velocity ω and is too fast to be perceived as other than a rapidly damped shivering motion.

This is an elementary example of the rotation of a rigid body about an axis which is not a principal axis of inertia, though very close to one, and the conical variation of the axis of rotation is typical of the more complex behaviour that one may find. We shall conclude this brief introduction to the vast field of rigid body dynamics with a more detailed discussion of this point.

(7) *Free rotation of a body whose principal moments of inertia are all different.* As implied by (6.12) we choose as ξ, η and ζ axes, fixed in the body, the principal axes of inertia. Then if the body has momentary angular velocity $\boldsymbol{\omega}$, we may write the components of $\boldsymbol{\omega}$ and $\mathbf{L}$ along the axes in the body:

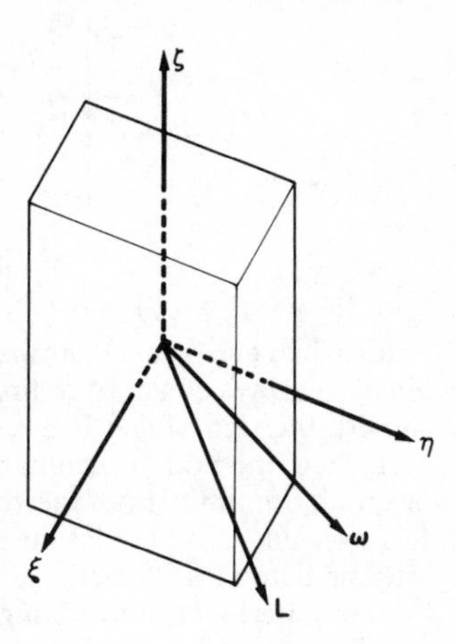

$$\boldsymbol{\omega} = (\omega_\xi, \omega_\eta, \omega_\zeta),$$

and

$$\mathbf{L} = (A\omega_\xi, B\omega_\eta, C\omega_\zeta); \qquad (6.15)$$

the axis of $\mathbf{L}$ is not parallel to $\boldsymbol{\omega}$. As the body spins, $\mathbf{L}$ must remain fixed in space while the axes spin around the vector $\boldsymbol{\omega}$. This means in general that $\mathbf{L}$ cannot maintain a constant direction relative to the principal axes, and therefore $\boldsymbol{\omega}$ cannot either. To find how $\boldsymbol{\omega}$ changes relative to the axes in the body, we need only note that as the body rotates infinitesimally about $\boldsymbol{\omega}$, while $\mathbf{L}$ stays fixed in space, an observer in the body would see $\mathbf{L}$ rotating at angular velocity $-\boldsymbol{\omega}$ about the axis of $\boldsymbol{\omega}$. To such an observer, then,

$$\dot{\mathbf{L}} = -\boldsymbol{\omega} \wedge \mathbf{L} = (\overline{B - C}\,\omega_\eta\omega_\zeta, \overline{C - A}\,\omega_\zeta\omega_\xi, \overline{A - B}\,\omega_\xi\omega_\eta) \text{ from (6.15).}$$

But

$$\dot{\mathbf{L}} = (A\dot{\omega}_\xi, B\dot{\omega}_\eta, C\dot{\omega}_\zeta),$$

so that

$$\left.\begin{aligned} A\dot{\omega}_\xi &= (B - C)\,\omega_\eta\omega_\zeta, \\ B\dot{\omega}_\eta &= (C - A)\,\omega_\zeta\omega_\xi, \\ \text{and} \quad C\dot{\omega}_\zeta &= (A - B)\,\omega_\xi\omega_\eta. \end{aligned}\right\} \qquad (6.16)$$

These three equations determine the changes of magnitude and direction of $\boldsymbol{\omega}$ in the body and hence, through (6.15), the motion of $\mathbf{L}$ which, relative to the body, will be seen as a vector of constant length but variable direction. By arranging that in real space $\mathbf{L}$ always points in the same direction, the complete motion of the body can in principle be mapped out step by step.

We shall only proceed as far as studying the implications of (6.16) regarding the motion of $\boldsymbol{\omega}$ in the body. We have chosen the axes so that $A > B > C$, and can see then that $\dot{\omega}_\xi$ has the same sign as $\omega_\eta\omega_\zeta$, $\dot{\omega}_\zeta$ the same sign as $\omega_\xi\omega_\eta$, but $\dot{\omega}_\eta$ the opposite sign to $\omega_\xi\omega_\zeta$. This singles out the intermediate axis as different from the other two. Now as $\mathbf{L}$ remains constant in magnitude, $L_\xi^2 + L_\eta^2 + L_\zeta^2$ is constant and this, according to (6.15), causes the end of the vector $\boldsymbol{\omega}$, if drawn from a fixed origin, to move on the ellipsoid

$$A^2\omega_\xi^2 + B^2\omega_\eta^2 + C^2\omega_\zeta^2 = \text{constant}. \qquad (6.17)$$

Consider first the motion on this ellipsoid, starting from a point close to the

ξ-axis, so that ω_η and ω_ζ are small. Then ω_ξ, taken as positive, stays sensibly constant, while ω_η changes with a sign opposite to ω_ζ, and ω_ζ changes with the same sign as ω_η. This enables us to sketch the sequence of ω_η and ω_ζ as a closed loop encircling the point where the ξ-axis cuts the ellipsoid. The same

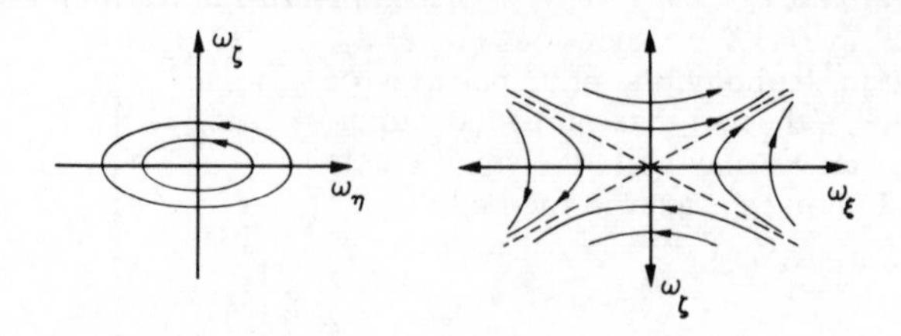

holds when $\boldsymbol{\omega}$ is nearly parallel to the ζ-axis, but with $\boldsymbol{\omega}$ nearly parallel to the η-axis, ω_η stays constant to first order while ω_ξ changes with the sign of ω_ζ and ω_ζ with the sign of ω_ξ. The resulting trajectories are hyperbolic.

If, then, the body is spun about an axis nearly parallel to that of maximum or minimum moment of inertia, its free motion involves a slight wobble about this axis but is otherwise stable. But if it is spun about an axis close to the intermediate axis of inertia, it straightway runs away from this axis of rotation, and performs a tumbling motion in which its rotation axis changes over a wide range of directions relative to the principal axes of inertia. One can sketch the trajectories of $\boldsymbol{\omega}$ on the ellipsoid (6.17), and the general pattern is

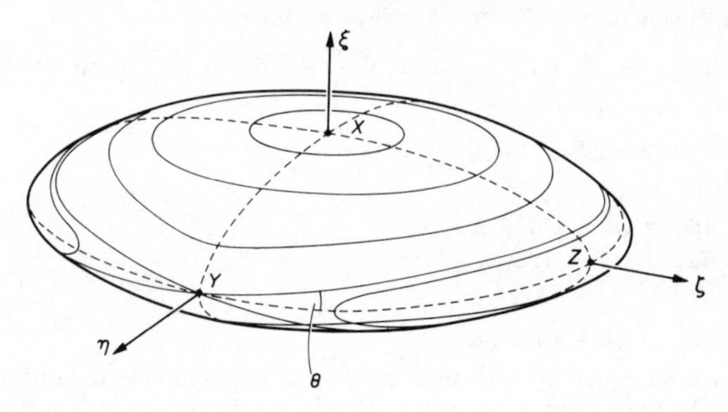

made clear by plotting out two particular trajectories, defined by the intersections of the ellipsoid with the planes passing through the η-axis and making an angle θ with the $\xi\eta$-plane, where $\omega_\zeta/\omega_\xi = \tan\theta = \{[A(A-B)]/[C(B-C)]\}^{1/2}$. On these planes, as the first and third equations of (6.16) show,

$$\dot{\omega}_\zeta/\dot{\omega}_\xi = \omega_\zeta/\omega_\xi.$$

The meaning of this equation is that on each of these planes ω_ζ and ω_ξ change so as to keep in the same proportion, and therefore $\boldsymbol{\omega}$ moves along a trajectory that runs from Y to Y' (on the opposite side of the ellipsoid from Y). These particular trajectories are the asymptotes of the hyperbolas we found in the vicinity of Y. Any other trajectory is a closed loop that encircles X or Z, but never Y.

Cavendish Problems: 3–16, 23–31, 35–39, 44–51, 108–110, 165.

READING LIST

Tidal Friction: H. JEFFREYS, *The Earth*, 4th edn., Ch. 8, Cambridge U.P.
Statistical Mechanics: F. REIF, *Statistical Physics* (Berkeley Physics Course, Vol. 5), McGraw-Hill; E. A. GUGGENHEIM, *Boltzmann's Distribution Law*, North-Holland.
Waves: C. A. COULSON, *Waves*, Oliver and Boyd; R. V. SHARMAN, *Vibrations and Waves*, Butterworth; H. N. V. TEMPERLEY, *Properties of Matter*. University Tutorial Press.
Elasticity: C. J. SMITH, *Properties of Matter*, Arnold; A. H. COTTRELL, *The Mechanical Properties of Matter*, Wiley.
Fluid Mechanics: H. N. V. TEMPERLEY, *Properties of Matter*, Arnold; L. PRANDTL, *Essentials of Fluid Dynamics*, Blackie.
Rigid-body Dynamics: R. HILL, *Principles of Dynamics*, Pergamon.
Engineering Structures: E. W. PARKES, *Braced Frameworks*, Pergamon.
Gyroscopes: R. N. ARNOLD and L. MAUNDER, *Gyrodynamics*, Academic Press.

7

Units and dimensions

It is convenient at this stage to pause and take a general look at a question that concerns all scientific procedures in which numbers are ascribed to phenomena, with a view to using those numbers in arithmetical calculations. Basically the question is one of discipline: what rules must be observed in quantifying the observations? The answer depends on the use to which the numbers are to be put.

Let us dismiss from our minds any idea that it is only scientists who are interested in numbers, but at the same time recognize that different people have different ends in view. A librarian is much concerned to assign a number to a book, and the way he does it depends on what use he makes of it. He may, for example, use a decimal classification system in which the number tells him the subject of the book and its year of publication; and he may arrange the books on the shelf in accordance with this number so that the casual reader, browsing through, is confronted with an array of more or less related works. But if he is short of space he may be forced to house them in groups of the same size, and put them on the shelves as they are acquired, without any thought for the subject; and this is perfectly acceptable if any reader is required to have a book fetched by an assistant. The class-mark then has purely geographical significance, and the librarian will usually have to keep a catalogue in which the number according to the decimal subject-classification is translated into a geographical class-mark. This catalogue, in which one set of numbers is put into a one-to-one relationship with another set, is simply the tabulation of a mathematical function of one variable—an eccentric function that exists only for integral values of the variable and has a highly erratic form. But as the sole use of this function is to relate one number to another, and no one has any thought of performing mathematical operations on it, its eccentricity is irrelevant. On the other hand, the geographical class-mark is something a little more significant in that the number is related by a comparatively simple rule to the position of the book on the shelves. In a well-ordered library, it should be possible, when told the location of two books of given class-mark, to guess fairly well where to find a book with an intermediate class-mark. That is to say, one may perform a simple arithmetical operation on the numbers and expect the result to be meaningful. But no one, told that books marked 32 were about kinetic theory and books marked 52 about solid state physics, might be reasonably expected to infer that books marked 42 were about electron optics*. The absence of logic does not, however, incommode the library user in any way, provided there is a catalogue.

Physics is, above all others, the science that concerns itself with quantifiable phenomena. Having experimented and observed, the physicist may give in the first instance a qualitative description of what he has seen, but he will be unhappy if he cannot follow it up by a quantitative description which enables another physicist to understand precisely what he wishes to convey. Often enough he will not be interested in aspects of the phenomena that are not quantifiable, though to others it may be those very aspects that are overwhelmingly important. The student of physics, faced with a problem in statics that begins 'A picture, 2 m × 1 m, is hanging from a picture hook etc.', is taught not to ask whether the picture is by Velasquez. Similarly, in a library, the librarian who most resembles a physicist is not the man who can advise a reader on the best books to consult, but the man who can find a specified book by looking up its class-mark and who is interested in devising a classification system that will enable him to do so most economically, without any regard to its quality as a book. The essentials, to a physicist, are those aspects of a phenomenon that can be characterized by numbers, and he is interested in the relations between the different numbers that arise in different circumstances. It is the latter point which really provides the starting point for our discussion of units.

There are infinitely many ways in which we can assign numbers to observable effects, if we can assign numbers at all. Consider, for example, the mass of a body. In our discussion of acceleration and force, and of gravitational effects, the only property of a body that really mattered was its mass—its colour, composition, temperature etc. were regarded as irrelevant, and two bodies having the same mass could be regarded as

* Example taken from the library of the Cavendish Laboratory, Cambridge.

interchangeable in any of the experiments. That being so, we may group together all bodies of the same mass and assign to them a number chosen at will. By doing this for every body, we provide them with a catalogue reference number having the one essential property, that all bodies with the same number are equivalent in certain dynamical experiments. In a similar fashion we might assemble a selection of forces (bent bars, stretched springs, weights on strings etc.) and number them all in such a way that equal forces had the same number. We should then find that when we carried out experiments on the acceleration of bodies by forces the number, a, measuring the acceleration was related uniquely to the mass and force numbers, m and F, in the sense that every body of a given m subjected to any force of a given F experienced the same acceleration a, i.e.,

$$a = f(m, F);$$

but we might also discover that the form of f was exceedingly complicated. An even simpler experiment, in which two bodies m_1 and m_2 in the same scale pan were counterpoised by a third body m_3 would reveal that m_3 was a unique function of m_1 and m_2, but it might not be easy to discover the form of the function except by a very large number of experiments.

The point of this fanciful excursion into what might have happened, but in fact did not, is that we would have done nothing wrong to have indulged our caprice in numbering masses and forces, but simply have made discovery of new laws much more difficult. There is no ambiguity in the numbering process suggested, but it is not as useful as it might be. By carrying out more experiments first, and studying the accelerations and counterpoising of different masses, we are enabled to replace a capricious scheme by one which reveals more readily the relations between the numbers; the functional form of $f(m, F)$ turns out to be (indeed is defined to be) F/m in the Newtonian system of enumeration, and m_3 is just $m_1 + m_2$—a law which is not hard to discover empirically with a small number of observations. There are a great many systems of enumeration used in everyday life that are quite unsuitable for revealing laws of physics, and which survive because they are not meant for this purpose but for something quite different. As examples we may cite:

the Beaufort scale of wind-speeds,
the Regulo number for the temperature of an oven,
the Standard Wire Gauge, and similar systems for specifying the sizes of drills, bolts, shoes, clothes, knitting needles etc.,
the Mohs and Brinell scales of hardness,
the magnitude of a star,
I.Q.,

and the reader will think of other examples for himself. To come nearer home, the assignment of a number to the temperature of a body by marking the position of the mercury in a thermometer at the boiling and freezing point, and subdividing the interval into 80, 100 or 180 degrees according to taste, is closely analogous to the arbitrary scales just mentioned. And as experiments became more refined, it appeared that the perfect gas had a peculiar advantage over other thermometric fluids in that the numerical values of temperature on this scale were related with high precision and by means of very simple formulae to other measurable effects. In this particular case, a system of labelling observations with numbers began as a qualitative description—a larger number meaning 'hotter', without any exact concept of quantity of hotness—and was gradually refined to the point where the numbers could be correlated with other measurements, and hotness (or temperature) took on an exact meaning in addition to its qualitative, or intuitive, meaning. In an exactly similar fashion, the magnitude of a star was first conceived as a means of describing its brightness in relation to other specified stars, and only later was it found convenient to replace the original set of discrete integers by a continuous measure based on the logarithm of the energy received.

All the examples cited above belong to the category of numbering systems applied not capriciously, but as a means of designating order rather than magnitude. Some (Beaufort scale, Regulo number, Standard Wire Gauge) could be replaced at a moment's notice by a quantitative measure if it were needed. There is no fundamental reason why wind-speed should not be measured in metres per second, but there is a sound practical reason if the majority of weather reports are to come from observers without instruments at their disposal. These systems, then, abandon niceties in the interest of convenience, and set up ordered and numbered arrays as examples for comparison; there is no meaning to be attached to intermediate numbers—no manufacturer markets a drill number 11·3, and no specification has been laid down by which he would be able to know exactly what to make if a special order for such a drill arrived. On the other hand, the Brinell hardness and the I.Q. are numbers in a continuum, in the sense that the specification of how they are to be determined allows for any figure to result from the measurements; it may be that the measurements are sufficiently imprecise as to make it unnecessary to use other than integers, but in principle a continuous range of numbers is conceivable. Here we have examples of systems where the number ascribed is to be found by a rigidly specified procedure, and where the process resembles strongly the process of measurement in a physical experiment. The only reason for denying exact comparison with a physical measurement is that for all their apparent exactitude, these systems of numbers were basically devised as a way of placing the observations in a recognizable order, rather than of producing numbers meaningful in themselves.

To reach the point, then, after so much preliminary discussion, there is a real difference between the statement that the mass of a body is 17 kg and the statement that a certain person's I.Q. is 126. When one states the mass of a body to be 17 kg, one states not only that it will behave in certain circumstances in the same way as all other bodies of 17 kg, and that it will weigh down the scale pan when counterpoised against any body of less than 17 kg; but, very significantly, one can also add such information that if it is cut into two, the masses of the two parts will add up to 17 kg. By contrast, a man whose I.Q. is 126 will perform in certain circumstances (i.e., during an I.Q. test) in the same way as others of I.Q. 126, and he will perform better than all those of lower I.Q. But one cannot add anything about the quantitative meaning of the number 126—there is no arithmetical rule, for example, by which one can decide what the I.Q.s of two men must be for them together to perform the task that one of I.Q. 126 could do in the same time.

In physical measurements we are concerned with ascribing numbers that not only put observations into a logical order but are capable of arithmetical processing in a rigidly defined manner to yield other meaningful numbers. There are various means of ascribing numbers, of which some are peculiarly important and underlie the following discussion of units and dimensions. Others, no less logical, appear in practice only rarely and must be excluded from the discussion; examples will be given. The simplest process that concerns us may be described as 'stepping out'. If we wish to measure the distance between two points we take a standard of measurement, e.g., a metre stick, and see how many times it must be put end to end to traverse the interval to be measured. If we imagine a lot of replicas laid out in a line, the analogy of this process to the determination of mass and time becomes obvious; having chosen a standard of mass and made many copies, we find how many are needed to build a body having the same dynamical behaviour as that whose mass is to be measured; in practice we do the comparison on a balance. And similarly the oscillations of a pendulum, a quartz crystal, or a caesium atom are to be thought of as replications of a standard time-interval. In each of these measurements we may choose to ascribe the number unity to the standard with which the unknown is compared, so that the number of stepping-off processes in the measurement is the number ascribed to the quantity measured. This number is the *measure* of the quantity in terms of the standard *unit*, the latter being chosen arbitrarily in accordance with what seems at the time a convenience. As the art has developed, rough and ready units such as the foot and the hand, whose convenient ubiquity outweighed their variability, were replaced by more permanent standards, bars of platinum-iridium alloy etc., which could be exactly copied and treasured for centuries; and more recently these too are giving way to standards supplied by nature, the wavelength of a certain spectral line, the vibration frequency of an atom etc. It does not concern us to go into the technical details of what constitutes a good standard, nowadays a very specialized and subtle branch of physics—it is enough to realize that the choice is free and that different people may make different choices without throwing the system into disarray, so long as they are prepared to compare their standards with one another.

The examples of stepping-off procedures, based on arbitarily chosen units, may be multiplied considerably. There were until 1948 a set of

International Units for electrical purposes, related to the resistance of a tube containing mercury, or the voltage across the terminals of a Weston cell; and there is a standard of pressure, the atmosphere, defined as the pressure required to raise a mercury barometer to 760 mm. At this stage, however, we must note that the stepping-off procedure is slightly varied without being altered in any essential principle. To find the pressure of the air, we do not pile up little bits of mercury until they balance the pressure, but simply measure the barometric height in millimetres and divide by 760 to give the answer as a multiple of the standard atmosphere. This exemplifies the second basic procedure for assigning numbers to physical quantities, which is to perform a set of operations that yield a number, and manipulate the numbers in a well-defined manner. Thus to ascribe a number to the speed of a body, we determine how far it goes in a given time, and divide the measure of the distance by the measure of the time. Symbolically

$$v = l/t,$$

and one should recognize that this equation is only a statement of the rule by which the numbers measuring length and time are to be manipulated to yield a number measuring speed. A similar procedure is involved in measuring force in terms of its accelerative properties; if in an experiment a body of mass m (m here is simply the symbolic representation of the number determined in the way already described) suffers an acceleration a, we define the number measuring the force F to be the product of the numbers m and a,

$$F = ma.$$

It may be noted that while the equation $v = l/t$ is little more than a definition of how we assign a number to the situation presented by a moving body, the equation $F = ma$ implies much more, as we have already discussed in detail in Chapter 3; it contains the results of observations that lead us to believe that a number ascribed to a force in this way will be consistent and meaningful.

It should not be thought that there is a unique way of assigning numbers to a physical quantity. There is nothing wrong with using a spring as a method of measuring force, and defining the measure of force in terms of the measure of the extension of the spring. It then becomes a matter for experiment to see what relation these two measures have to one another. According to Hooke's law they turn out to be simply proportional.

The examples quoted were typical of procedures in which the measure was derived from the observations by the arithmetic operations of multiplication and division. Extensions of the procedure are obvious, e.g., the kinetic energy of a body is given a numerical measure by calculating $\frac{1}{2}mv^2$. Sometimes, however, other operations are involved, as for example when the acidity or alkalinity of a solution is measured by its pH, defined as minus the logarithm, to base 10, of the hydrogen-ion concentration in grammes per litre; the magnitude of a star and the loudness of a sound in decibels are measured in similar logarithmic fashion. There is absolutely nothing wrong in this; quite the reverse, it is a very convenient system for widely varying quantities. But we exclude these cases from our analysis because they are comparatively unusual and need a variant of the standard treatment—they can always be dealt with as special cases when we understand the more usual situation.

The simplicity of systems of measurement that employ for the assignment of measure only multiplicative processes, i.e., the multiplication of powers (positive, negative and fractional) of the numbers derived from experiments involving stepping-off processes or their equivalents, lies in the way in which the measures change when we decide to change our standard units. Any measured quantity that depends, like length, on the stepping-off of a standard, clearly has the property that the size of the standard and the measure of a given distance vary inversely; the length of a table may be 5 feet or 60 inches—reduction of the unit of length by a factor 12 obviously enlarges the measure by the same factor. If, then, we define the process by which a number is to be assigned to speed by the equation $v = l/t$, a change in the standards of length and time that results in the measure of length being multiplied by α and the measure of time by β necessarily involves the measure of speed being multiplied by α/β. And just as multiplication of the measure of length by α is the consequence of division of the standard by α, so we can interpret the change in the measure of speed as a consequence of the division of its standard by α/β. All speeds are affected in the same way, and the same is true for any measure which is derived by an arithmetical process involving combinations of powers of the measured quantities. Thus if in addition to the changes of the standards of length and time we divide the standard of mass by γ, the measure of mass is multiplied by γ, and the measure of kinetic energy by $\gamma\alpha^2/\beta^2$. Again, the measure of the kinetic energy of every body is changed by the same factor, and we may speak of the standard of kinetic energy as having been divided by $\gamma\alpha^2/\beta^2$. On the other hand, if we were to define the pH of a solution as minus the logarithm of the H-ion concentration, measured in terms of the prevailing standards of mass and length, a change in these would change the measure of H-ion concentration by a factor γ/a^3, and the pH would not be multiplied by a factor but would simply be increased by the addition of a constant, $-\log(\gamma/a^3)$. The introduction of the logarithm into the arithmetical process alters the way in which the measure is affected when the standards are changed. There is no unit of pH in the sense that there is a unit of length.

We now restrict our attention to physical quantities whose measure is defined by processes involving the arithmetical operations of multiplication and division, and not by such functional operations as taking logarithms. The conventional way of writing the rule governing the change of units of derived quantities is to write the formula to be applied without troubling to include numerical constants or any other parts which are unaltered by changes of variables, and to put the measured quantities in square brackets to show that such constants are missing and the equation is meant only to specify the rule for changes of units. Thus we write

$$[V] = [L]/[T] \qquad \text{and} \qquad [E] = [M][V]^2 = [M][L]^2[T]^{-2}.$$

The first of these *dimensional equations* relates the *derived* unit $[V]$ to the *basic* units $[L]$ and $[T]$; the second relates the derived unit to basic units which are either $[M]$ and $[V]$, or $[M]$, $[L]$ and $[T]$. There is nothing sacred about the basic units—they are only those that we elect to vary at will. The first equation tells us that we cannot arbitrarily alter the units of v, l and t simultaneously and still maintain the arithmetical truth of the equation $v = l/t$, but it does not tell us that the unit of velocity must always be subservient to the units of length and time. When we speak of the distance of a star as being so many light-years, we imply that we have chosen the unit of time to be a year and the unit of speed to be the speed of light; the unit of length is then related to them in accordance with the rule

$$[L] = [V][T]$$

Note that we do not imply that the unit of length is *necessarily* the distance travelled by light in unit time, though in this case it happens to be so. To relate the absolute values of the units, as distinct from the way they change relative to each other we must know the numerical constants. Thus the kinetic energy of unit mass moving with unit velocity is half a unit of energy, since the formula applied is $\frac{1}{2}mv^2$.

To sum up the argument to this point:

The dimensional formula for a physical quantity expresses the factor by which the units for this quantity must be multiplied when the basic units occurring in the formula are multiplied by arbitrary numbers. Exactly the same statement holds for the arithmetical measure of a quantity and the way it changes when arbitrary changes of the basic units alter the measures of the quantities appearing in the formula.

EXAMPLE

In the c.g.s. system the unit of force is a dyne, in the m.k.s. system it is a newton. How many dynes are there in a newton?

The dimensions of force are $[M][L][T]^{-2}$. In going from c.g.s. to m.k.s. the unit of mass is multiplied by 10^3 and the unit of length by 10^2, that of time being unaltered. Then the unit of force is multiplied by $10^3 \times 10^2$, i.e., 10^5, therefore

$$10^5 \text{ dynes} = 1 \text{ newton}.$$

PROBLEM

How many kilowatt-hours are there in a therm? A therm is 10^5 British thermal units (Btu); 1 Btu is the heat required to raise the temperature of a pound (454 g) of water through 1°F; 1°F is $\frac{5}{9}$°C; 1 calorie (=4·186 J) is the heat required to raise the temperature of a gramme of water through 1°C. [Answer: 29·3]

In the light of this statement it should be clear that the Theory of Dimensions does not conceal any subtle hints on the ultimate nature of things. This warning is necessary because of the rather widespread view that the dimensions of a physical quantity in some way express its essential quality divorced from the crude numerical value, as if leaving out the constants in an equation and enclosing all the variables in square brackets was all that was needed for the attainment of true enlightenment. It would be nearer to the truth to assert that the dimensions tell us nothing of the quality of an experience but only something about its quantitative measure, but we shall not debate the point at length. It is enough to assert that any deep significance claimed for the theory should be explicitly displayed before one allows oneself to believe too firmly in its existence. If we adopt this view we need not be dismayed to find that 'physically meaningless' fractional powers sometimes appear, or that the same quantity may have different dimensions in different circumstances. On this latter point, we shall see in due course that the number of basic units may be as many or as few as we please, and it follows that as we change the number the dimensions of derived quantities will necessarily alter.

There is, however, one set of physical quantities whose special significance is revealed by dimensional analysis, and these are the dimensionless numbers, whose value is independent of any choice of units and which may therefore be said to have a meaning that transcends merely human choice. It is amusing that the velocity of light should be within 0·1% of the simple number 3×10^8 m s^{-1}, but the fact has no significance unless one believes, against all probability, that it was to this end that the French savants who defined the metre were led by Providence. On the other hand, it is no accident of choice in the matter of units that confers on the proton a mass 1836·1 times that of the electron. This number must be accepted either as an arbitrary choice on God's part at the creation of the world, or as something still hidden in the theory of fundamental particles, yet awaiting to be revealed as the inevitable corollary of some act of divine will* more basic than merely inventing a number. This particular number is one that belongs to a well-defined branch of physical theory, the theory of fundamental particles, whose

* I purposely adopt this language to make clear that I should think it arrogant to rule out the possibility that we were set in our course and are maintained in it by a divine power, simply because we do not see the Hand of God continually interfering in the workings of the physical world.

practitioners are reasonably confident of ultimately finding an explanation when the theory has been sufficiently developed. There are other numbers, however, whose existence is tantalizing in the extreme. We have noted, in Chapter 4, that the Coulomb force between electrons is 4×10^{42} times stronger than the gravitational force. This is a dimensionless number whose explanation demands a theory linking the two basic forces, electric and gravitational, and it is a formidable task to devise a unified theory that can have among its consequences two effects, apparently similar in kind, yet so vastly different in magnitude.

How many basic units?

The short answer is—as many as you like. In the foregoing discussion, force was defined in terms of mass and acceleration by prescribing a routine for determining what number to assign to a given force, but other procedures are just as acceptable. We might devise a standard spring or cantilever (specifying its exact shape and material) and denote as unit force that which would cause a certain displacement; the measure of another force would be defined as a number proportional to the displacement it produced. In an acceleration experiment we should then find that the product of mass and acceleration was proportional to force,

$$ma = CF, \tag{7.1}$$

a law of nature involving in its exact statement a constant, C. Because C changes its magnitude when the units of mass, length, time and force are changed, C has dimensions

$$[C] = [M][L][T]^{-2}[F]^{-1}, \tag{7.2}$$

and is called a *dimensional constant*. There is no objection to proceeding in like manner with every sort of physical quantity; indeed the old international system of electrical units was like this. Sometimes it happens by accident, because the law of nature that could be used to avoid defining a separate unit has not yet been discovered. Long before the work of Rumford, Joule and other pioneers of thermodynamics, the concepts of heat (Q) and work (W) were well enough understood to be reduced to quantitative measurement, the former in terms of the calorie, the latter in terms of mechanical units. The first law of thermodynamics then arrived to relate the two by an equation

$$W = JQ, \tag{7.3}$$

in which J, the mechanical equivalent of heat, is a dimensional constant

$$[J] = [W][Q]^{-1}. \tag{7.4}$$

When one recognizes that two physical quantities are uniquely related by a law of nature, it often seems an unnecessary luxury to carry independent units for both, though it may be a great convenience as the survival of the calorie testifies. By adopting the same unit, the dimensional constant is made to take a defined value (usually unity) and being thereafter invariant has immediately become dimensionless. Alternatively stated, we may define constants such as C in (7.1) and J in (7.3) to take the value unity, whereupon the requirement that they be invariant when units are changed implies, through (7.2) and (7.4) that

$$[F] = [M][L][T]^{-2},$$

and

$$[Q] = [W];$$

that is, the units of force and heat are now derived instead of basic units.

The suppression of dimensional constants by the use of laws of nature is something that has happened on many occasions without conscious effort. It is probably not generally appreciated how the isotropy of space has made it possible to maintain one standard of length, against a prevalent tendency to think of lengths measured in different directions as different in kind. As evidence for what may seem a wild statement, one may cite the cases of angles and gradients whose numerical measure may vary in spite of their apparent dimensionlessness. A right angle is said by some to be $\pi/2$ and by others 90°; a hill may have a gradient of 1 in 20 (i.e., 1/20) or 264 ft/mile, the latter representation being usual if the gradient is measured off a map. The possibility of variation lies in the dimensions of angle being $[L_t]/[L_r]$, tangential and radial lengths being measured in different systems of units. There is nothing wrong in such systems, but they are inconvenient in mathematics because one has to carry a dimensional constant (e.g., $\pi/180$ radians/degree) through all manipulations; it is therefore customary to suppress one of the length units by insisting on angles being measured in radians.

Of considerably greater interest is the tacit suppression of a conceivable dimensional constant relating gravitational to inertial mass. In this case it happened that Newton saw inertia and gravitation as separate manifestations of the same property or matter, and later philosophers and experimenters have agreed with him; but it need not have been so. It would have been quite reasonable to define a derived unit of gravitational mass, m_g, by specifying that two bodies of unit gravitational mass shall exert unit force on each other when placed unit distance apart, or, in general

$$|F| = m_{g1}m_{g2}/r^2. \tag{7.5}$$

Remembering that F has dimensions $[M_i][L][T]^{-2}$, where $[M_i]$ represents inertial mass, we see that in this system gravitational mass has the dimensions $[M_i]^{1/2}[L]^{3/2}[T]^{-1}$. In fact, of course, the law of force was written in the form

$$|F| = Gm_{i1}m_{i2}/r^2,$$

where G is a dimensional constant having dimensions $[M_i]^{-1}[L]^3[T]^{-2}$.

We have here a situation where the recognition of a universal law of nature connecting two apparently different quantities, inertial and gravitational mass, is not used to suppress one of the units. It would be perfectly justifiable to recognize (7.5) as the law of gravitation and then to lay down, once and for all, that m_g/m_i was to be unity. This is the same as defining G as unity and depriving it of dimensionality. The implication is that $[M]^{-1}[L]^3[T]^{-2}$ is dimensionless, or that, for example,

$$[M] = [L]^3[T]^{-2}.$$

The unit of mass is then to lose its basic status and be derived from the units of length and time.

EXAMPLE

In the above system, what is the measure of a 1 kg mass in terms of the metre and the second, given that G is experimentally found to be $6{\cdot}67 \times 10^{-11}$ $kg^{-1}\,m^3\,s^{-2}$?

The statement of the magnitude of G may be rewritten as $1\ kg = 6{\cdot}67 \times 10^{-11}\,G^{-1}\,m^3s^{-2}$ so that if $G = 1$ it follows that $1\ kg = 6{\cdot}67 \times 10^{-11}\,m^3s^{-2}$.

Alternatively, if we multiply the unit of mass by α, the measure of mass of a given body is reduced by α, and since G has dimensions $[M]^{-1}$ in mass, the measure of G is multiplied by α. If then we choose α to be $10^{11}/6{\cdot}67$, i.e., $1{\cdot}5 \times 10^{10}$, leaving length and time unaltered, the measure of G becomes unity, as required. The unit mass in this system is therefore $1{\cdot}5 \times 10^{10}$ kg. Correspondingly unit force, having dimensions $[M][L][T]^{-2}$, is $1{\cdot}5 \times 10^{10}$ N and two unit masses ($1{\cdot}5 \times 10^{10}$ kg each) placed 1 metre apart attract one another with unit force ($1{\cdot}5 \times 10^{10}$ N).

The example makes clear that the unit of mass defined in this way would be highly inconvenient for everyday purposes, but there is a much more serious objection to it, that it is extremely hard to realize in practice. To determine the mass of a body in terms of this 'natural' standard would involve comparing it with a mass (sub-standard) that had been used in a gravitation experiment; the comparison might be done by using a balance, with a precision of 1 part in 10^9 or better, but the absolute determination of the sub-standard would involve performing one of the determinations of G (e.g., Heyl's) which are capable of an accuracy of 1 in 10^3 at best (though there is work in progress that may improve the determination by at least a factor of 10). In this situation it is far better to forget the idea of relating all masses to length and time by making G equal unity, and carry on in the old way of relating them to an arbitrarily selected standard mass and leaving G as a dimensional constant to be determined as well as possible. This is an especially strong argument when one realizes how rarely the accurate value of G enters into any calculation; the gravitational effects that need to be known very well are terrestrial effects for which g is needed to high accuracy, while in cosmic calculations involving G high precision is as yet hardly relevant.

A closely related case where suppression of dimensional constants is feasible is Coulomb's law (4.2) which can be written in the form

$$|F| = q_1q_2/r^2,$$

by adopting a unit for electric charge which is related dimensionally to those of mass, length and time exactly as was the unit of gravitational mass; of course the magnitude is different since γ (see equation 4.1) and ε_0 are different in magnitude. Here there is no experimental reason for rejecting this system as there was in the gravitational case, and in fact this is the absolute electrostatic system which could in principle be used with any set of mechanical units, but which in practice has always been associated with the c.g.s. units.

PROBLEM

The electronic charge is $4{\cdot}8 \times 10^{-10}$ in the electrostatic system based on c.g.s. units. What would it be in a hypothetical 'British electrostatic' system based on yards (91·4 cm) pounds (454 g) and seconds?

[Answer $2{\cdot}6 \times 10^{-14}$]

There are still many physicists whose early conditioning makes them feel that $4{\cdot}8 \times 10^{-10}$ is the most comfortable number to associate with the electronic charge and who are resistant to SI units, but international usage has swung decisively in favour of an independent unit of charge, the Coulomb, with the consequence that the Coulomb law must contain a dimensional constant as in (4.2).

Finally we may note the existence of many other constants of nature, which, like G, are so unchanging as to deserve to be given simple values and to justify suppression of independent units. The velocity of light, c, pervades relativistic theories and cosmology just as Planck's constant $\hbar$ pervades atomic and nuclear physics. What could be better than to define G, c and $\hbar$ to be unity and in this way eliminate all artificial units of mass, length and time?

PROBLEM

In the system where $G(6{\cdot}67 \times 10^{-11}\ \text{kg}^{-1}\ \text{m}^3\ \text{s}^{-2})$, $c\ (3{\cdot}00 \times 10^8\ \text{m s}^{-1})$ and $\hbar\ (1{\cdot}05 \times 10^{-34}\ \text{J s})$ are all set equal to unity, show that the unit of mass is $2{\cdot}2 \times 10^{-8}$ kg, the unit of length $1{\cdot}6 \times 10^{-35}$ m and the unit of time 5×10^{-44} s.

When one sees these figures, one appreciates that the elegance of suppressing constants of nature may be bought very dearly. In the end, it is essential to recognize that the units must be chosen for convenience and reproducibility, and that whatever a theorist may do in the way of selecting rational systems of units to give a tidy appearance to his mathematics, he must not attempt to impose his criteria on practical physicists and engineers, but leave that side of it to the experts.

We have learnt, however, by this discussion, that there is no magic number of basic units, and that the theory of dimensions is essentially a practical matter whose philosophical overtones are so slight as to be safely ignored by the majority. The general agreement to accept three basic units, of mass, length and time, for mechanical purposes and to supplement these by a unit of current for electrical purposes reflects the fact that these units can be constructed, copied and compared with great accuracy. In due course, other and better systems may be discovered, but until that time the agreed set of units offers the best chance for physicists all over the world to make useful comparisons among themselves of the numbers they attach to their observations.

Dimensional analysis of physical problems

We now proceed to the application of dimensional analysis, and may conveniently start with an example to show the method in action—the velocity of a wave in a deep ocean. We begin by listing the parameters that might influence the velocity, bearing in mind that the experiment is not limited to oceans of water nor to the surface of this planet; that is to say, the density of the liquid, ρ, and the gravitational acceleration g, may be taken as variables. We also assume that the velocity may depend on the wavelength, λ. If we stop there, our task is to construct a combination of ρ, g and λ that shall have the dimensions of velocity. This may be achieved by writing the dimensions of everything involved in terms of some suitably selected basic units, such as mass, length and time:

$$[v] = [L][T]^{-1}$$
$$[\rho] = [M][L]^{-3}$$
$$[g] = [L][T]^{-2}$$
$$[\lambda] = [L].$$

Since $[M]$ appears only in ρ, ρ must be absent from the combination, which is now seen to be $(g\lambda)^{1/2}$. Thus, we argue, the velocity of a deep ocean wave is expressed in the form, $v = C(g\lambda)^{1/2}$, in which C is a dimensionless number, concerning whose magnitude the analysis is powerless to speak.

This example raises three questions which we shall discuss separately: (1) why should the equation have the same dimensions on both sides, (2) why did we only write down a limited number of parameters, and what would have happened if we had thought more were involved, and (3) if the answer contains an undetermined constant is not the whole process a bit pointless?

(1) *Dimensional congruence of equations.* It may seem obvious that the expression for v should have the same dimensions as v, for otherwise the equation could be true only in one particular set of units. Remember,

however, before taking this for granted, that there are many useful relations which are true only in one set of units. A housewife who cooks her joint of beef according to the rule '15 minutes to the pound and 15 minutes over' is applying a formula,

$$T = 15(W + 1),$$

which is valid only if W is measured in pounds and T in minutes. So too the formula,

$$♀ = \tfrac{1}{2}♂ + 7,$$

which represents the ancient recipe that the girl's age at marriage should be half the boy's plus seven, is applicable, if at all, only to ages measured in years. We must not dismiss these 'rules of thumb' as invalid or unscientific just because they are dimensionally inhomogeneous, but should explore the difference between them and what we regard as a more physical formula.

The essential difference is that the rules of thumb are approximate statements relating to exceedingly complex situations, and valid only within a restricted range of circumstances. Thus a truly scientific formula for cooking would be a formula for cooking any sort of meat to be fit for the table of any sort of man, and would involve as variables the detailed characteristics of the proteins to be found in mouse and mammoth, and psychological factors underlying the taste of Earl and Eskimo. The rule of thumb valid for beef would then be seen as an approximate expression for one small part of the complete solution of the problem. In just the same way we may save a clockmaker the trouble of remembering that the time of swing (half-period) of a pendulum is $\pi(l/g)^{1/2}$ by telling him that a grandfather clock with a one-second pendulum will gain 43 seconds a day if the pendulum is shortened by one millimetre. This statement is true only for a one-second pendulum on the Earth, but after all that is the usual context in which it is likely to be needed.

The question to ask, then, is why certain equations are not approximations of this sort and have the property of holding in any set of units. The answer is to be found by considering how one would derive from first principles such an equation as that giving the velocity of a water wave. We should write down a number of basic statements relevant to the motion of the water, such as:

(a) If the density is ρ and the velocity at a point $\mathbf{u}$, then the force $\mathbf{F}\,\mathrm{d}V$ acting on any elementary volume $\mathrm{d}V$ is related to the acceleration by the equation $\mathbf{F} = \rho\dot{\mathbf{u}}$.

(b) The force $\mathbf{F}$ acting on unit volume is related to the variation of pressure P by the equation $\mathbf{F} = -\operatorname{grad} P$.

(c) There is a gravitational force ρg per unit volume everywhere.

(d) If the liquid is incompressible, $\operatorname{div} \mathbf{u} = 0$.

From these statements one proceeds by mathematical manipulations, which in this case are fairly lengthy, to derive a differential equation for $\mathbf{u}$

whose solutions turn out to be wavelike, and so the wave velocity is determined. The point to observe in all this is that the basic statements are all dimensionally homogeneous, either because like (b) and (d) they are statements in which each term is of the same kind (e.g., in (d) $\partial u_x/\partial x$ has the same form as $\partial u_y/\partial y$ etc.; in (b) we are in effect adding up a number of forces acting all round a volume element and equating the resultant to a force) or because like (a) and (c) they involve a law (N2) which is itself the means whereby the dimensions of force are established. The analysis of the problem from first principles thus starts with a set of dimensionally homogeneous equations, and from then onwards the mathematical operations are applied to each term in an equation in the same way, and do not destroy homogeneity.* This explains why an analysis from first principles may be assumed to yield a dimensionally homogeneous answer, something of whose structure may be inferred from that very fact, even though the numerical constants cannot.

(2) The second question concerned the number of parameters on which we assumed the solution to depend. Here experience plays a considerable role, but one may in principle analyse the situation by going back to the basic equations again. To take the wave problem once more, in the end we hope to derive a differential equation in which z_0, the vertical displacement of the surface, is related to x, y and t, and we can see that the physical parameters that define the problem are g and ρ; the variables $\mathbf{F}$ and P are introduced as intermediates in the calculation and will be eliminated during the manipulations leading to the differential equation. The trial solution which we substitute into the equation itself contains three parameters, one expressing the amplitude of the wave, A, with dimensions $[L]$, and the others the wavenumber, k ($= 2\pi/\lambda$), and angular frequency, ω. We might for instance look for a solution in which the wave travels in the x-direction, and hazard the guess that $z_0 = A \cos(kx - \omega t)$; this would prove to be not quite good enough and a more general function $Af(kx - \omega t)$ would be needed. The function f, like cosine, must be regarded as dimensionless, in the sense that it must operate on a dimensionless variable (both kx and ωt are dimensionless), and have its value unaffected by changes of units. When this trial solution is substituted in the equation the process of differentiation of a dimensionless function does not leave it dimensionless since $\partial f/\partial x$, for example, has the dimensions $[L]^{-1}$ as a consequence of dividing a little dimensionless δf by a little length δx. The dimensional coefficient introduced by differentiation is, of course, provided by the parameters in the trial solution; thus $(\partial/\partial x)[\cos(kx - \omega t)] = -k \sin(kx - \omega t)$, the sine function

* One can, of course, destroy homogeneity by adding two equations which have different dimensions, but this is a futile operation which does not help the analysis and leaves one in the end with an equation which can be dissected into separate homogeneous equations, each correct in itself. In complex algebra, the symbol i may be thought of as something like a dimensional symbol that keeps incongruent elements of the equation apart; in the end, however, the real and imaginary parts have to be separated for numerical purposes.

being dimensionless and k having dimensions $[L]^{-1}$. We therefore expect each term in the equation, after substituting the trial solution, to have as coefficient some combination of the physical parameters g, ρ, A, k and ω. The condition that the trial solution shall satisfy the equation leads to some relation between the coefficients and it is this that gives us, say, the frequency ω as a function of g, ρ, A and k, and the velocity ω/k as a function of the same four variables.

This analysis shows that we could reasonably have expected v to depend on four independent parameters defining the system, yet in practice we chose only three, g, ρ and k. There was, indeed, no justification for dropping the amplitude A, except the knowledge, based on experience, that many oscillations and waves of small amplitude are not amplitude dependent in their properties, and one might hope that this would prove to be yet another example, as indeed experiment verifies, if A is small compared with λ.

To avoid the reproach of having inserted prior knowledge of the solution into the analysis, let us repeat it with A as a potentially significant variable. We now find that the form of the solution is much less well defined, since there is an infinite number of combinations of g, A and λ that have the dimensions of velocity (A, having dimensions $[L]$, does nothing to bring ρ back into the calculation). The function $(g\lambda)^{1/2}$, with the dimensions of velocity, may be multiplied by any power of the dimensionless combination A/λ, or indeed by any mathematical function of A/λ without losing this property, and the most general solution can be cast in the form

$$v = (g\lambda)^{1/2}\phi(A/\lambda), \tag{7.6}$$

where ϕ is any function.

We can look at this from a slightly different point of view by thinking once more of the hypothetical differential equation into which we substituted a trial solution. The condition for the solution to be valid was the satisfying of a certain dimensionally homogeneous equation in which combinations of the parameters appear as coefficients of dimensionless functions. If we divide through by one of these coefficients we arrive at an equation in which all coefficients and the arguments of all functions are dimensionless combinations of the physical parameters. Representing these dimensionless combinations by α, β, γ, δ etc., we may schematically write the solution of the equation in the form,

$$\alpha = F(\beta, \gamma, \delta \ldots),$$

though until we have set up and solved the equation we have no idea of the form of the function F. Now not all these dimensionless combinations need be independent variables; it may happen, for example, that δ is just $\beta\gamma$ or α/β^2, in which case it need not be displayed explicitly, but absorbed into the structure of F, which now depends on as many dimensionless combinations of parameters as cannot be derived from the others by multiplication etc. In the problem of the wave we can form two independent dimensionless combinations of which $v(g\lambda)^{-1/2}$ and A/λ are a typical pair ($v(gA)^{-1/2}$ is an equally valid substitute for either of these but is not independent since it can be formed by dividing the first by the square root of the second). The most general statement we can make is therefore that $v(g\lambda)^{-1/2}$ depends in a perfectly unique manner on A/λ, but we cannot at this stage say anything about the form of the dependence, i.e.,

$$v(g\lambda)^{-1/2} = \phi(A/\lambda),$$

which is the same as (7.6).

The question then arises, how many independent dimensionless parameters are there? We answer this by writing the dimensional formulae for all the variables known to affect the problem, and thence find the condition that an algebraic combination shall be dimensionless. Now

$$[v]^a[\rho]^b[g]^c[\lambda]^d[A]^e = [M]^b[L]^{a-3b+c+d+e}[T]^{-a-2c}$$

Any dimensionless combination must therefore have $b = 0$, $a - 3b + c + d + e = 0$, and $a + 2c = 0$. These three equations with five unknowns reduce to the form $d + e - c = 0$, and any two of these three can be chosen arbitrarily, after which the rest are fixed. If we decide to make d and e the arbitrary variables, there are two elementary choices from which all others can be generated by linear combinations, and these choices may be taken to be $d = 1$, $e = 0$ and $d = 0$, $e = 1$ giving as independent dimensionless variables $\alpha = g\lambda/v^2$ and $\beta = gA/v^2$. Any other dimensionless variable formed by putting $d = n$, $e = m$ is just $\alpha^n\beta^m$ and is therefore not independent. The general rule then follows by an obvious extension—the difference between the number of physical variables and the number of basic units involved in writing their dimensions gives the number of independent dimensionless parameters.

It can now be seen that the closeness to which the form of the solution of a problem can be specified by dimensional analysis is enhanced by keeping as many basic units as possible. This does not mean that one can improve matters by the arbitrary introduction of extra independent units, for this simultaneously increases the number of physical variables. Thus if we decide to have force as a standard as well as mass, length and time, we have to write N2 in the form (7.1) and C appears as a further dimensional quantity; we are no better off. On the other hand, although we may know that position and time coordinates are in principle connected by the velocity of light, we should not use this fact to eliminate the unit of length or time unless we are sure that relativistic effects are involved. But in this case we need to keep the velocity of light in the equation, so that we neither lose nor gain by eliminating a unit, since simultaneously c, being equal to unity, has disappeared as a variable. The whole conclusion, then, is that we can only get the fullest information about a system by dimensional analysis if we take pains not to introduce any irrelevant physical variable. To finish with a last look at the ocean wave, if we have reason to believe, perhaps by experiment, that the amplitude plays no significant part, we should leave it out and reach thereby the most complete answer we can ever expect, that $v = C(g\lambda)^{1/2}$. But if we have any doubt, we should leave it in and be content with (7.6). It may well prove, when experiments are performed, that even this does not describe the facts; in that case we must look back at our analysis to see what significant factor (surface tension?) was left out.

(3) The third question concerned the value of an admittedly incomplete solution; the question has acquired more point by what we have just discussed, in that we have found that the incompleteness may extend from unknown constants to unknown functions. If the problem is readily solved, there is indeed little value in the dimensional analysis except as a check of the form of solution obtained. It is quite another matter if the exact solution is unknown, for then the dimensional analysis may enable experiments to be carried out economically and the results presented cogently so as to display in empirical form the complete solution.

EXAMPLE

The flow of liquid in a smooth circular pipe. The theory of streamline flow of a viscous liquid leads to Poiseuille's formula, which we shall simply quote since a detailed understanding is unnecessary for present purposes. If the mean velocity of flow is v, the radius of the pipe r, the pressure gradient along the pipe grad P and the viscosity η, then for streamline flow

$$v = r^2 \operatorname{grad} P/(8\eta).$$

Writing the dimensions,

$$[v] = [L][T]^{-1}, \ [r] = [L], \ [\operatorname{grad} P] = [M][L]^{-2}[T]^{-2}$$

and

$$[\eta] = [M][L]^{-1}[T]^{-1},$$

we see that this result is dimensionally correct and in fact takes the only possible form. But if the speed is increased, turbulence sets in with the formation of eddies whose production and subsequent behaviour are affected by the inertial properties of the fluid; in streamline flow, where the fluid flows in straight lines without acceleration, there is no need to consider its density, but when turbulence appears ρ also must enter. We can now construct two independent dimensionless parameters out of v, r, grad P, η and ρ, and those conventionally chosen are Reynolds' number, Re, defined as $2\rho vr/\eta$, and a parameter λ, defined as $4r \operatorname{grad} P/(\rho v^2)$. Dimensional analysis tells us that all experiments on the flow of incompressible fluids through smooth pipes (these two requirements obviate the need to insert into the theory the bulk modulus of the fluid and a further parameter specifying the roughness of the walls) should give numerical results for Re and λ that lie on a single curve,

$$\lambda = \lambda(Re).$$

For Poiseuille flow $\lambda = 64/Re$, but when Re exceeds about 2000, turbulence sets in and, as the experimentally determined curve shows, the function $\lambda(Re)$

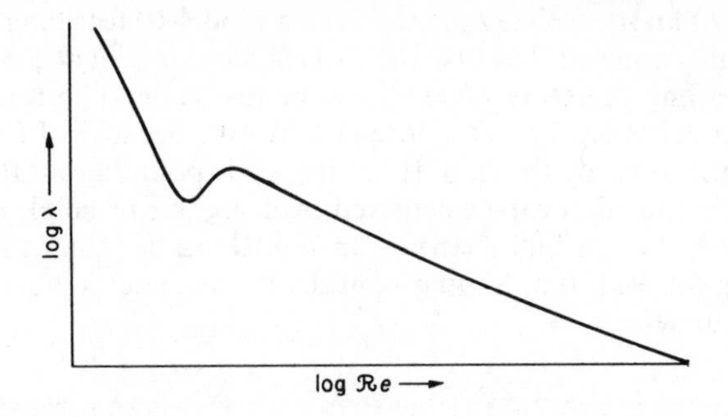

is rather extraordinary. Although theories of its form when *Re* is very large have been given, they are by no means as simple as the theory of Poiseuille flow, and no adequate theory exists for the intermediate region where the curve kinks.

The practical value of dimensional analysis in such a situation is that it enables experimental data to be presented with economy on a single curve. It is no longer necessary to have a curve of flow resistance for every liquid separately—all lie on the one curve when the dimensionless parameters are employed. If there were three independent dimensionless parameters the universal function relating them would be a surface in a three-dimensional plot; more complicated, but nevertheless a more manageable representation than a whole bookful of uncoordinated data on different materials. A good idea of the practical value of dimensional analysis may be obtained from any standard work on the engineering aspects of fluid flow and heat transfer, which depend on dimensionless charts for the presentation of design data obtained empirically and without the support of any but the sketchiest of theories. This is indeed a simple technique for bringing within the scope of rational examination problems so complex that conventional mathematical analysis is almost useless.

Very commonly it is not necessary to determine by experiment the complete form of the function relating the dimensionless variables, since perhaps only a single point is needed. Consider, for example, the problem of finding what resistance a ship will experience. If we expect that the major source of resistance will be the waves it sets up, rather than the eddies breaking away from its underwater surface, it is likely that the viscosity of the water will play only a small part compared with its density and the gravitational acceleration. The retarding force F on a ship of given shape is then tentatively supposed to depend on ρ, g, the velocity v, and a single length l specifying the size of the ship, and dimensional analysis enables one to write

$$\frac{F}{\rho g l^3} = f\left(\frac{v^2}{gl}\right).$$

If we wish to find what force a ship 1000 feet long will experience at a speed of 20 knots, we may instead take a model 10 feet long and move it through the water at 2 knots; the *Froude number*, $v^2/(gl)$, is the same in both cases and therefore $F/(\rho g l^3)$ will be the same. The force measured on the model must then be multiplied by 10^6, because of l^3, to give the force experienced by the ship. It hardly needs pointing out the enormous saving in money that can be achieved by using this principle of *dynamical similarity* to design hull shapes with models rather than using the full-sized ship for each test. Similar considerations underlie the use of models in wind-tunnels.

EXAMPLE

We shall not pursue this particular application any further, or discuss how viscosity can be brought into the model experiments, but as a last example of the use of dimensional analysis show how it was most elegantly applied by de Boer in 1949 to predict certain properties of the rare light isotope of helium before enough had been prepared to measure them directly. We begin by recalling that the equation of state of N molecules of a perfect gas may be written

$$PV = RT \quad \text{or} \quad Pv = kT,$$

where $v = V/N$ and is the volume per molecule, and k is Boltzmann's constant which takes the same value for all gases. If we believe that two molecules, r apart, interact with one another with a conservative force derived from a potential $\phi(r)$ which gives a weakish attraction at larger distances and a strong repulsion when they get too close, we must modify the equation of state to take this force into account. It happens that the general form of the curve is very similar for most simple molecules, but the depth ε of the minimum, and the radius σ at which the minimum occurs, vary from one to another. It is therefore quite a good approximation to characterize the individual gas by the two parameters σ and ε as well as the mass of the molecule, m, and write as a general form of the equation of state

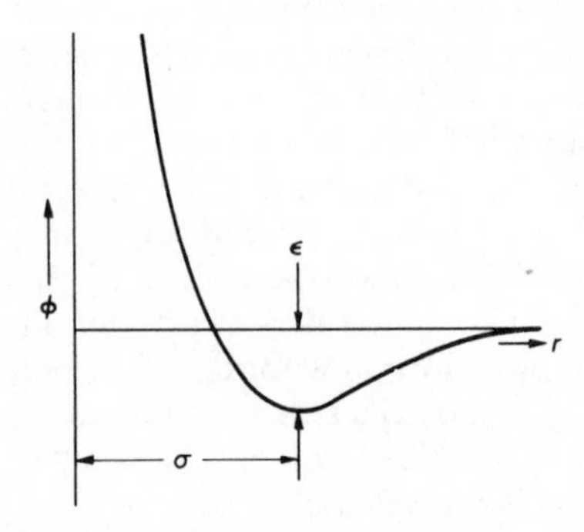

$$f(P, v, kT, \sigma, \varepsilon, m) = 0.$$

We now write this in dimensionless parameters, noting that P^*, defined as $\sigma^3 P/\varepsilon$, V^* defined as v/σ^3, and T^* defined as kT/ε are three independent parameters, and that no more can be constructed; m cannot be incorporated and must therefore be irrelevant. It follows that in dimensionless form the equation of state should be the same for all gases,

$$F(P^*, V^*, T^*) = 0; \tag{7.7}$$

and this is indeed quite a good approximation and is known as the *law of* *corresponding states*. The diagram shows schematically isotherms at various values of T^*, and there is one of particular interest, the *critical isotherm*, above which the liquid and vapour phases merge as P^* is changed, without any intermediate state of mixed phases. The dimensional analysis indicates that in so far as $\phi(r)$ has the same form for different molecules, so T_c^* should be the same. Now de Boer pointed out that the argument assumes classical, rather than quantal, behaviour on the part of the molecules, in that Planck's constant is not assumed to be a relevant parameter. Moreover this assumption is particularly dubious for the light gases at low temperatures, for the molecules have then a low momentum ($p = (2mE)^{1/2}$ if E is the kinetic energy, and therefore $p \propto (mT)^{1/2}$), and if the product $p\sigma$ is comparable with $\hbar$ the Uncertainty Principle leads us to expect that quantum mechanics will play a part in the collision process. It is in fact well known that the properties of helium near and below its critical temperature, and especially in the liquid state, are strongly influenced by quantal effects. At high temperatures, however, the classical approximation is fairly good, and de Boer used the high temperature departures from Boyle's law to estimate the relative values of σ and ε in different gases; this is done by finding how P, V and T have to be scaled so as to make all gases fit on the same set of isotherms. This enables one to determine, for example, relative values of ε for a number of gases, and so to check that the reduced critical temperatue T_c^*, i.e., kT_c/ε, is the same for all. The check is satisfactory for the heavier molecules, as expected, but marked discrepancies occur for the lighter molecules. At this stage de Boer introduced $\hbar$ into the argument, in the form of yet another dimensionless parameter Λ^*, defined as $\hbar/(\sigma^2 m\varepsilon)^{1/2}$, and pointed out that one would expect a more general equation of state, replacing (7.7), to take the form

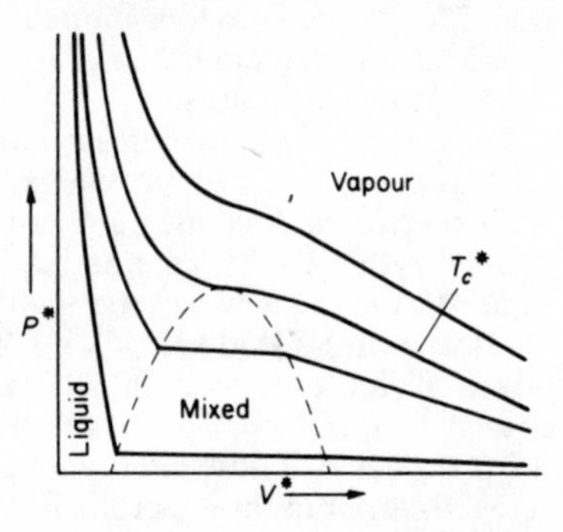

80

$$F(P^*, V^*, T^*, \Lambda^*) = 0.$$

This implies that each gas, characterized by its peculiar Λ^*, would exhibit

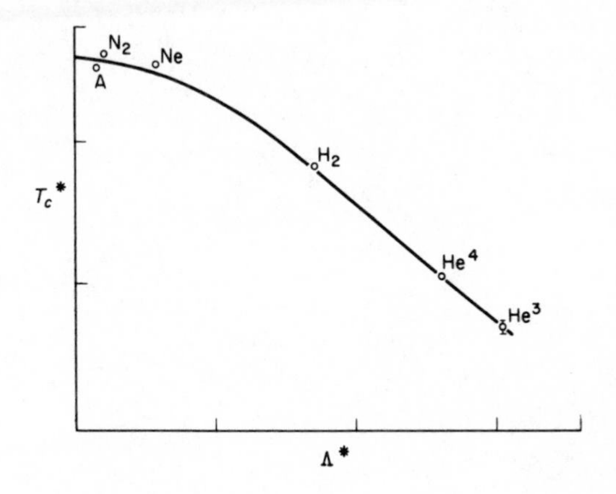

isotherms of different shapes, so that the reduced critical temperature would no longer be constant but would depend on Λ^*, according to a universal function $T_c^*(\Lambda^*)$. He therefore plotted T_c^* against Λ^* for a number of gases and found, as the diagram indicates, that a smooth curve could be drawn close to the experimental points.

The last step in the argument introduces the new isotope of helium, whose nucleus consists of two protons and only one neutron, instead of two as in ordinary helium. Its mass number is 3 instead of 4. Outside the nucleus, however, everything is the same, so that ε and σ are the same as for normal helium. Thus Λ^* for the new isotope is $(4/3)^{1/2}$ times the value for normal helium. De Boer extrapolated his $\Lambda^* - T_c^*$ curve by eye until he reached the appropriate Λ^* for He^3, read off T_c^* and converted it by means of ε into a real critical temperature, which he gave as 3·3 K with an estimated error of ±0·2 K. When, a little later, enough He^3 had been prepared in a nuclear reactor to carry out the experiment of condensing it, it was found that the critical temperature was right in the middle of de Boer's predicted range. He went further than this and predicted the vapour pressure as a function of temperature, and again scored full marks as a prophet. This is a very fine example of the use of dimensional analysis to achieve practical results that are quite beyond the reach of the most elaborate existing theories.

Cavendish Problems: 98, 99, 101, 102, 117, 131.

READING LIST

Flow in Pipes: J. M. Kay, *Fluid Mechanics and Heat Transfer*, Cambridge U.P.

Hydraulic Models: J. Allen, *Scale Models in Hydraulic Engineering*, Longmans.

8

The electrostatic field: charges and conductors in free space

In this chapter we develop the consequences of Coulomb's law in terms of the field configurations produced by various arrangements of charge and how these are modified by the presence of conductors. This leads to the introduction of the concept of capacitance, which is especially important in electrical circuits. From this we proceed to the law of conservation of energy and its formulation when electric fields are present, and to the forces experienced by conductors in an electric field. This prepares the way for a discussion, in Chapter 9, of the behaviour of the molecules of a gas or a liquid in the vicinity of charged bodies and thence to an explanation in microscopic terms of how the characteristic effects of a dielectric medium arise.

Let us begin by drawing the equipotentials and field lines produced by simple charge distributions, starting with the simplest of all, a single

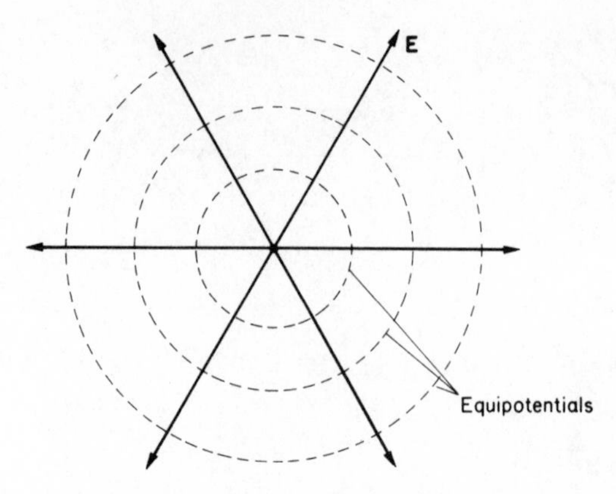

isolated charge q. The equipotentials are spheres, whose radius r is related to the potential ϕ by the equation

$$\phi = q/(4\pi\varepsilon_0 r). \tag{8.1}$$

The field lines are the orthogonal trajectories of the equipotentials and are radial.

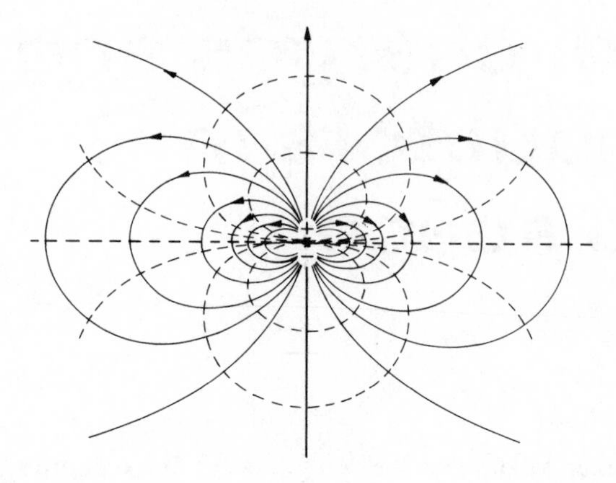

Next, the dipole, consisting of charges q at $\mathbf{a}$ and $-q$ at $-\mathbf{a}$, and possessing a dipole moment $\mathbf{p} = 2q\mathbf{a}$. If $a \ll r$, the distances $|\mathbf{r} \pm \mathbf{a}|$ may be approximated by $r \pm (\mathbf{r}.\mathbf{a})/r$, so that

$$\phi(\mathbf{r}) \approx \frac{q}{4\pi\varepsilon_0}\left[\left(r - \frac{\mathbf{r}.a}{r}\right)^{-1} - \left(r + \frac{\mathbf{r}.\mathbf{a}}{r}\right)^{-1}\right] \approx \frac{q\mathbf{r}.\mathbf{a}}{2\pi\varepsilon_0 r^3},$$

if terms in $(a/r)^2$ are neglected. This gives for the potential of an ideal dipole (one in which a has been shrunk to zero while the dipole moment $2qa$ is kept constant):

$$\phi(\mathbf{r}) = (\mathbf{p}.\mathbf{r})/(4\pi\varepsilon_0 r^3). \tag{8.2}$$

Then we write for the field surrounding an ideal dipole:

$$\begin{aligned}\mathbf{E}(\mathbf{r}) &= -\operatorname{grad}\phi \\ &= \frac{1}{4\pi\varepsilon_0}\left[\frac{3\mathbf{p}.\mathbf{r}}{r^4}\operatorname{grad} r - \frac{1}{r^3}\operatorname{grad}(\mathbf{p}.\mathbf{r})\right] \\ &= \frac{1}{4\pi\varepsilon_0}\left[\left(\frac{3\mathbf{p}.\mathbf{r}}{r^5}\right)\mathbf{r} - \frac{1}{r^3}\mathbf{p}\right], \text{ by XI.}\end{aligned} \tag{8.3}$$

In the vertical direction of the diagram, with **r** parallel to **p**, the first term in the square brackets dominates the second and **E** has the same sense as **p**; but in the broadside position, $\mathbf{r} \perp \mathbf{p}$, only the second term contributes and **E** is oppositely directed, as the diagram makes clear.

PROBLEM

A uniform gas of molecules each carrying the same dipole **p** (all dipoles parallel) is contained in a sphere. Show that they produce no resultant field at the centre of the sphere.

This result was used by Lorentz in his analysis of the influence of polarized molecules on each other in a dielectric medium.

It will be observed that the potential of a charge falls off as $1/r$ while that of a dipole falls off as $1/r^2$. We calculated the potential of the dipole by taking ϕ_q, the potential due to a charge q, and subtracting from it the potential due to the same charge at a neighbouring point, $-\mathbf{p}/q$ away. We could just as well have taken the difference in potential due to a single charge **q** at two points $\mathbf{p}/q$ apart,

$$\phi_p = (\operatorname{grad}\phi_q).\mathbf{p}/q,$$

which is readily seen to give (8.2). The change from $1/r$ to $1/r^2$ arises from the operation grad applied to $1/r$. In the same way, if we take two equal and opposite dipoles and displace them slightly we create a quadru-

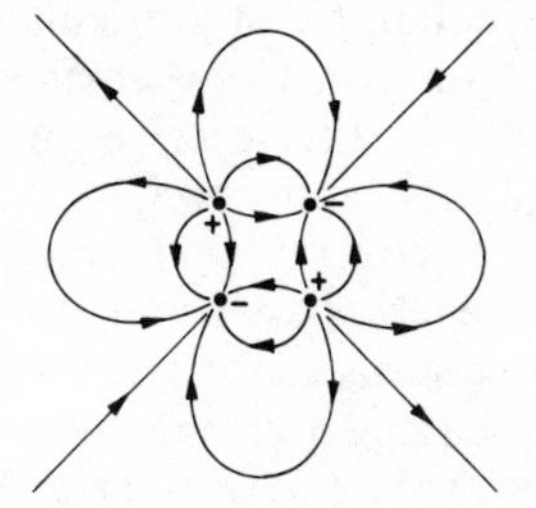

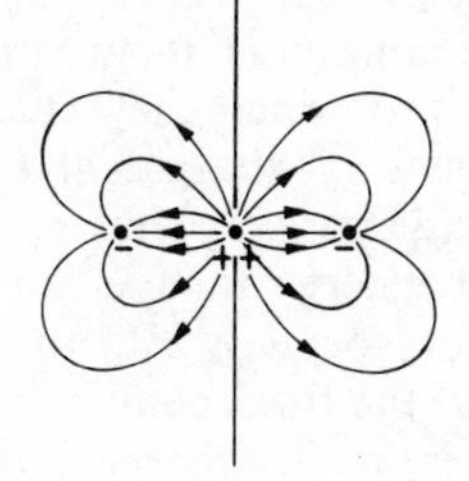

pole, whose potential decays as $1/r^3$ and field as $1/r^4$; two examples are illustrated. And we may continue the process to create octupoles, 16-poles, etc., whose field patterns are more and more elaborately lobed, and die away with steadily increasing inverse powers of r; a 2^n-pole has a field that decays as $r^{-(n+2)}$.

If we have an arbitrary assembly of charges q_i occupying a limited region of space, the field close to the charges is likely to have a very complex form, but at greater distances it becomes simpler. This is because we can always dissect the charge distribution into a charge, a dipole, a quadrupole etc. This is illustrated by the following dissections of linear charge distributions:

$$\begin{array}{ccclrrrl} \cdot & \cdot & \cdot & \text{is the superposition of} & 0 & 1 & 0, & \text{a charge} \\ 5 & -3 & -1 & & 6 & -6 & 0, & \text{a dipole} \\ & & & & -1 & 2 & -1, & \text{a quadrupole} \end{array}$$

$$\begin{array}{ccclrrrl} \cdot & \cdot & \cdot & \text{is the superposition of} & 11 & 0 & -11, & \text{a dipole} \\ 16 & -10 & -6 & & 5 & -10 & 5, & \text{a quadrupole} \end{array}$$

At great distances, the field of the first distribution will be dominated by that of the charge because it falls off more slowly than that of the dipole, even though the latter dominates close in. The second distribution has no net charge, and thus at great distances has a dipole-like field, decaying as $1/r^3$.

PROBLEM

Sketch the field lines for these two distributions.

The calculation of the field due to a given system of charges is obviously a matter of routine computation—the only purpose of mathematical analysis is possibly to shorten the labour. It is not so when conductors are present, for the routine solution of the problem may be so excessively lengthy as to make analysis well worth while. The essential difficulty posed by conductors is that they permit the free movement of charge, so that the actual charge distribution is not given when the problem is set. The task is to find how the charges distribute themselves on the conductors so as to be in equilibrium. Since any electric field within a conductor causes charge to move, equilibrium is only attained when every point within a given piece of conductor is at the same potential. In particular, the surface of a conductor is an equipotential surface, from which field lines leave or into which they enter normally. If therefore we are given an arrangement of charges at specified points, and specified conductors whose total charge is known, we have to discover a configuration of field lines that emerge in due number from each charge and run to other charges, or normally into the conducting surfaces, or right away to infinity; and the total number entering or leaving a given conductor is determined by its net charge. Alternatively we may have an arrangement of conductors at specified potentials, and the problem is solved when we have discovered the family of equipotential surfaces of which the conductors are specified members.

We shall first derive, by use of Gauss' theorem, some simple general results about conductors, and shall proceed from there to synthesize some field configurations involving conductors, so as to illustrate the characteristic form of the field lines. This will lead us to consider the general problem and to outline some of the methods, mathematical and experimental, available for finding solutions.

The following corollaries of Gauss' theorem may be stated and demonstrated in a few words each, and are the consequence of the absence of any electric field in a conductor:

(1) The interior of a conductor is electrically neutral, for if it were not a surface could be drawn around any excess charge through which the total induction of **E** would not vanish. Any charge on a conducting body is therefore carried on its surface.

(2) The electric field, which is normal to the surface, is related to the local surface charge density σ (charge per unit area of surface) by the equation $E_n = \sigma/\varepsilon_0$. This is proved by applying Gauss' theorem to the pillbox in the diagram.

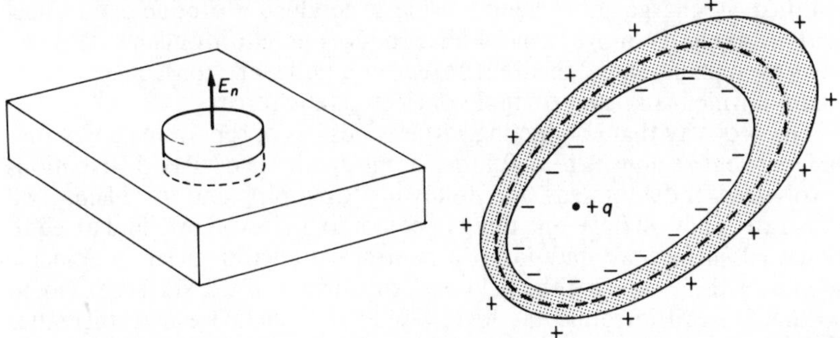

(3) If a closed surface is drawn which lies wholly within the material forming a hollow conducting shell, the total charge within this surface is zero. Hence any charge q placed in the interior space must result in such movement of charge across the shell as to establish a charge $-q$ spread over the interior surface of the shell. If the shell is neutral, there must then be a surface charge q spread over the outside. The disposition of the surface charges is such as to neutralize the field due to the original q at every point within the conductor.

(4) If the shell contains no interior charge there is no electric field anywhere inside it. For the inner surface is an equipotential, and if there exists another equipotential surface, at a different potential, within the interior space it cannot touch the surface and so must be wholly enclosed by the shell. The electric field described by such a pair of closed equipotential surfaces runs from one to the other in the same sense at all points on each surface, and therefore has a non-vanishing induction, contrary to the initial statement that there is no interior charge. Hence no inner equipotential exists, and the whole space is field-free and at constant potential.

55

(5) Movement of charge outside a closed conducting shell produces no effect within. For if the interior is empty, it is field-free, whatever happens outside, and the superposition principle then ensures that the fields due to charges distributed in the interior are not affected by movement of other charges. Thus a conducting shell may be used to isolate a part of space from everything outside, and it is sometimes convenient to imagine any arrangement of charges, etc., under discussion to be enclosed in a very large conducting surface. It is readily shown that if we let the size of the surface tend to infinity and ascribe to it zero potential, the potential at any interior point $\mathbf{r}$ due to enclosed charges q_i at $\mathbf{r}_i$ is $\sum_i q_i/(4\pi\varepsilon_0|\mathbf{r} - \mathbf{r}_i|)$. Such a surface is an *ideal earth*; the surface of the Earth may often be regarded as an adequate approximation to an ideal earth, even though it does not conduct well and does not usually enclose the experimental arrangement; this is because all that is usually required of it, if it is to perform the function of an ideal earth, is that its potential shall not be affected by charge movements during the experiment, and this it achieves merely by virtue of its bulk.

(6) Just as charge movement outside a conductor produces no effect within, so charge movement within produces no effect outside. This can be proved from Gauss' theorem or seen as a necessary consequence of (5) and N3 which is known to apply to electrostatic forces.

The property that conducting surfaces possess of separating space into independent regions is a useful tool in the synthesis of field distributions involving conductors, as the following examples and problems will illustrate. They all have one thing in common, that when we find an equipotential surface we may insert a neutral conducting sheet to coincide with it, without altering the field pattern; further, if the surface is closed so that we can introduce a closed conducting shell, we may thereafter shift charges around outside the shell without altering the field inside, and vice versa.

EXAMPLES

(1) *A conducting sphere carrying charge q.* Consider first the field of an isolated charge q, for which the equipotentials are spheres. We may build up a neutral conducting sheet to coincide with one of the spheres without disturbing the field. According to (3) above, the sheet, though remaining neutral of course, will carry $-q$ on its interior and q on its exterior surface. If we bring the original charge q to the interior surface to neutralize the $-q$ already there, leaving no charge within, the exterior field is unchanged, by (6). We have thus synthesized a situation in which a conducting sphere carries charge q, and have seen that the field produced outside the sphere is the same as if we had a point charge at its centre. This is of course a trivial example and is given simply to expound the synthetic process. Since we are concerned with synthesis we describe the processes as they are carried out, and only at the end formulate the problem whose solution we have discovered. In this case we started with the field of a point charge and discovered the solution to the problem of a charged conducting sphere.

(2) *Two equal and opposite charges*, q at $x = a$ and $-q$ at $x = -a$, produce field lines and equipotentials as shown, of which one in particular, the equipotential $\phi = 0$, is the plane $x = 0$. If then we divide space into two halves with a neutral conducting plane, and move the left-hand charge $-q$ on to it, we have synthesized a situation in which a charge q is placed at a distance a from a plane conductor carrying charge $-q$, and we know that the resulting field lines on the right-hand side are exactly as if we had no plane but only an

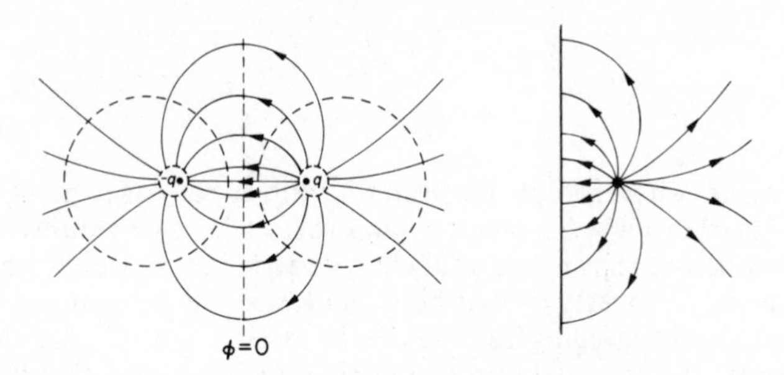

image charge $-q$ at $x = -a$. It might be thought that it would be difficult to create the situation in practice in which the plane carried exactly $-q$, but this is not so. For the equipotential which we covered with a neutral conducting sheet was the zero equipotential, which may be connected to earth without causing any change. The problem we have thus solved is the field distribution due to *a charge in the vicinity of an earthed conducting plane.*

PROBLEMS

(1) Determine the distribution of charge on the surface of the conducting plane, and by integration verify that it amounts to $-q$ in all.

(2) Consider the equipotentials of a ring of $2n$ charges, alternately $\pm q$, evenly set around a circle, and hence show how to calculate the field lines for a charge placed symmetrically in the wedge formed by two earthed conducting planes set at an angle π/n.

> *Comment*. The synthetic method works for integral values of n but not otherwise, even though the form of the field lines does not depend on this fact. The method is indeed powerful in such cases as it works at all, but is otherwise helpless; it is useful for compiling a catalogue of solutions, but not normally as an approach to a new problem.

(3) A uniform field $\mathbf{E}$, parallel to the x-axis, is described by a potential $\phi = -Ex$. Place a dipole $\mathbf{p}$, also parallel to the x-axis, at a point where $x = a$, and show that the equipotential $-Ea$ is a sphere. Hence show that a neutral conducting sphere of radius r placed in a uniform field acquires the potential of the point where its centre now lies, and distorts the field outside in the same way as would a dipole $4\pi\varepsilon_0 r^3\mathbf{E}$ at the centre.

180

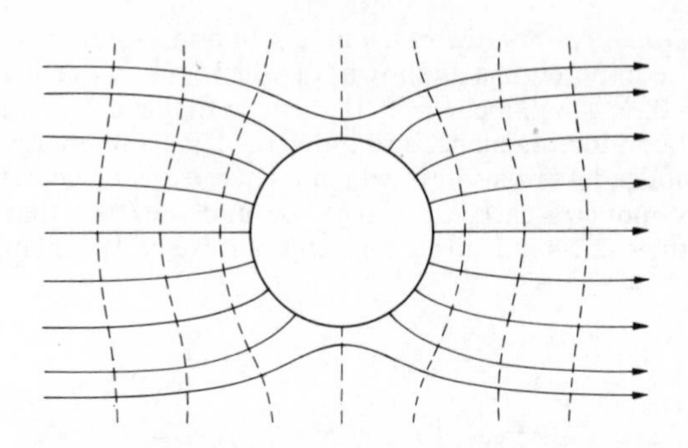

Comment. This example illustrates a simple, but basic, point which must be thoroughly appreciated in order to have an intuitive grasp of how fields behave. The total charge carried by an isolated body is an intrinsic property of that body, and can only be changed by the positive act of bringing up or taking away electrons or other charged particles; the potential of a conducting body, on the other hand, is determined not only by the total charge it carries but by its situation with respect to other charges. In the above problem, the sphere was neutral, but if moved in the field **E** its potential would alter so as to match the potential of the point which its centre occupied at any instant. In principle one can determine the charge on a body by taking it to pieces and counting the particles of either sign, and the body can be taken away to a convenient place for the analysis; but determination of the potential must be done *in situ*, in principle by integrating $\int \mathbf{E}.d\mathbf{r}$ along any line running from the body to the ideal earth at infinity.

(4) Two charges, q_1 and q_2, produce a potential proportional to $q_1/r_1 + q_2/r_2$ at a point distant r_1 from q_1, r_2 from q_2. Take $q_1 = 5, q_2 = 1$

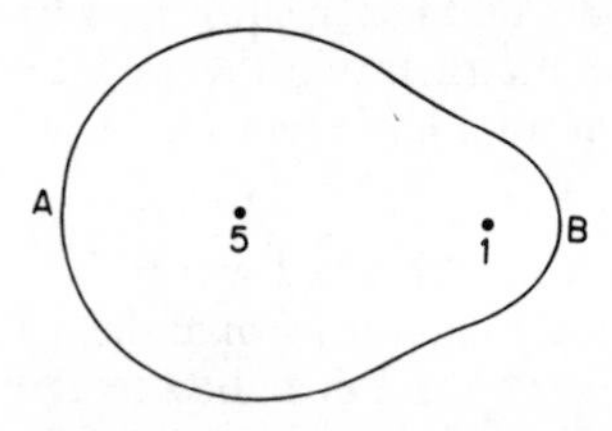

and put them 10 units apart; construct the pear-shaped equipotential for which $q_1/r_1 + q_2/r_2 = 0{\cdot}75$. Show that the field strength at B is 1·6 times that at A.

Comment. There is a concentration of field on pointed parts of charged conductors, relative to the more rounded parts. The same effect can be seen by using the solution of problem (3) to draw the field lines emanating from a charged infinite conducting plane to which is attached a hemispherical box. The field at the pole of the hemisphere is three times as strong as the distant field. It is this field concentration at points which is used in lightning conductors to ensure as far as possible that the electrical discharge occurs where it will do no damage.

In problem (4) the field concentration was demonstrated explicitly, but it is possible to see by a qualitative argument that it must occur. Suppose we spread a uniform positive charge density over the surface of the pear-shaped body; it will no longer be an equipotential, since the sharper point at B ensures that the centroid of the charge is further from B than from A. Hence B will have a lower potential than A, and if the charge is now freed it will move towards B to raise its potential; the greater charge density implies a stronger field.

It would be a mistake to exaggerate the degree of field enhancement. In the example of problem (4) the fields of B and A are in the ration 1·6 which is rather less than the inverse ratio, 1·9, of the radii of curvature at the ends. Nevertheless by sharpening up the point one can achieve a considerable concentration of field, even though the greater part of the charge is still carried on the large rounded surfaces.

In the field-ion microscope a very fine needle of a refractory metal such as tungsten lies at the centre of a sphere, so that by establishing a potential difference between the needle (+) and the sphere (−) a very strong electric field is produced at the surface of the tungsten, just at its tip. On the atomic scale there are field variations between one atom and the next, since the surface is bumpy where the atoms are, and there is field enhancement at the bumps. So strong is the field that when a very small quantity of helium gas is admitted, individual atoms may actually lose an electron as they approach one of the tungsten atoms on the surface; being then positively charged they are attracted strongly to the sphere, which they strike sufficiently energetically to make a flash of light in the phosphor that coats its inside surface (like the coating of a cathode-ray tube). Now the helium ion travels almost exactly along the radial field line which starts from the point where it was ionized, and the pattern of light on the screen thus images, at enormous magnification, the regions of the tungsten tip where ionization is most readily achieved. As the photograph shows, individual atoms show up clearly, and the instrument provides a direct picture of the atomic arrangement on the tungsten surface.

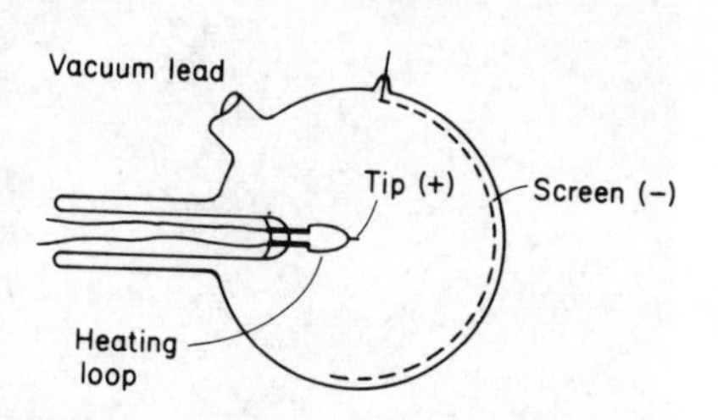

With these examples in mind we proceed to examine the general problem of charges and conductors, which may be formulated in terms

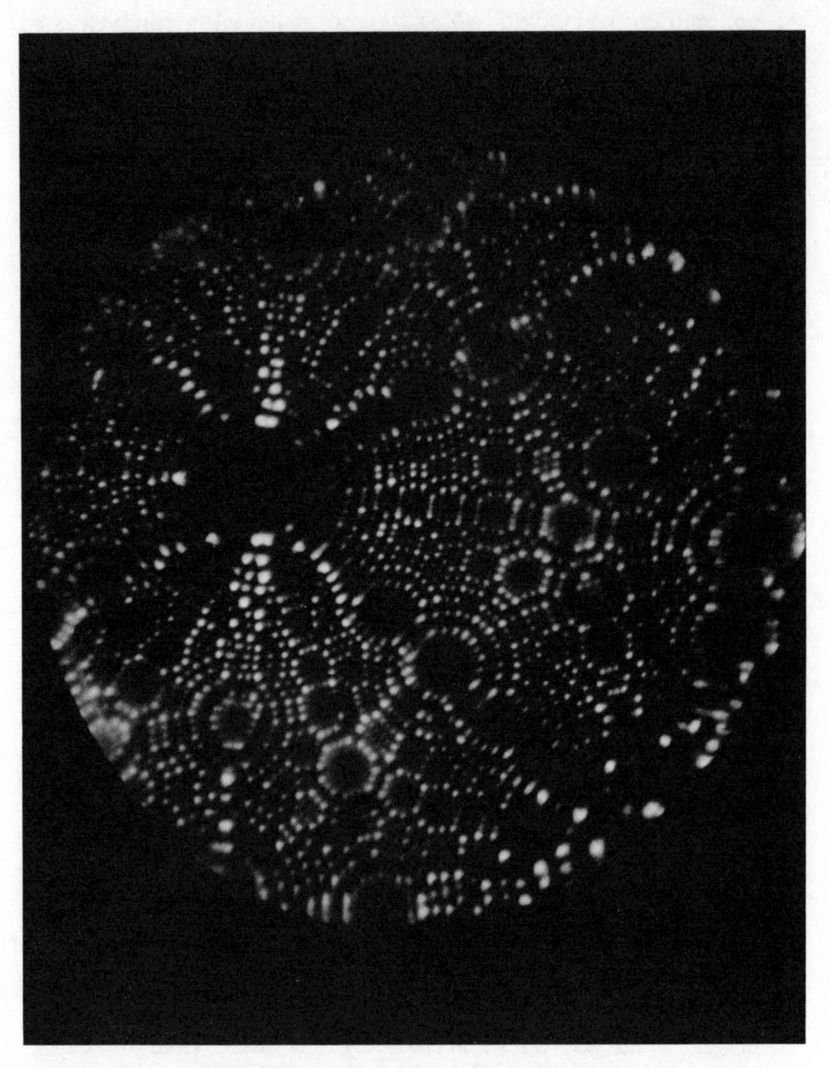

The atoms on the surface of a tiny spherical tip of tungsten, as revealed by the field-ion microscope. (Photograph supplied by the Department of Metallurgy and Materials Science, Cambridge University)

of the partial differential equation governing the potential in the free space between the conductors, and the boundary conditions imposed by the free charges and the conductors themselves. The differential equation may be derived by remembering that the divergence of a vector is the limit of the flux of the vector through a surface bounding a small volume element, divided by the volume δV of the element. Now if we have a continuous distribution of charge, of density $\rho(\mathbf{r})$, the charge contained in δV

is $\rho\,\delta V$, so that the flux of $\mathbf{E}$ through the surface is $\rho\,\delta V/\varepsilon_0$, by Gauss' theorem. It follows that

$$\text{div}\,\mathbf{E} = \rho/\varepsilon_0. \tag{8.4}$$

Further, since the potential is defined by the equation grad $\phi = -\mathbf{E}$, we have that

$$\text{div grad}\,\phi \equiv \nabla^2\phi = -\rho/\varepsilon_0. \tag{8.5}$$

This is Poisson's equation relating the potential to the charge distribution. In any region of space where ρ vanishes, the potential obeys Laplace's equation,

$$\nabla^2\phi = 0. \tag{8.6}$$

The examples that we shall discuss will be typically the fields in the space surrounding conductors, without any free charges except on the surfaces of the conductors. We shall thus be concerned with Laplace's equation, which we have already met and which, occurring as it does in almost every branch of physics, has been very systematically studied. This is not to say that all its solutions are known—far from it—but methods are available for computing solutions for a wide variety of boundary conditions, probably indeed for all boundary conditions, given enough computer time. It is not necessary to discuss either mathematical or computational methods here, since there are many specialized texts which enter fully into these things. It is, however, worth noting the occurrence of the equation in different branches of physics, for this not only gives us insight into its meaning, and some feel for the character of its solutions, but we find as well that solutions discovered experimentally in one field may be transferred by analogy to others where the experiments would be harder to perform.

Laplace's equation characteristically turns up to describe stationary flow when the substance in motion is conserved. By *stationary flow* we mean a flow pattern that remains unchanged as time proceeds, and can be described by a velocity vector $\mathbf{v}(\mathbf{r})$, or something analogous, which is dependent on $\mathbf{r}$ but not t. If the pattern does not change, and if there are no sources or sinks introducing or extracting the substance, the fact that it is conserved means that the surface integral $\int \mathbf{v}.\text{d}\mathbf{S}$ must vanish over all closed surfaces, and that div $\mathbf{v} = 0$. The 'substance' involved may be an incompressible fluid, electrons in a metal whose motion produces a current, heat etc., and the corresponding analogues of velocity are velocity itself, $\mathbf{v}$, for an incompressible fluid, current density $\mathbf{J}$ in a metal ($\mathbf{J}$ is the current crossing unit area normal to $\mathbf{J}$), and heat flux $\mathbf{Q}$ (heat crossing unit area normal to $\mathbf{Q}$ in unit time). All these flows are non-divergent in a large variety of physical situations. In order that the flow pattern shall be controlled by Laplace's equation, it is necessary that the flow itself at any point shall be strictly proportional to the gradient of a scalar, since if div $\mathbf{v} = 0$ and $\mathbf{v} = A$ grad ϕ, div grad $\phi = \nabla^2\phi = 0$. We already know that under stationary conditions the electric field can be written as the gradient of a scalar potential, so that we only need the conductivity, relating $\mathbf{J}$ to $\mathbf{E}$ by the equation $\mathbf{J} = \sigma\mathbf{E}$, to be independent of position for the potential in a conducting medium to satisfy Laplace's equation. A similar condition on the thermal conductivity, relating $\mathbf{Q}$ to the gradient of a scalar temperature by the equation $\mathbf{Q} = -\kappa$ grad T, ensures that $\nabla^2 T = 0$ in stationary heat flow. As for the flow of an incompressible fluid, this we havc already discussed briefly to explain that it is not in general described by a scalar potential, because of the onset of turbulence; but a special case of interest here is the percolation of fluid through a uniform porous bed, such as gravel, where the flow velocity is determined by the local pressure gradient, $\mathbf{v} = -C$ grad P, leading to $\nabla^2 P = 0$ if C is independent of position.

102

These are all examples of Laplace's equation in three dimensions, which are hard enough to study experimentally whether for their own sake or as analogues of the problem whose solution is really wanted. In two dimensions, however, there exist comparatively simple techniques for the experimental solution of Laplace's equation, some of them deriving from a problem already analysed in Chapter 6, the normal displacement of a stretched membrane or soap film, which can be used to solve two-dimensional electrostatic problems or, more realistically, problems involving cylindrical conductors of constant cross-section. If, for example, we wish to know the field distribution around a channel girder inside a cylindrical tube and raised in potential relative to the tube, we must solve Laplace's equation $\nabla^2\phi = 0$, in this case $\partial^2\phi/\partial x^2 + \partial^2\phi/\partial y^2 = 0$ since there is no variation along z. The boundary condition is that ϕ is zero round the inner circle and takes some constant value ϕ_0 at the surface of the channel. Now this is formally the same problem as pushing a similar channel on to a large uniform circular membrane, so as to displace it by a constant amount h_0 round the edge of the channel. If we carry out the experiment and determine the contours of constant height on the membrane we have also a map of the equipotentials in the electrostatic problem. Or, if we are interested in a quick rough answer, we can form a soap film between the end of a piece of channel and a circular ring, and look for regions of high and low gradient which correspond to regions of strong and weak field. It is difficult to take a convincing photograph of the film, but the experiment is so easy to carry out that the reader should try it for himself, noting how as the end of the channel is drawn out of the plane of the ring, the film inside the angle stays flat (indicating very little field in this region) but leaves

101

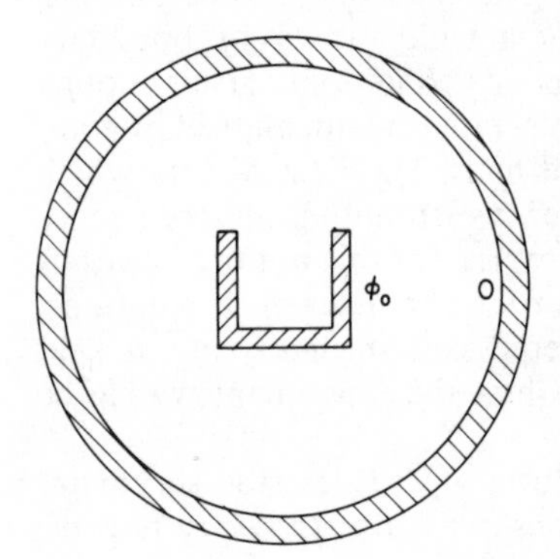

the sharp external corners at a steep angle (indicating field concentration at these points).

Another convenient way of solving two-dimensional electrostatic problems is to make use of the electric current analogy, and determine by a Wheatstone's bridge the potential distribution on a uniform conducting sheet. Such sheets are commercially available,* covered with graphite to give a remarkably uniform conducting layer. Electrodes of tin foil may be pressed on to the surface to act as equipotential electrodes, but it is better to use conducting paint. The bridge circuit is balanced only when the potentials at P and C are equal, and the locus of P is an equipotential

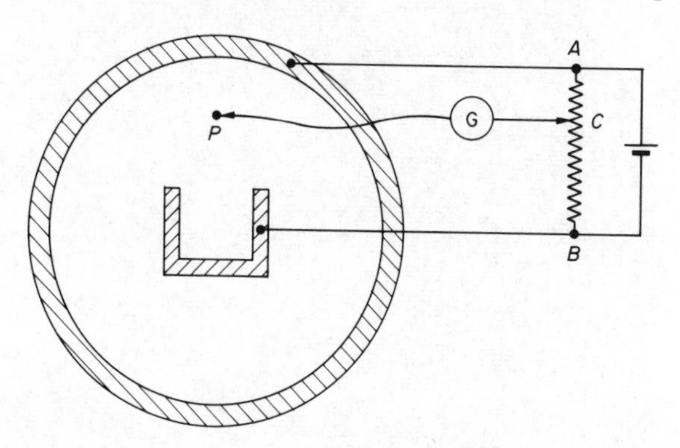

whose actual potential is known from the position of the tapping-point C on the resistor AB. If the probe P is a pointed stylus, it may be pressed into the paper wherever a balance point is found, and the equipotentials drawn on the sheet, as shown in the photograph (p. 158) of the result of such an experiment. Note again how close the equipotentials run round the corners, where the field is strong, and how weak the field is within the channel. The two parallel sets of electrodes above and below the test sample were painted on to measure the resistivity of the sheet. One of them gave 1930 Ω and the other 1990 Ω for the resistance between opposite sides of a square cut from the sheet (all squares cut from uniformly resistive sheet have the same resistance). This measurement indicates how uniform the sheet is, with only 3% difference between two pieces a foot apart.

Analogue methods of obtaining the solution of Laplace's equation have become especially important in the design of electrode systems to act as lenses in electron microscopes and other instruments depending on the precise manipulation of electron beams. The techniques for determining equipotentials, especially with electrolytic tanks to replace the rather crude conducting paper just described, have been refined to a considerable degree and are well described in specialist works.

225

* E.g., tele delta paper from Sensitized Coatings Limited, Croydon, Surrey, England.

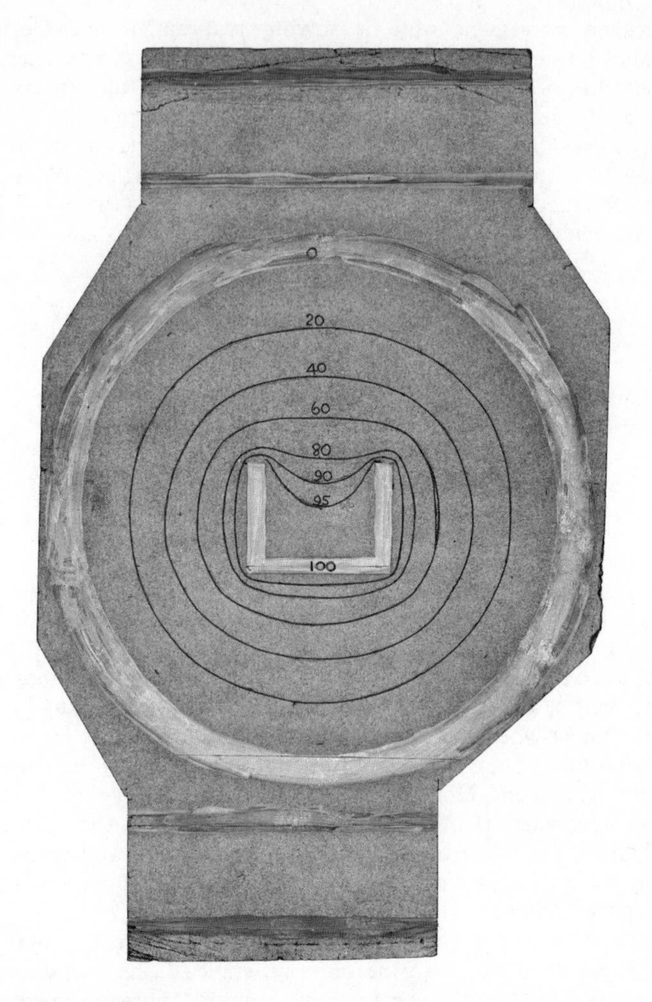

Capacitance

Two closely related conceptions of the meaning of capacitance should be familiar from elementary treatments of electrostatics. The commercial condenser (*capacitor*), consisting of two metal foils separated by a thin insulating layer and rolled compactly into a cylinder, exemplifies the *mutual capacitance* between two conductors which enables them to carry equal and opposite charges on their surfaces at the expense of a potential difference between them. The ratio of the charge q on either of the conductors to the potential difference V is the magnitude of the mutual capacitance, taken as positive, i.e., $C = |q/V|$. There is also the self-capacitance possessed by a single conductor, which expresses the fact that if charge q is brought to that conductor its potential changes; as before, the measure of self-capacitance is the ratio of the added charge to the resulting change of potential. Since the Earth (or a conducting sphere at infinity, or any other suitable reference conductor) is taken as fixed in potential, the self-capacitance may be thought of as the mutual capacitance between the conductor and earth.

EXAMPLE

Mutual capacitance between two concentric spheres. If the inner sphere has radius a and carries charge q, its potential is $q/(4\pi\varepsilon_0 a)$ and the potential of the outer sphere, of radius b, is $q/(4\pi\varepsilon_0 b)$. Then the potential difference is $q(1/a - 1/b)/(4\pi\varepsilon_0)$, and the mutual capacitance is $4\pi\varepsilon_0/(1/a - 1/b)$. If we let the radius of the outer sphere tend to infinity, its potential tends to zero, and the self-capacitance of the now-isolated inner sphere is $4\pi\varepsilon_0 a$. 218

In a conventional circuit diagram mutual capacitances are represented by a symbol $\dashv\vdash$, and we shall show what this means by analysing the particular problem presented by two conducting bodies isolated from any other influences, and charged to potentials ϕ_A and ϕ_B. The equivalent circuit shown illustrates the existence of a mutual capacitance C_{AB}

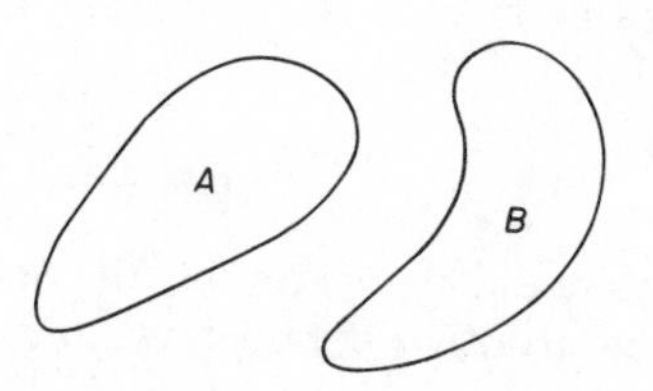

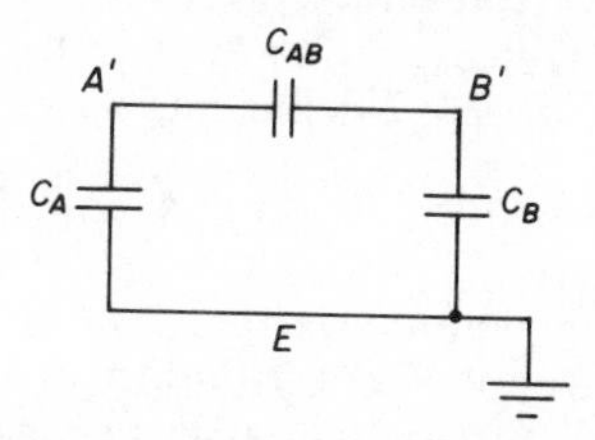

between A and B, and self-capacitances C_A and C_B for each, which are here represented by mutual capacitances to earth ($\phi = 0$). Now the idealized mutual capacitances of the diagram have the property that they carry charges $\pm q$ on their plates with a potential difference q/C between them, the positively charged plate being at the more positive potential. If we imagine the diagram to be a picture of a construction made of metal, the thin wires carry no charge and each separate piece of metal is at a constant potential. The diagram states that the relations between the total charges on the bodies, Q_A and Q_B, and their potentials, ϕ_A and ϕ_B, may be exactly reproduced in the wire and plate construction, with the wire A' and its attached plates simulating A, B' simulating B, and E simulating the conducting sphere at infinity. Since the positive charge on the left-hand plate of C_{AB} is $C_{AB}(\phi_A - \phi_B)$, and that on the upper plate of C_A is $C_A\phi_A$, we have that in the model

$$Q_A = (C_A + C_{AB})\phi_A - C_{AB}\phi_B;$$

Similarly

$$Q_B = -C_{AB}\phi_A + (C_B - C_{AB})\phi_B. \qquad (8.7)$$

It is not surprising that charges and potentials are related by linear equations since the field due to any charge is proportional to the magnitude of the charge,* but it is noteworthy that the linear equations are not the most general:

$$\left.\begin{aligned} Q_A &= c_{AA}\phi_A + c_{AB}\phi_B \\ Q_B &= c_{BA}\phi_A + c_{BB}\phi_B. \end{aligned}\right\} \tag{8.8}$$

The most general pair of equations has four independent coefficients, while (8.7) has only three and requires that c_{AB} and c_{BA} shall be the same. It is therefore not obvious that the circuit model can always be assumed to describe the general behaviour of charged conductors. Nevertheless it is so, as can be proved by demonstrating that $c_{AB} = c_{BA}$ or, more generally, that if the charges on a system of conductors are related to their potentials by a matrix of *capacity coefficients*,

$$Q_i = \sum_j c_{ij}\phi_j, \tag{8.9}$$

then $c_{ij} = c_{ji}$, i.e., the matrix c_{ij} is symmetric. The proof of this important result is most readily given by considering the energy of the system, and will be deferred until we have discussed energy further.

The relation of the capacity coefficients to the conventional circuit capacitances can be derived by rewriting (8.9) in terms of the potential differences V_{ij}, defined as $\phi_i - \phi_j$:

$$Q_i = -\sum_j c_{ij}V_{ij} + \phi_i \sum_j c_{ij}.$$

In a conventional circuit we should write the charge on the ith conductor as the sum of contributions $C_{ij}V_{ij}$ due to other conductors, and a contribution from the self-capacitance:

$$Q_i = \sum_j C_{ij}V_{ij} + \phi_i C_i,$$

so that the mutual capacitance C_{ij} is just $-c_{ij}$ and the self-capacitance C_i is $\sum_j c_{ij}$.

In the light of this analysis we may safely use the circuit diagram to represent the behaviour of real systems and as an aid in calculating what will happen when potentials and charges are altered. For numerical calculations it is necessary, of course, to know the values of the capacitances, and this information may be obtained either by calculation or by suitable

* If we imagine the two bodies to be uncharged and then attach charge Q_A to A, the potentials of both A and B will alter to values proportional to Q_A. The same thing will happen if we start again and attach Q_B to B. If both charges are present together, superposition ensures that the final potentials are the sum of those due to Q_A and Q_B separately. Thus we know that there exist coefficients p such that

$$\phi_A = p_{AA}Q_A + p_{AB}Q_B, \quad \phi_B = p_{BA}Q_A + p_{BB}Q_B,$$

and solution of these equations for Q_A and Q_B gives a linear dependence of charge on potential as exemplified by (8.8).

experiments. It is not, however, quite obvious how one should proceed to get the required data economically—a single calculation or a single experiment is not enough. To take the case of the two bodies A and B just discussed, if we assign arbitrary potentials ϕ_A and ϕ_B, and calculate or measure the resulting equipotentials, we may deduce from them the local surface-charge density and by integration the total charge carried by each. This provides, by use of (8.8), two equations involving the three unknowns c_{AA}, c_{BB} and $c_{AB}(= c_{BA})$. A second arbitrary choice of ϕ_A and ϕ_B gives two more equations, of which one may be used to provide the third equation and determine the capacitances, and the other serves as a check. In general, with n conductors, the number of unknown capacitances is $\frac{1}{2}n(n + 1)$, and for each arbitrary assignment of potentials n equations are obtained; at least $\frac{1}{2}(n + 1)$ separate measurements or calculations are therefore needed. Alternatively, given a capacitance bridge or other instrument for measuring the capacitance between any two of the conductors, measurements may be made with the conductors connected together or to earth in different ways. But here each measurement gives one item of information only, and as many separate measurements are needed as there are capacitances in the circuit.

Instead of measuring the capacitance between two of the conductors, it may prove more convenient to use the analogue of conductivity, immersing the conductors in an electrolyte and measuring the resistance between any two of them.

PROBLEMS

(1) The capacitance between two metal objects is C. Show that when the whole system is immersed in an electrolyte of conductivity σ, the resistance between them is $\varepsilon_0/(C\sigma)$.

(2) In the experiment described on p. 157, the resistance between the outer circle and the channelled inner conductor was 348 Ω. Taking the mean resistivity of the sheet to be 1960 Ω, as measured, calculate the capacitance per metre between the channel girder and the round tube, of which the diagram is a cross-section.

(3) In a separate experiment to determine the capacitance per unit length between the channel and the tube, a rubber membrane stretched over the end of a piece of the tube ($6\frac{5}{8}$ inches in diameter) was mounted horizontally so that a squared-off piece of the channel could be pressed into it. By mounting the channel on a balance, it was arranged that its end remained parallel to the plane defined by the edge of the membrane, and it was possible to weigh the force needed to deflect the membrane by a given amount. The deflection was found to be proportional to the force for deflections up to 6 mm, the weight of 100 g causing a deflection of 5·3 mm. When the channel was replaced by a circular tube 1·32 inches in diameter the weight of 70 g sufficed for a deflection of 4·9 mm. Calculate the capacitance per metre between the channel and the tube, and compare with

the estimate in the preceding problem to get some indication of the reliability of the simple methods used.

(4) If the cross-sectional dimensions of the conductors are scaled by a factor A, how is the capacitance per metre affected?

It is important to recognize that the self-capacitance of a given body is not an intrinsic property, but is affected by the surroundings. Thus an isolated sphere (A) of radius a has self-capacitance $4\pi\varepsilon_0 a$, but if it is enclosed in a conducting shell (B) its self-capacitance vanishes. For then the charge it carries depends solely on the potential difference between itself and the shell, $Q_A = C(\phi_A - \phi_B)$; comparing with (8.7) we see that this implies that $C = C_{AB}$ and $C_A = 0$.

PROBLEM

Two conducting spheres, each of radius a, have their centres b apart ($b \gg a$). If one carries charge Q and the other is neutral, show by considering the potentials, field strength and resulting polarizations how to calculate the first terms in the expansions of ϕ_1 and ϕ_2 as power series in a/b. Hence show that the spheres have self-capacitance $4\pi\varepsilon_0 a(1 - a/b + \cdots)$, and mutual capacitance $4\pi\varepsilon_0 a^2/b(1 + a^2/b^2 + \cdots)$.

Before leaving this topic we note that the symmetry of the matrix of capacity coefficients, $c_{ij} = c_{ji}$, implies a *reciprocity* of charge and potential in the following sense: if a charge Q deposited on a conductor A raises the potential of another conductor B by ϕ, then a charge Q deposited on B will raise the potential of A by ϕ. This is true for any two conductors in any assembly, however complicated, and is a simple consequence of the symmetry of c_{ij} or, what is equivalent, of p_{ij}; for Q deposited on A raises the potential of B by $p_{BA}Q$, and Q deposited on B raises the potential of A by $p_{AB}Q$. An application of this theorem will be given later. Similar theorems are found in many other branches of physics, for example in the theory of elastic structures; if a force F applied at a point A causes the structure to be displaced by a distance x at B, then force F at B causes displacement x at A. This result is due to Maxwell who also enunciated reciprocity theorems in thermodynamics. The method of proof in all cases depends on being able to define the energy stored in the system concerned. There are some reciprocity relations, notably those discovered by Onsager which apply to irreversible flow, which are not immediately related to an energy function. But these require the most subtle analysis and are quite outside our scope. Here we are more interested in such reciprocity relations as are intimately linked to energy conservation, to which topic we now direct our attention.

Energy

Since the Coulomb force, being central, is conservative, we have reason to suspect that the concept of energy as a conserved quantity can be extended to electrostatic systems. We shall not, however, take this for granted but demonstrate it from first principles. The argument will hinge on showing that the work needed to assemble any system of charges, bringing them from infinity, is independent of the order of assembly. Once we have shown this, we may ascribe potential energy U to the assembly, with the property that in any change the work done to effect that change is equal to the increase in the value of U.

Believing that we can achieve this programme, we shall start by assembling the charges in a particular way to make the calculation of the work easy, and shall afterwards show that all ways would have given the same result. If the final arrangement is specified by the charge density $\rho(\mathbf{r})$, we assemble the charges from infinity bit by bit, in such a way that at an intermediate stage the pattern is the same but less dense; that is, an intermediate charge density is $\alpha\rho(\mathbf{r})$, with α taking the same value at all points. The process of assembly involves letting α increase from 0 to 1. Now if the potential due to $\rho(\mathbf{r})$ is $\phi(\mathbf{r})$, the potential due to $\alpha\rho(\mathbf{r})$ is $\alpha\phi(\mathbf{r})$. Consider then a volume element $\mathrm{d}V$ situated at $\mathbf{r}$, where the potential is $\alpha\phi(\mathbf{r})$; when α increases by $\mathrm{d}\alpha$, the charge brought to this element is $\rho(\mathbf{r})\,\mathrm{d}\alpha\,\mathrm{d}V$, and the work that must be done on it to overcome the Coulomb forces of the other charges is $\alpha\phi(\mathbf{r}) \times \rho(\mathbf{r})\,\mathrm{d}\alpha\,\mathrm{d}V$, i.e., $\phi\rho\,\mathrm{d}V \times \alpha\,\mathrm{d}\alpha$. As α changes from 0 to 1, $\phi\rho\,\mathrm{d}V$ is constant for this volume element, and the total work is $\phi\rho\,\mathrm{d}V \int_0^1 \alpha\,\mathrm{d}\alpha$ i.e., $\frac{1}{2}\phi\rho\,\mathrm{d}V$. Integrating over all space we have that the work needed to assemble the complete system is $\int \frac{1}{2}\phi\rho\,\mathrm{d}V$.

The next stage in the argument is to show that this result is not peculiar to the method chosen to assemble the charges. We *define* a function U as $\int \frac{1}{2}\phi\rho\,\mathrm{d}V$; since for any pattern of charge density $\rho(\mathbf{r})$, $\phi(\mathbf{r})$ is determined uniquely, the value of U is fixed by the form of $\rho(\mathbf{r})$. If we can show that in any infinitesimal variation of $\rho(\mathbf{r})$ the work needed is given by the change in U, it follows that the same must be true for all finite variations, which are made up of a sequence of infinitesimal variations. Consider therefore a variation $\delta\rho(\mathbf{r})$, in which the potential changes by $\delta\phi(\mathbf{r})$. The resulting change in U is given by

$$\delta U = \frac{1}{2}\int (\phi\delta\rho + \rho\delta\phi)\,\mathrm{d}V,$$

while the work done in effecting the variation, which involves as a typical process bringing $\delta\rho(\mathbf{r})\,\mathrm{d}V$ from infinity to a point where the potential is $\phi(\mathbf{r})$, is given by

$$\delta W = \int (\phi\,\delta\rho)\,\mathrm{d}V.$$

If we can show that, in any infinitesimal variation, $\phi\,\delta\rho = \rho\,\delta\phi$, the theorem is proved, since $\delta W = \delta U$.

This is readily shown by writing these quantities explicitly from Coulomb's law. The charge $\rho(\mathbf{r}')\,\mathrm{d}V'$ in a volume element $\mathrm{d}V'$ at $\mathbf{r}'$ produces at $\mathbf{r}$ a contribution to the potential

$$\rho(\mathbf{r}')\,\mathrm{d}V'/[4\pi\varepsilon_0|\mathbf{r}' - \mathbf{r}|],$$

so that

$$\phi(\mathbf{r}) = \frac{1}{4\pi\varepsilon_0}\int\frac{\rho(\mathbf{r}')\,\mathrm{d}V'}{|\mathbf{r}' - \mathbf{r}|}.$$

When the change $\delta\rho(\mathbf{r})$ is imposed,

$$\delta\phi(\mathbf{r}) = \frac{1}{4\pi\varepsilon_0}\int\frac{\delta\rho(\mathbf{r}')\,\mathrm{d}V'}{|\mathbf{r}' - \mathbf{r}|}$$

Therefore

$$\int(\rho\,\delta\phi)\,\mathrm{d}V = \frac{1}{4\pi\varepsilon_0}\iint\frac{\rho(\mathbf{r})\,\delta\rho(\mathbf{r}')\,\mathrm{d}V'\,\mathrm{d}V}{|\mathbf{r}' - \mathbf{r}|}.$$

Similarly

$$\int(\phi\,\delta\rho)\,\mathrm{d}V = \frac{1}{4\pi\varepsilon_0}\iint\frac{\rho(\mathbf{r}')\,\delta\rho(\mathbf{r})\,\mathrm{d}V'\,\mathrm{d}V}{|\mathbf{r}' - \mathbf{r}|}.$$

Since these double integrals are taken over the same range (i.e., over all space) for both $\mathrm{d}V$ and $\mathrm{d}V'$, and differ only in interchange of primes in the integrand, they are equal, and we have proved that an energy conservation theorem holds, with U playing the part of potential energy.

The form $\frac{1}{2}\int\rho\phi\,\mathrm{d}V$, in which U has been defined is not usually the easiest to apply, and we shall now manipulate it into alternative forms. First, consider the case when all charges are carried on the surfaces of conductors. Then the volume integral can be reduced to a sum over all conductors, on each of which ϕ is constant:

$$U = \tfrac{1}{2}\sum_i Q_i\phi_i,$$

in which Q_i is the charge on the surface of the ith conductor whose potential is ϕ_i.* From (8.9) we see that this takes a very symmetric form

$$U = \tfrac{1}{2}\sum_i\sum_j c_{ij}\phi_i\phi_j. \tag{8.10}$$

* Now that we know that the energy U exists as a well-defined function, we may write the change in energy due to infinitesimal variations of the Q_i in the form

$$\mathrm{d}U = \sum_i \phi_i\,\mathrm{d}Q_i$$

Then

$$\partial U/\partial Q_i = \phi_i \quad \text{and} \quad \partial^2 U/\partial Q_i\partial Q_j = \partial\phi_i/\partial Q_j = p_{ij}$$

Similarly $\partial^2 U/\partial Q_j\partial Q_i = p_{ji}$. Since for a well-behaved function the order of differentiation is irrelevant, $p_{ij} = p_{ji}$, and the symmetry of the matrices p_{ij} and c_{ij} is proved.

EXAMPLE

The energy stored in a charged capacitor. The self-capacitance is assumed negligible compared to the mutual capacitance.

(1) First we give the standard derivation. Imagine the plates of the capacitor to be isolated and to carry such charges $\pm Q$ that there is a potential difference ϕ between them, Q being $C\phi$. If we detach an element of charge $\mathrm{d}Q$ from the plate carrying $-Q$ and transfer it to the other plate, we have to carry it across a potential difference ϕ and do work $\phi\,\mathrm{d}Q$ on it, i.e., $C\phi\,\mathrm{d}\phi$. In the whole process of raising the potential difference from 0 to ϕ then, the work required is $\int_0^\phi C\phi\,\mathrm{d}\phi$ or $\frac{1}{2}C\phi^2$.

(2) We may reach the same answer, rather long-windedly, by use of (8.10). Labelling the plates by subscripts 1 and 2, we observe that the self-capacitance of each, $c_{11} + c_{12}$ and $c_{22} + c_{12}$ in terms of capacity coefficients, is taken as zero, so that $c_{11} = c_{22} = -c_{12}$. Also the mutual capacitance is $-c_{12}$ and this must be equated to C. Thus

$$c_{11} = c_{22} = C \quad \text{and} \quad c_{12} = -C.$$

If the plates are at potentials ϕ_1 and ϕ_2, according to (8.10)

$$U = \tfrac{1}{2}C(\phi_1^2 - 2\phi_1\phi_2 + \phi_2^2) = \tfrac{1}{2}C(\phi_1 - \phi_2)^2, \text{ the same as before.}$$

We may also express U in terms of the electric field strength at all points by use of Gauss' theorem in the form (8.4). From the definition of U,

$$U = \tfrac{1}{2}\int \phi\rho\,\mathrm{d}V = \tfrac{1}{2}\varepsilon_0 \int \phi \operatorname{div} \mathbf{E}\,\mathrm{d}V.$$

Now consider the function div $(\phi\mathbf{E})$, which by writing in Cartesian coordinates is easily seen to equal $\phi \operatorname{div} \mathbf{E} + \mathbf{E}.\operatorname{grad}\phi$, or $\phi \operatorname{div} \mathbf{E} - E^2$. We therefore write $\phi \operatorname{div} \mathbf{E}$ as $E^2 + \operatorname{div}(\phi E)$ and

$$U = \tfrac{1}{2}\varepsilon_0 \int E^2\,\mathrm{d}V + \tfrac{1}{2}\varepsilon_0 \int \operatorname{div}(\phi\mathbf{E})\,\mathrm{d}V. \tag{8.11}$$

If we suppose, as is necessary to keep U finite, that the charges responsible for the field are all contained within a finite volume of space, we may enclose the volume in a surface S and evaluate the integrals over the enclosed volume, V, afterwards letting S expand in the assurance that the expression will tend to a finite limit which will be the true value of U. The second term in (8.11) can be transformed by use of IX into a surface integral over S:

$$U = \lim_{V\to\infty}\left(\tfrac{1}{2}\varepsilon_0 \int_V E^2\,\mathrm{d}V + \tfrac{1}{2}\varepsilon_0 \int_S \phi\mathbf{E}.\mathrm{d}\mathbf{S}\right).$$

Now as S expands to the point when the surface is a great distance, R say, from the centroid of the charges, ϕ ultimately varies as $1/R$ and E as $1/R^2$. The second integral thus involves an integrand varying as $1/R^3$ to be integrated over a surface which increases only as R^2, so that the integral goes to zero as $1/R$. The first integral is convergent when E varies as $1/R^2$, and the energy is therefore simply obtained by integrating $\frac{1}{2}\varepsilon_0 E^2$ over all space. In so far as the work needed to assemble the charges can be imagined as stored potential energy in some sense like the work involved in compressing a spring is stored as elastic energy, $\frac{1}{2}\varepsilon_0 E^2$ can be thought of as an energy density associated with the field, which can be transformed from its potential form to mechanical work, kinetic energy, heat etc. when the field is reduced by any operation. We need not, as the exponents of the aether theory of electromagnetism did in the 19th century, believe in this energy as really stored elastically—if we prefer a severer philosophy we may insist that it is only a mathematical construction. Nevertheless, it is often of great help to the imagination to ascribe reality to the stored energy and to visualize its being transferred like so many bales of merchandise when the field changes its strength and configuration, and such aids to thought must not be refused on doctrinaire grounds, unless they prove positively mischievous in closing our minds to alternative and more productive viewpoints. In this case, however, it seems to be common ground among physicists that the pretence that energy is as real as matter is so valuable a fiction that it may prove in the end to be nearly enough true. And certainly the Einstein relation between mass and energy, $E = mc^2$, encourages one to adopt the same attitude to both—to accept both as real, thereby showing sturdy common sense, or to treat both as mathematical constructions, thereby showing delicate appreciation of philosophical niceties.

266

A third form in which U may be expressed is of value when the system is an assembly of charged particles. If these are treated as point particles, the field energy of each is infinite (see the following problem (2)), but is at any rate independent of the position of the particles, so that we can ignore it when calculating the extra work needed to assemble the particles. Since this is $\frac{1}{2}\sum_i q_i\phi_i$, where ϕ_i is the potential at the ith particle due to all others and neglecting its own infinite contribution, we have

$$U = \tfrac{1}{2}\sum_i \sum_j q_i q_j/(4\pi\varepsilon_0 r_{ij}), \qquad 45$$

in which r_{ij} is the distance between the ith and jth particles. This particular form of U is useful when the energy of an assembly of particles is written down as the first step in a quantum-mechanical treatment of their behaviour; the theories of atomic structure and of covalent chemical bonds are examples of problems where the interaction of charged particles plays a fundamental role.

PROBLEMS

(1) Show that the field energy in a parallel-plate and a concentric spherical condenser is $\frac{1}{2}CV^2$.

(2) Imagine an electron to be a sphere of radius r carrying its charge on the outside; calculate the field energy and, by use of Einstein's relation,

determine what r must be for the mass associated with this energy to be equal to the electronic mass.

The answer, $1{\cdot}4 \times 10^{-15}$ m, is not much smaller than typical distances between particles in the nucleus and although electrons are not found as such in the nucleus it is an indication that electrostatic energy is not a negligible contribution to the dynamical behaviour of fundamental particles.

EXAMPLE

A large parallel-plate capacitor has its plates, which are a distance a apart, joined externally by a conducting wire. A charge q between the plates moves a distance δx normal to the plates. Show that $q\,\delta x/a$ flows in the wire from one plate to the other.

We shall give four different solutions to this problem, to illustrate the application of four ideas we have met so far. The first two are trick solutions in the sense that they give the answer very elegantly for a parallel-plate condenser, but are difficult to extend to any other geometry. The last two are immediately extendable and may be considered better methods to remember as likely to be useful in a variety of circumstances.

(1) *Superposition principle.* If a point charge gives a certain charge flow when moved, a sheet of charge will give the same result as if each element were moved separately. Consider then a sheet of charge at x, with charge density σ per unit area. Let the total charge on the plates be zero, and at any instant let the charge densities on them be $\pm\sigma_1(x)$. If the fields in the various regions are E_0 to E_3, as shown, Gauss' theorem tells us that

47

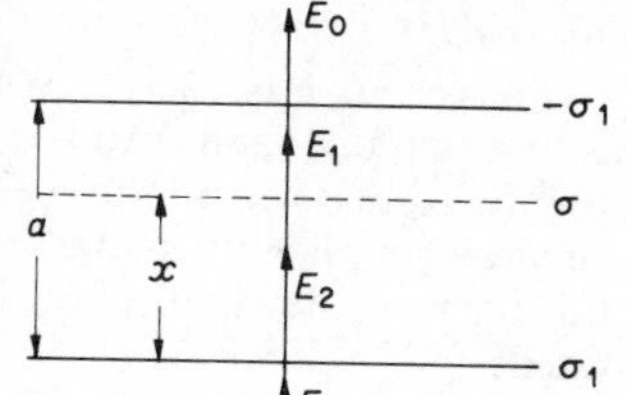

$$\varepsilon_0(E_0 - E_1) = -\sigma_1, \quad (8.11)$$

$$\varepsilon_0(E_1 - E_2) = \sigma, \quad (8.12)$$

$$\varepsilon_0(E_2 - E_3) = \sigma_1. \quad (8.13)$$

Now the values of E_0 and E_3 are determined by σ and by any charges external to the capacitor, and do not change with x. Differentiating (8.11), we thus have that $\varepsilon_0\,\delta E_1 = \delta\sigma_1$ when movement of the charged plane alters the fields. But $E_2 x + E_1(a - x) = 0$ since the plates are at the same potential, and this may be rewritten, using (8.12), in the form

$$\sigma x - \varepsilon_0 E_1 a = 0,$$

i.e.,

$$\sigma\,\delta x - \varepsilon_0 a\,\delta E_1 = 0,$$

i.e.,

$$\sigma\,\delta x - a\,\delta\sigma_1 = 0;$$

so that $\delta\sigma_1 = \sigma\,\delta x/a$, which is equivalent to the answer given.

(2) *What happens in a closed conducting box does not affect the outside world.* 150

Since the plates are large and connected by a wire, we shall not alter the behaviour by joining them all round the edge to form a conducting pillbox. When charge q is moved through δx the effect is the same as adding a dipole consisting of $-q$ at x and q at $x + \delta x$, with moment $q\,\delta x$. Such a dipole would produce a dipole field at great distances and contradict the principle of isolation enunciated at the heading; the box itself must therefore produce an opposing dipole by transferring $q\,\delta x/a$ from one plate to the other.

162 (3) *Reciprocity theorem.* Let one plate be earthed and the other isolated, and insert a very small metallic sphere S between the plates. When a charge q is deposited on A, its potential rises to q/C, and the potential of the sphere to $x/a \times q/C$, since it acquires the potential of the point where it is placed. Hence, by the reciprocity theorem, if instead we start with A uncharged and deposit q on S, the potential of A rises to $xq/(aC)$. Now connect A to the earthed plate; charge xq/a will flow as its potential falls to zero. Thus inserting a charge q at x into a capacitor with its plates earthed causes charge $-xq/a$ to flow to one plate and $-(a - x)q/a$ to flow to the other; as x is changed, $q\,\delta x/a$ is transferred.

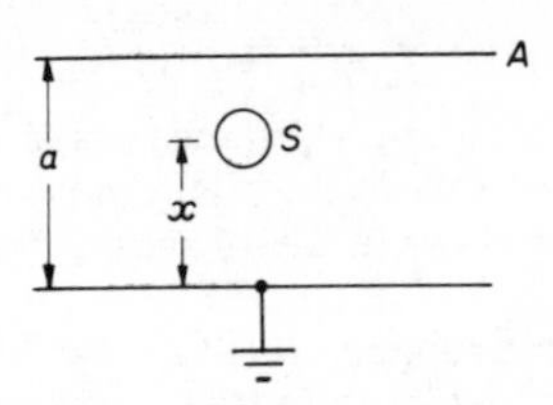

This method, unlike the first two, is immediately generalized, and the reader may easily supply the proof of the following statement: define the potential inside a capacitor of any form by $\phi(\mathbf{r})$, where ϕ is normalized to take the value zero on one plate and unity on the other (i.e., $\phi(\mathbf{r})$ is the potential distribution with one volt between the plates). Then if the plates are connected and a charge q moved from $\mathbf{r}_1$ to $\mathbf{r}_2$ the charge flowing in the connecting wire is $q[\phi(\mathbf{r}_2) - \phi(\mathbf{r}_1)]$.

166 (4) *Conservation of energy.* We now imagine that we have a cell in the connecting circuit, and can adjust the potential difference ϕ of the cell at will. We perform the following operations and change the energy in the way stated:

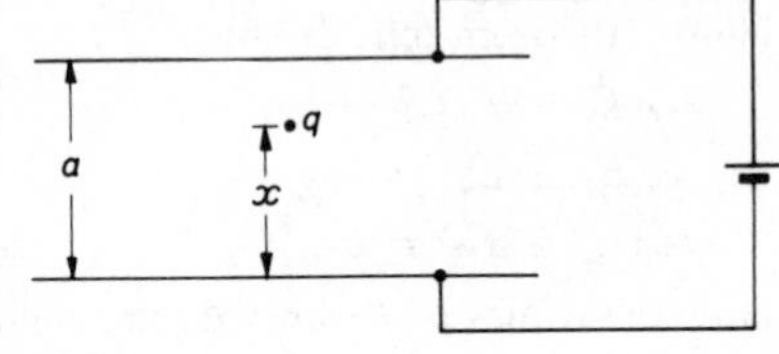

(a) with $\phi = 0$, insert q into the capacitor; the field energy is $\frac{1}{2}\,\varepsilon_0 \int E^2\,\mathrm{d}V$, which we call U_0.

(b) Raise ϕ to ϕ_0 holding q in position; the charge that flows is independent of the presence of q, and the work done by the cell is $\frac{1}{2}C\phi_0^2$.

(c) Move q by a distance δx in the direction of E and suppose that as a result $\alpha q\,\delta x$ flows through the cell; then work $\alpha q\phi_0\,\delta x$ is done by the cell and $-qE\,\delta x$ by the force holding q in position.

(d) Reduce ϕ_0 to zero; the work done by the cell is $-\frac{1}{2}C\phi_0^2$.

As a result of all these processes, q has been moved through δx and the final energy U_1 has the form:

$$\begin{aligned} U_1 &= U_0 + \tfrac{1}{2}C\phi_0^2 + \alpha q\phi_0\,\delta x - qE\,\delta x - \tfrac{1}{2}C\phi_0^2 \\ &= U_0 + (\alpha q\phi_0 - qE)\,\delta x. \end{aligned}$$

Now the field E acting on q is made up of ϕ_0/a due to the potential difference between the plates, and any field, E', produced by the images of q in the plates. The latter is itself proportional to q and makes a contribution to the work which is proportional to q^2; also U_1 and U_0, being proportional to the square of the field due to q, are quadratic in q. We may therefore separate linear and quadratic terms in the energy balance, writing

$$(\alpha - 1/a)q\phi_0\,\delta x + (U_0 - U_1 - qE'\,\delta x) = 0.$$

Since this must be true for all q, the linear and quadratic terms vanish separately; i.e., $\alpha = 1/a$, which shows that $q\,\delta x/a$ flows in the external circuit. The vanishing of the second term indicates that the change in field energy due to q itself is matched by the work done on q by the forces of attraction to the capacitor plates.

This method also can be generalized like the previous one.

The voltaic cell; contact potentials

We have just made use of a voltaic cell as an aid to solving a problem, and it is convenient to discuss now, though only superficially, the processes that go on in such a cell. The basic phenomenon is one of great generality, that when two systems can exchange charged particles across the interface separating them, they can normally be in equilibrium only with a potential difference between them. To illustrate this point, by an example which is not closely connected with the cell, we shall consider the junction between two metals, across whose interface electrons may flow freely.

First we must discuss what happens in a single piece of metal, and here it is convenient to introduce a simplified model which is nevertheless of considerable value in illustrating many fundamental aspects of metallic behaviour. The reality is a regular lattice of metallic ions (e.g., sodium atoms each lacking one electron) forming a rigid background to the more or less free motion of the electrons liberated from the atoms when the ions were formed; the simplified (*free-electron*) model is a box within which the electrons move like atoms of a perfect gas. It is a peculiarity of the wave character of particles in quantum mechanics that the electrons can pass readily through the dense ionic lattice, much as light passes through glass, so that the model of an empty box is not at all ridiculous. The reader should be warned, however, that this account of electrons in metals glosses over a number of very subtle difficulties which are the special province of the expert in solid-state theory. The electrons cannot escape from the metal unless they are given a considerable extra kinetic energy; at a high temperature ($\sim$2000 K), a few acquire enough by accidental collisions to be able to leave, and this is the phenomenon of thermionic emission, used in the filaments of radio valves; or irradiation by light can cause photo-electric emission. It is clear from this that the potential energy of an electron is lower inside the metal than outside, and we may perhaps visualize why this is by imagining an electron wandering through the positive ions of the lattice, as one particle in a gas of electrons. The other electrons are repelled from any one of their number, so that on the average it finds itself in an environment that is somewhat positive rather than neutral; any electron coming in from outside therefore experiences an attractive force towards this positive environment, which is another way of saying that the potential energy is lower inside than out.

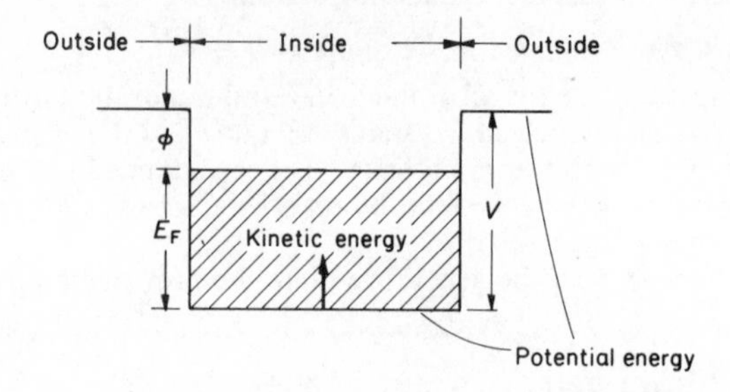

At this stage we must take notice of another peculiarly quantal phenomenon, the *Pauli Exclusion Principle*, which forbids two electrons from ever being in exactly the same state. There is no need to elaborate here on this statement, which can in fact be made quite precise in meaning, since all we need recognize is that even at the absolute zero of temperature the electrons are not permitted to come to rest; there are in the lowest permitted energy state electrons moving at all speeds from zero to the *Fermi velocity*, different for different metals but of the order of 10^6 m s^{-1}, corresponding to a kinetic energy (*Fermi energy*) of several electron volts. Any electron seeking to enter the metal must find an unoccupied state, i.e., it must enter with kinetic energy at least as great as the Fermi energy. We may thus picture an electron with the Fermi energy E_F as possessing, relative to an electron just outside, potential energy $-V$ and kinetic energy E_F, the total energy $E_F - V$ being still negative, $-\phi$ (ϕ is called the *work function*). It is necessary to give an electron at least ϕ extra energy before it can escape as a thermionic or photoelectron. This discussion is summed up in the energy diagram which shows how the different contributions to the total energy vary with position, and in particular exhibits the potential energy discontinuity at the boundary of the metal, and the kinetic energy states of electrons from 0 to E_F, all filled at 0 K.

We now see what happens when two different metals are joined together. If they are so arranged that an electron has the same potential energy just outside each, as illustrated in (a), their work functions being in

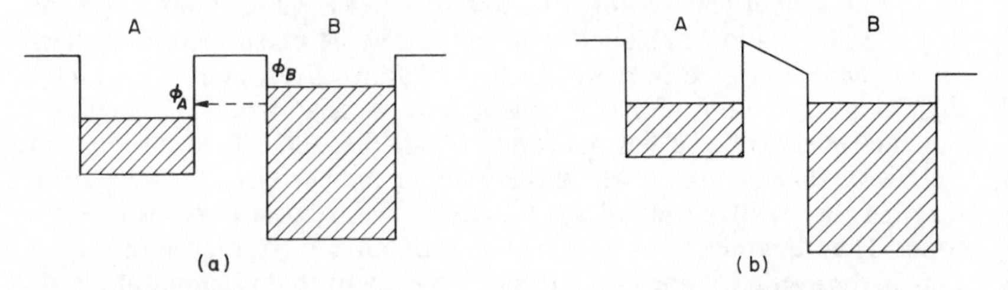

general different, the total energy at the Fermi level will differ by $\phi_A - \phi_B$. It is now possible for electrons to pass from B to A, and having got there, collide and lose energy till they sink down to the Fermi level of A. The reverse process being prohibited by the Exclusion Principle, A will acquire an excess of electrons, and become negatively charged. This will raise the potential energy of all electrons in A, and lower it in B, so that the diagram will come to resemble (b), at which point no further change is possible and equilibrium has been attained. The sloping line between A and B reflects the change of potential with position, i.e., the presence of an electric field between the metals, sufficient to create a potential energy difference $\phi_A - \phi_B$ for an electron situated just outside the two metals. When the metals are joined together with no more than a few atomic layers of mixed composition as interface, the field in this region is very strong indeed, and there is a potential drop of perhaps 1 V in a distance of 10^{-9} m, i.e., 10^9 V m^{-1}. It is this field that accelerates or decelerates electrons leaving one metal so that they slide comfortably into the environment provided by the other. It should be noted that in the transition region the concentration or deficiency of electrons that is needed to set up the required field is very small indeed compared to the number of electrons available in the metals, and there is normally no need to consider where they are to come from. The same is not true for semiconductors and insulators, but we shall not pursue this point.

The *contact potential* just described leads to the presence of an electric field in the vicinity of two different metals that are joined together, for each metal surface in itself is an equipotential, but not the same for both. This was observed as long ago as 1860 by Kelvin, who made a ring half of copper and half of zinc, and hung over it on a torsion wire an insulated

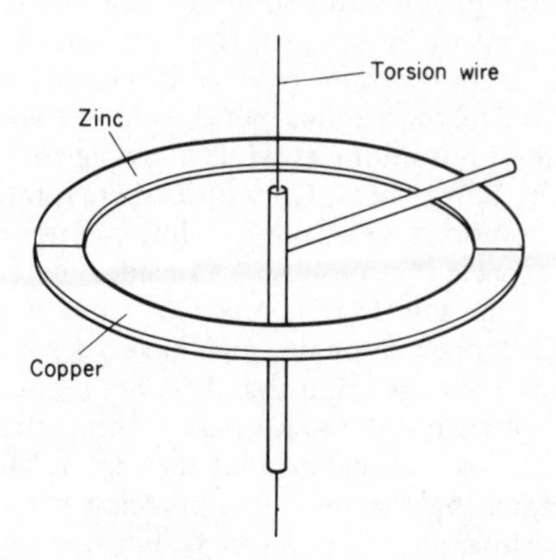

metal pointer; when it was charged electrically it swung towards one or other of the metals, depending on the sign of the charge, and showed that a field existed outside the ring. He later separated the two metals and arranged to vary the potential difference between them by means of a cell, finding that with an additional difference of just under 1 V the pointer remained unmoved when charged. In this way he determined the contact potential between copper and zinc and showed, moreover, that it was very sensitive to the physical and chemical condition of the surfaces. Nowadays in fact, any valid measurements of this sort are performed under strict vacuum conditions with scrupulous attention to details of surface preparation.

It might be supposed that the contact potential could be used like a voltaic cell as a continuous source of electrical power, but a moment's thought shows that this is not so. If two wires of copper and zinc are joined, there does indeed exist a potential difference between the free ends, but of such a magnitude as to ensure that when those ends are touched together there is no need for any current to flow to establish equilibrium, and the same holds however many different wires are joined end to end.

We turn now to the voltaic cell, where a similar contact potential exists, but with the technically important difference that it is useful. There are many varieties of cells employing different chemical reactions, and we shall treat one only, the classic Daniell cell. Inside a copper vessel stands a porous earthenware pot, the space outside the pot being filled with concentrated copper sulphate ($Cu\,SO_4$) solution and the inner space with acid zinc sulphate ($Zn\,SO_4$) solution in which stands a zinc electrode. A potential difference of 1·1 V is developed between the copper (+) and zinc (−) electrodes. When current flows, electrons in the external circuit go to the copper electrode where they combine in pairs with the doubly-ionized copper ions, which then leave the solution and are deposited as neutral metallic copper. At the zinc electrode, zinc atoms from the metal may be imagined as giving up two electrons, which travel into the external circuit while the zinc ions go into solution. The migration of Zn^{2+} and Cu^{2+} in the direction of $Zn \rightarrow Cu$ in the solution, and of SO_4^{2-} in the opposite direction, is the mechanism whereby current passes through the liquid. The function of the porous pot is simply to prevent accidental mechanical mixing of the solutions while allowing free passage of ions.

We have here three interfaces, Cu with $CuSO_4$ solution, $CuSO_4$ solution with $ZnSO_4$ solution, and $ZnSO_4$ solution with Zn; potential differences may be expected across each. Consider for example the Cu—$CuSO_4$ interface. In the solution the ions Cu^{2+} and SO_4^{2-} (which exist in that form in solid copper sulphate, and have such a strong Coulomb attraction that they hold the solid together) are separated, for the high dielectric constant of water (about 80) weakens the attraction of positive and negative ions to such an extent that thermal agitation is enough to prevent them sticking together for any appreciable fraction of the time. We may therefore imagine a Cu^{2+} ion within one or two atomic diameters of the electrode pulled towards the metal by those (essentially quantum-mechanical) bonding forces that make copper the strong material it is, and in the opposite sense by those forces that cause the ions in copper sulphate crystal to prefer to dissolve in water rather than remain bonded to their parent crystal. As soon as a copper rod is dipped in copper sulphate solution these forces move the ions until such a distribution of charge is established at the interface, in the form of a double layer of positive and negative charge excess, that the electrical force on the ions cancels any tendency for further movement. We may thus imagine the metal and the solution to form two equipotential regions, with a potential difference between them, and we would expect to be able to detect the fields in the vicinity by an experiment analogous to Kelvin's. A similar situation exists round the zinc electrode, with a different contact potential between metal and solution.

As for the interface between the solutions, here again we may expect a potential difference, though in fact it is usually much smaller. The mechanism is somewhat different, for to a good approximation the individual ions are not subject to any translational force, but migrate freely through the water surrounding them, largely oblivious of the presence of other ions. However, this individual migration is enough to ensure that in a concentration gradient there is a tendency for ions to move from stronger to weaker concentration—it is just that more ions are available to wander from regions of high to regions of low concentration than are available to wander in the reverse direction. Where there is a rather sudden change from copper sulphate to zinc sulphate solution, Cu^{2+} are migrating towards the zinc solution, and Zn^{2+} in the opposite sense, while if the SO_4^{2-} concentration is not constant there will be yet another diffusion process taking place. In general we may expect the diffusion currents not to balance each other, so that a separation of charge will proceed until the electric field so produced causes an organized ionic current that matches the diffusion current. A state of quasi-equilibrium is then established (not true equilibrium since the solutions continue to diffuse into one another) with a potential difference between the solutions.

Adding all these effects together, we can expect the copper and zinc to be at different potentials and because these are controlled by ionic, not electronic, processes the difference need not be the same as the contact potential difference of electronic origin between copper and zinc. It is from this that the usefulness of the cell derives, for if the copper and zinc are now touched together, their Fermi levels will not be the same, and a current will flow, and continue to flow so long as the chemical processes of dissolution of zinc and precipitation of copper can be maintained. To relate the useful e.m.f. of the cell to the various interface potentials we have considered, it is convenient to connect a copper wire to the zinc electrode, so that a voltmeter placed across the cell is connected to copper on both terminals and has no contact potential troubles. Then if the potential difference between Cu and $CuSO_4$ solution is V_{Cu}, between Zn and $ZnSO_4$ solution V_{Zn}, between the $CuSO_4$ and $ZnSO_4$ solutions V_s, and between metallic copper and zinc V_m, the measured e.m.f. of 1·1 V is $V_{Cu} + V_s - V_{Zn} - V_m$.

When a resistance is connected across the cell and a current flows, the potential distribution is disturbed at all interfaces, and it is this that upsets the balance and causes copper ions to be deposited more rapidly on the electrode than they leave into the solution, and zinc ions to be dissolved more rapidly. The movement of ions under the disturbed potential distribution in the solutions causes the interface between the solution to shift, and the relative proportions of $CuSO_4$ and $ZnSO_4$ shift in favour of the latter. The total energy of the components in their various forms is altered; with less Zn and more $ZnSO_4$, more Cu and less $CuSO_4$, the energy is lowered, and the loss of stored chemical energy is matched by the heat generated in the resistance and in the cell itself by the passage of current. The ideal cell is therefore to be thought of as a 'black box' inside which many complex processes are going on, but which can be treated in its relations to the rest of the world as a pair of terminals between which a constant e.m.f. is automatically maintained. An electric field exists in the space around the terminals. When charge Q flows through the cell, in the normal discharge direction, the stored chemical energy is reduced by VQ, and this energy appears in the external circuit as heat, mechanical energy in a motor, stored electrical energy in a condenser, or in a variety of other forms.

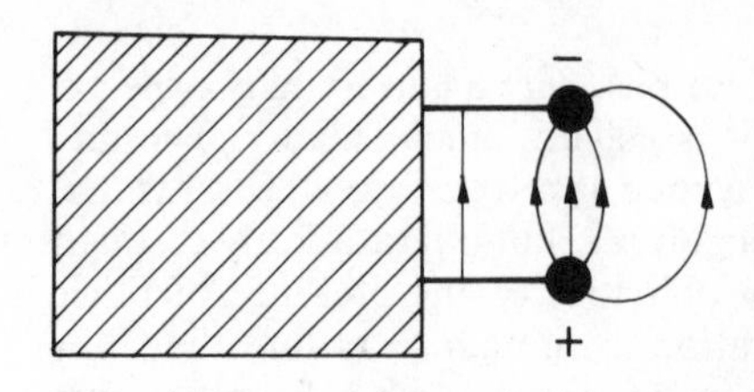

EXAMPLE

Comparison of energy content of a voltaic cell and various other systems. A typical lead-acid battery weighs $3\frac{1}{2}$ kg and can discharge at 1 A and about 2 V for 30 h, thus delivering about $2{\cdot}2 \times 10^5$ J. This is a great deal of energy, enough for example to raise a block weighing one ton (10^3 kg) through a height of 22 m, or to accelerate it from rest to a speed of 21 m s^{-1} (47 mile/h). Chemical energy is indeed very concentrated, as is also clear from the distance a heavy car can be propelled, against the frictional drag of air, tyres and brakes, on a gallon of fuel; or we may note the speed of a bullet driven by a small cartridge, or the destruction produced by a high-explosive shell. The concentration of energy in a battery is not less than that in an explosive by as large a factor as one might at first think—it seems less potent because it is released gently over a long period. A mixture of acetylene and oxygen gases is possibly as devastating an explosive mixture as any, and can release $2{\cdot}2 \times 10^5$ J from about 20 g of gas; it is therefore only some 200 times more concentrated than a lead-acid cell, and the factor is considerably less if we compare with a more modern battery using lighter ingredients than lead.

Other instructive comparisons of the chemical energy of a battery may be noted:

(1) $2{\cdot}2 \times 10^5$ J $= 5 \times 10^4$ calories, so that the battery, discharged through a resistor, could be used to bring $\frac{1}{2}$ kg of water from 0°C to 100°C.

(2) A commercial electrolytic condenser of capacity 160 μF, with a maximum working voltage of 350 V, can store 10 J in a volume of 10^{-4} m^3. To store in this way as much as the lead-acid battery would require a volume of $2{\cdot}2$ m^3.

Comment. It is interesting to look at the large density of energy in a chemical store from the atomic point of view. Consider the relative inefficiency of storage in a condenser; basically the reason is that the field strength that can be maintained by the dielectric insulation between the plates is not very large compared with the fields in atoms and molecules. The best one can expect of a dielectric is that it can withstand a few hundred thousand volts across a gap of 1 mm, say $E = 3 \times 10^8$ V m^{-1}. Such a field in free space would possess energy density, $\frac{1}{2}\varepsilon_0 E^2$, of 4×10^5 J m^{-3}, four times more than what we have already found for an electrolytic condenser, which has much of its volume taken up by metals in which there is no field and no stored energy. The field strength quoted here should be compared with a typical field in an atom, say the field acting on an electron in the lowest Bohr orbit, $5{\cdot}3 \times 10^{-11}$ m from a proton; this is [75] $6{\cdot}5 \times 10^{12}$ V m^{-1}, giving an energy density of 2×10^{14} J m^{-3}, 5×10^8 times as great. It will be realized, of course, that most of the electrostatic energy within a molecule is due to the field which is inseparably attached to the charged particle. The potential energy of an electron in a Bohr orbit is negative, and can be attributed to the diminution, as a result of bringing the charged particles together, of the distant fields they generated when far apart. We cannot therefore expect to release more than a small fraction of the electrostatic energy by a rearrangement of the charged particles.

It is more significant to consider the kinetic energy of the electrons, which we have seen to be a fair measure of the binding energy in the Bohr atom. Applying this to the acetylene-oxygen mixture, we may neglect all but the outermost shell of electrons, the inner electrons being too tightly bound to make much of a contribution; this leaves four electrons in each carbon atom, one in each hydrogen, and six in each oxygen, i.e., forty electrons in the interaction of one molecule of acetylene C_2H_2 with five oxygen atoms to give carbon dioxide and water. These electrons make up one part in 5000 of the total mass, and they are moving with speeds around 3×10^6 m s^{-1}. If we were to suppose that all their kinetic energy were to be released in a reaction and used to drive a bullet equal in mass to the reacting chemicals, it would be accelerated to a speed $(5000)^{-1/2}$ times the electronic speed, or $4{\cdot}3 \times 10^4$ m s^{-1}, 130 times the speed of sound in air. Comparing this with a typical rifle bullet, moving at 1·5 times the speed of sound when driven by propellant weighing half as much as itself, and making allowances for the greater potency of the acetylene-oxygen mixture (the reader is not encouraged to test this for himself), we still find a discrepancy indicating that only about one part in 1000 of the electrons' kinetic energy is made available in the explosion.

Summing up, one can appreciate that the density of chemical energy storage, compared with storage in a condenser or a spinning fly wheel, arises from the much greater electric fields and electronic speeds in atoms and molecules than are to be found in any macroscopic system. An intermediate storage density is provided by thermal storage, where the characteristic speed which is the reservoir of kinetic energy is the molecular speed, typically 10^3 m s^{-1} and rather larger than comfortable in fly wheels. At the other extreme from mechanical storage, we have nuclear energy, the most concentrated of all, which depends on the binding energy of protons and neutrons in the nucleus, of the order of millions of electron volts per particle compared with about one electron volt per atom in a chemical combination.

Force on a conductor in an electric field

The energy conservation law may be used to derive the tractive force exerted by an electric field on the surface of a conductor, as may be seen by analysing a special case. Consider a parallel plate condenser whose plates have area A and are a small distance x apart. If charges $\pm Q$ are deposited on the plates, which are then isolated, a force F will be needed on each to hold them apart. The field strength E between the plates is $Q/(\varepsilon_0 A)$, and this remains unchanged when the separation is altered. When x is increased by δx, an additional volume $A\ \delta x$ is created in which the energy density is $\frac{1}{2}\varepsilon_0 E^2$ and the work needed, $\frac{1}{2}\varepsilon_0 E^2 A\ \delta x$, must be provided by the force F moving a plate through δx. The total attractive force between the plates is therefore $\frac{1}{2}\varepsilon_0 E^2 A$, and the force per unit area is $\frac{1}{2}\varepsilon_0 E^2$, the same as the energy density.

PROBLEM

Instead of depositing $\pm Q$ on the plates, put a cell across them and repeat the argument, now taking account of the variation of E with x and the work done by or on the cell. You should of course get the same result for the force.

The force can also be derived directly, in such a way as to show that the result is very general. Consider a small element of a conducting or non-conducting sheet which has a different field on its two sides; the element is small enough to be considered plane. Then the external field distribution can be dissected into a symmetrical part (S) in which the field is the same on both sides, and an antisymmetrical part (A) in which it takes

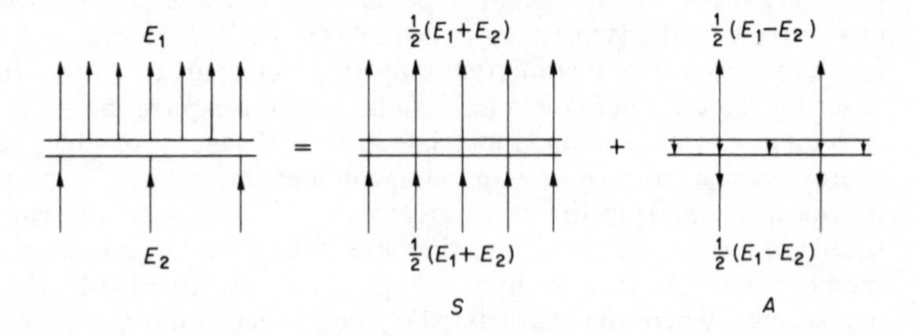

equal and opposite values on the two sides. We do not need to know what happens inside the sheet; it is enough to make the field S continue straight through and to make any necessary corrections to the internal field by choosing the internal field in A appropriately. As drawn, the sheet is taken to be a conductor, with zero internal field, but this is not a necessary assumption. Now the field S clearly does not depend on the presence of the sheet, and must originate from charges outside the surface element, while A is generated by surface (and possibly internal) charges whose total surface density is, by Gauss' theorem, $\varepsilon_0(E_1 - E_2)$. To derive the force acting on the surface element we need only note that by N3 the charge distribution in A cannot of itself produce a resultant force, and therefore the force is the result of the field in S acting on the charge in A. The force per unit area is $\frac{1}{2}\varepsilon_0(E_1 + E_2)(E_1 - E_2)$ upwards, and can be thought of as a tractive force $\frac{1}{2}\varepsilon_0 E_1^2$ on the upper surface and a tractive force $\frac{1}{2}\varepsilon_0 E_2^2$ on the lower surface. This is consistent with our analysis of the condenser. Note that the direction of the force is always tractive and is not related to the direction of the field.

50

PROBLEM

A sphere is charged uniformly with surface charge density σ. By direct application of the inverse square law show that the force per unit area is $\sigma^2/(2\varepsilon_0)$, consistent with the result obtained directly from the field strength just outside.

Cavendish Problems: 166, 196, 197, 199–203, 207, 208, 212, 218, 241, 242, 265.

READING LIST

Field-ion Microscope: R. Gomer, *Field Emission and Field Ionization*, Harvard U.P.

Metals: J. M. Ziman, *Electrons in Metals*, Taylor and Francis.

Experimental Solution of Laplace's Equation: W. J. Karplus and W. W. Soroka, *Analog Methods* (Ch. 11), McGraw-Hill.

General: E. S. Shire, *Classical Electricity and Magnetism*, Cambridge U.P.

9

Dielectric materials

In the last chapter we considered electric fields in free space, and we shall now show how the molecular conception of matter enables one to understand, by simple extension of what we have already discovered, some of the well-known effects that occur when dielectric materials are present. In particular we shall demonstrate the reasons for the following experimental observations:

(1) When two charged and isolated conductors have the space around and between them filled with a non-conducting (*dielectric*) fluid, the force between them is reduced by a factor, ε, characteristic of the fluid and quite independent of the shape of the conductors.

(2) The capacitance between two conductors is increased by a factor ε on flooding the system with the dielectric fluid.

We shall further show how the differential equation for the electric field may be formulated when dielectrics are present, and derive the boundary conditions at the surfaces of dielectrics. From that point on it is usually unnecessary to think any more of the molecular constitution, and enough to solve any problem that arises as an exercise in differential equations, or in the physics of macroscopic bodies.

We shall regard a molecule as a neutral assembly of positive and negative charges which, though in continuous rapid motion, have a well-defined mean configuration. This may be spherically symmetrical as in single atoms, or axially symmetrical as in H_2, HCl, CO_2, or it may have more complicated form as in H_2O, NH_3, CH_4. Unless the symmetry is so high as to preclude it, we may generally expect the molecule, though neutral, to carry a dipole moment, the centroids of positive and negative charge being not coincident. Of those molecules mentioned above HCl, H_2O and NH_3 have dipole moments. The molecules are constantly turning in all directions and even if they possess dipole moments the time-average of the moment is zero, until an electric field is applied and causes them to spend longer with their dipoles pointing along the field than with them pointing against it.* An electric field can also distort a molecule and cause a slight charge separation where none existed before, thus creating a dipole moment. Whichever sort of molecule we consider, the effect of an electric field is to change its structure or motion so that on the average there is a dipole moment parallel to the field and proportional to it, $\mathbf{p} = \alpha\varepsilon_0\mathbf{E}$; α is the *molecular polarizability*.

PROBLEM

75

An electron in a Bohr orbit of radius a round a proton is also acted upon by a uniform electric field $\mathbf{E}$, in the plane of the orbit. Show that if $\mathbf{E}$ is not too strong, the new orbit is a circle with its centre displaced by a distance $\mathbf{r}_0$ from the proton, where $\mathbf{r}_0 = 4\pi\varepsilon_0 a^3\mathbf{E}/e$; show also that the electron traverses its new orbit at constant speed, so that the mean dipole moment is $e\mathbf{r}_0$, and $\alpha = 4\pi a^3$. The condition for this approximation to be valid is that terms in $(r_0/a)^2$ may be neglected.

The same result, $\alpha = 4\pi a^3$, is obtained for the polarizability of a Bohr atom when $\mathbf{E}$ is normal to the plane, and we may take this as a reasonable estimate of the polarizability of a real atom. This view is confirmed by another estimate based on a model that resembles more closely the modern view of atomic structure; a cloud of electrons, not localized in a definite orbit, may be thought of as not unlike a metallic

* This is a rather nice point. If we imagine a dipole spinning rapidly before the field is applied, the effect of the field is to cause it to go faster when $\mathbf{p}$ is along $\mathbf{E}$ and slower when $\mathbf{p}$ is directed opposite to $\mathbf{E}$, and therefore to spend longer pointing opposite to $\mathbf{E}$, contrary to what we have stated. There are, however, certain dipoles which happen to spin so slowly that application of $\mathbf{E}$ sends them into pendulum oscillations with a mean direction parallel to $\mathbf{E}$. At high temperatures, when there are dipoles present, spinning at all rates up to such rapidity that $\mathbf{E}$ leaves them virtually unaffected, the average effect due to fast and slow speeds cancels out; at lower temperatures an excess of slower rotation speeds leads to a resultant average direction along, rather than against, $\mathbf{E}$. Any given dipole, of course, is continually caused by collision to switch from one rotation speed to another, and the time-average of its behaviour favours the pendulum oscillations at low temperatures and gives the mean polarization along $\mathbf{E}$ that is actually observed.

sphere, in that the electrons are free to respond to an applied field and will move so as to exclude it if they can. If a is now interpreted as the radius of the equivalent metal sphere, we take over the result already proved that such a sphere acquires a moment $4\pi\varepsilon_0 a^3\mathbf{E}$ in a field, so that $\alpha = 4\pi a^3$ as before.

Yet another approach to the problem, and an especially interesting one because of its close connection with the real properties of atoms, is to relate the polarizability in a steady field to the frequency at which the atom can absorb and emit radiation. If we look on the atom as an electron of mass m bound by an elastic force to the centre so that for displacement x the restoring force is $-\mu x$, we know that ω_0, the resonant angular frequency of the resulting harmonic oscillator, is $(\mu/m)^{1/2}$; moreover in a steady field E the force on the electron, eE, produces a displacement eE/μ and a dipole moment e^2E/μ. Using these two expressions to eliminate μ, we write

$$\alpha = e^2/(\varepsilon_0 m\omega_0^2). \tag{9.1}$$

If the model has any validity, it explains why atoms like helium with high absorption frequencies (helium absorbs only in the far ultraviolet) have low polarizabilities.

Let us apply these results to the hydrogen atom. Using the Bohr theory we have found the lowest orbit, of radius a_0, to have a polarizibility of $4\pi a_0^3$. On the other hand (9.1) gives $(8/3)^2 \times 4\pi a_0^3$, 7·1 times as great, when ω_0 is assigned the frequency of the radiation that excites the atom from its ground state ($n = 1$) to the next highest state ($n = 2$). A complete calculation based on the quantum-mechanical formulation of the hydrogen-atom problem yields $4{\cdot}5 \times 4\pi a^3$. All the calculations give the same form of answer, differing only in numerical constants. We may dismiss the first result immediately, since it is well known that the quantum-mechanical calculation differs significantly from the Bohr theory in its description of the ground state; in particular the real ground state has zero angular momentum, while Bohr's circular orbit obviously has not.* The difference between the correct coefficient 4·5 and the value 7·1 obtained from (9.1) derives from the fact that the ground state is able to make transitions to many other levels above that for which $n = 2$, and the appropriate value for ω_0 is some average of all radiation frequencies that can excite the atom; this, being necessarily higher than what we assumed, will lead to a reduction of the coefficient. In fact (9.1), developed by quantum mechanics into a similar formula that takes account of all possible transitions, provides a precise description of the polarizability of atoms and molecules in terms of their spectra, without losing the essential simplicity that comes from treating the atom as if it were something like a harmonic oscillator.

* The conducting sphere of radius a_0 does not therefore represent the ground state of the hydrogen atom very well. It is distinctly better for atoms containing many electrons.

So much for the polarizability by distortion, to which we shall return later to see the implications of its magnitude in terms of the properties of real dielectrics. As for the polarizability resulting from alignment of already existing dipoles, this increases as the temperature is lowered since thermal agitation does not then so readily randomize the dipole directions; even at room temperature, molecules with a permanent dipole moment may exhibit considerably higher values of α than those without, and in the solid or liquid state the dipoles may play an important role in determining the arrangement of molecules. Thus in water and ice the dipoles tend to point in head-to-tail fashion, and are sufficiently strongly locked in this pattern that the polarization produced by an electric field is notably less than one would expect for freely rotating dipoles. These are very complicated problems which we must pass over, and we shall confine our attention to molecules for which α is small enough that each may be assumed to be influenced only weakly by its neighbours.

This is not to say that such influences are altogether negligible, but we shall suppose that they can be adequately taken into account by a modification of the field acting on the molecule. There are two effects to be considered, a macroscopic shape effect and a microscopic effect. The shape effect is simply that a lump of dielectric placed in a uniform field distorts it, so that it is no longer possible to assume that the field inside is related in a straightforward manner to the original applied field. We bypass this problem by defining the field inside, $\mathbf{E}$, as a local average in the following sense: if ϕ is the potential difference between two points separated by a distance $\mathbf{r}$, which is many molecular spacings, $\langle \phi/r \rangle$ is the component of $\mathbf{E}$ in the direction of $\mathbf{r}$. The brackets indicate that we take the average of ϕ/r for many samples of the same $\mathbf{r}$ differently situated with respect to the molecules, or keep $\mathbf{r}$ fixed and let the molecules move as they will, to give a time-average. Leaving aside for the moment the question of how we are to measure $\mathbf{E}$ in practice, we assert that $\mathbf{E}$ is a natural measure of the field responsible for the local polarization of the molecules. It should not be assumed that $\mathbf{E}$ is exactly the polarizing field, for it is the average field taken over all space, inside as well as outside the molecules, whereas a molecule is unable to sample the field so fully and knows nothing of the field inside another. Since the difference between the fields inside and outside a molecule is due to the polarization, we may be sure that the correction needed to $\mathbf{E}$ to give the effective polarizing field is proportional to the polarization. If there are n molecules per unit volume, each with moment $\mathbf{p}$, we write $\mathbf{P}$ for $n\mathbf{p}$, the dipole moment per unit volume, and suppose that there is a number λ, whose true value demands detailed analysis, such that the field acting on a molecule is $\mathbf{E} + \lambda\mathbf{P}/\varepsilon_0$. As a result the dipole moment acquired by each molecule is $\alpha(\varepsilon_0\mathbf{E} + \lambda\mathbf{P})$, so that

$$\mathbf{P} = n\alpha(\varepsilon_0\mathbf{E} + \lambda\mathbf{P}),$$

or

$$\mathbf{P} = n\alpha\varepsilon_0\mathbf{E}/(1 - n\alpha\lambda). \tag{9.2}$$

Usually λ is positive, and the mutual interactions of molecules enhance the polarizability of the material by a factor $(1 - n\alpha\lambda)^{-1}$. So long as $n\alpha\lambda$ does not approach unity, we may subsume this mutual interaction into a modified polarizability, $\alpha' \equiv \alpha/(1 - n\alpha\lambda)$, and otherwise forget about it. When $n\alpha\lambda$ approaches or exceeds unity, this approximation breaks down. It is now that the dipoles tend to order themselves in chains or other patterns, and simple treatments are without value.

After this preliminary discussion we may proceed to account for the observations quoted at the beginning of the chapter, starting with the second, the increase of capacitance when a fluid dielectric floods the space around and between the conductors. Let us deposit charges $\pm Q$ on the condenser plates (of any shape) when they are in a vacuum, and let the charge distribute itself with a surface density σ depending on position. The resulting field in the interspace is non-divergent. Now we freeze those charges in position and flood the system with dielectric, which means allowing a collection of polarizable molecules to fill the interspace more or less evenly. We shall assume, and justify later, that the effect of the dielectric is to change the field **E** at any point by a constant factor from the value that obtained before the dielectric was added. Then the resulting polarization **P** is $\kappa\varepsilon_0\mathbf{E}$, where κ is written for $n\alpha'$ and is called the *volume susceptibility*. Clearly div **P**, in parallel with div **E**, is zero. The significance of this may be seen by considering separately the movement of positive and negative charges in the molecules under the influence of **E**. The positive charges form an even distribution of points in space which, so far as their influence on external circuits is concerned, can be smeared out into a uniform charge density, ρ_0; similarly with the negative charges. In the absence of a field, these two uniform distributions $\pm\rho_0$ overlap completely to give no field outside or inside; but when the molecules are polarized, the distributions are shifted relative to each other by a distance $\boldsymbol{\xi}(\mathbf{r})$ determined by the requirement that at every point $\rho_0\boldsymbol{\xi}$ shall be the dipole moment per unit volume, **P**. The non-divergence of **P** therefore implies that div $\boldsymbol{\xi} = 0$—each charge distribution shifts like an incompressible fluid, and the balance of positive and negative charges is not dis-

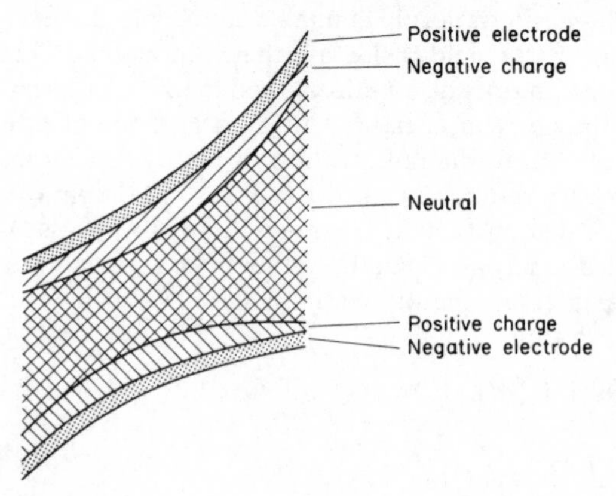

turbed inside. There is, however, a surface charge on any surface bounding the dielectrics, in this case next to the conductors, and the surface charge density due to the dielectric polarization has magnitude $\rho_0\xi$ i.e., P. It does not really matter that the dielectric is not made up of two interpenetrating continua of positive and negative charge as in our smeared-out model; in the molecular model the average charge density is also zero everywhere except very near the surface, and there instead of a continuous surface distribution we have a discrete pattern of charges whose average is still P per unit area. Thus we may treat the condenser as a pair of plates carrying surface charge density σ (provided by an external source) with another surface charge density $-P$ provided by the dielectric and located in the empty space immediately next to the metal surfaces. It is now easy to see that a consistent solution is obtained by supposing that P at the surface follows the form of σ exactly, $P = \beta\sigma$. The electric field between the plates has then exactly the same form as before the dielectric was added, but scaled by a factor $1 - \beta$. This holds right up to the surface of the plates, where the original field was σ/ε_0 and is now $\sigma(1 - \beta)/\varepsilon_0$. Consequently P at the surface is $\kappa\sigma(1 - \beta)$, and β is thus determined by the requirement that

$$P = \beta\sigma = \kappa\sigma(1 - \beta),$$

or

$$\beta = \kappa/(1 + \kappa).$$

The fact that β is constant shows that our supposition that P and σ follow each other is justified. The field within the plates does indeed keep the same form when the dielectric is added, as we assumed at the beginning. This was, to be sure, with the charge distribution clamped on the surfaces of the condenser plates, but the field we have just found, having the same form as the original field, has equipotentials coinciding with the plates. If then we unclamp the charges they will not move, and we have indeed found the correct solution to the problem.

A condenser, after filling with uniform dielectric fluid, thus has identical charge and field configurations, but the internal field produced by a given charge supplied from outside is multiplied by $1 - \beta$, i.e., $1/(1 + \kappa)$, as a result of the polarization of the dielectric. In consequence, the potential difference is multiplied by $1/(1 + \kappa)$ and the capacitance by $(1 + \kappa)$ for any shape of condenser. This factor is called the *relative permittivity* (or *dielectric constant*),

$$\varepsilon = 1 + \kappa, \qquad (9.3)$$

and ε is a property of the dielectric.

This completes our discussion of the capacitance change, but we may interrupt the argument for a moment to estimate the magnitude of ε from our knowledge of atomic polarizability. We have found that for atoms or molecules carrying no permanent dipole moment α', which is near enough to α for present purposes, is not far from $4\pi a^3$, where a is the atomic radius. Then κ may be taken as $4\pi na^3$ or $3\,nV$ if V is the atomic volume defined as $\frac{4}{3}\pi a^3$. Now nV is simply the fraction of the available space actually occupied by atoms, which in a condensed solid or liquid may be taken to be about one-half. We therefore expect κ for condensed materials 275 to be of the order of unity, and ε to be 2 or more, especially more if there are dipoles still able to rearrange themselves when a field is applied. In fact there are very few solids with ε as low as 2; hydrocarbons and many plastics range between 2 and 3, glasses from 4 to 8. In some solids, the local interactions may modify α' to a much higher value, producing, as in titanium oxide and a number of titanates, values of ε ranging up to 1000, or maybe leading to a catastrophe in which the solid settles down to a state of permanent polarization even in the absence of a field (*ferro-electricity*). This is a difficult matter, and all we need note here is that simple atomic models correctly predict the order of magnitude (about unity) for the dielectric susceptibility of simple forms of condensed matter.

To see what the dielectric does to the force between the plates we need only apply energy conservation. Any operation involving changing the separation between charged and isolated plates, if performed first with no dielectric and then with dielectric filling all space around and between them, results in exactly the same changes of field configuration but on a different scale in the two cases. The stored energy, $\frac{1}{2}Q^2/C$, is smaller by a factor ε when the dielectric is present, and therefore the work needed to effect the change, and the force between the plates, are reduced by ε. It should be noted that this argument depends on the dielectric being fluid, for otherwise the plate separation could not be changed. It is in fact a much more troublesome matter to discuss the forces in solid dielectrics; one must always verify that the question is meaningful. For example, we cannot say what happens to the force between parallel condenser plates when a solid dielectric slab is inserted unless we know whether it slips in with room to spare (in which case the force is unchanged) or whether it is forced in with a mallet. This may sound like a trivial and irrelevant issue, but in fact it is not, as the following discussion should reveal.

The energy argument certainly produces the required answer for the effect of a fluid, but it is not quite obvious what causes the force to be less. If we picture the charge σ on the condenser plates creating a field σ/ε_0 just outside, in the atomically small space between the metal and the dielectric molecules the tractive force of the field is exactly the same as if no dielectric were present, and one must look to the direct mechanical influence of the molecules in exerting a bigger pressure on the interior than on the exterior faces. There is in fact a force drawing the molecules towards regions of high field strength, and when

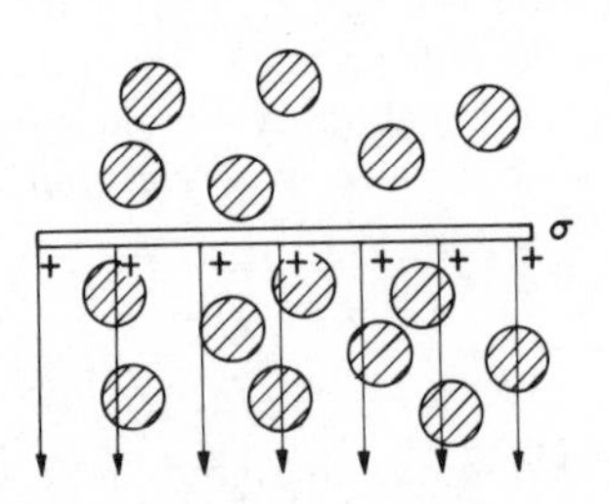

we calculate its effect on the pressure in the fluid we shall find that it exactly accounts for the reduction in force between the condenser plates. 278

The diagrams illustrate pictorially how a molecule, considered here as a polarizable sphere, is drawn towards the higher field regions. In the left-hand diagram, the increase in field from left to right is

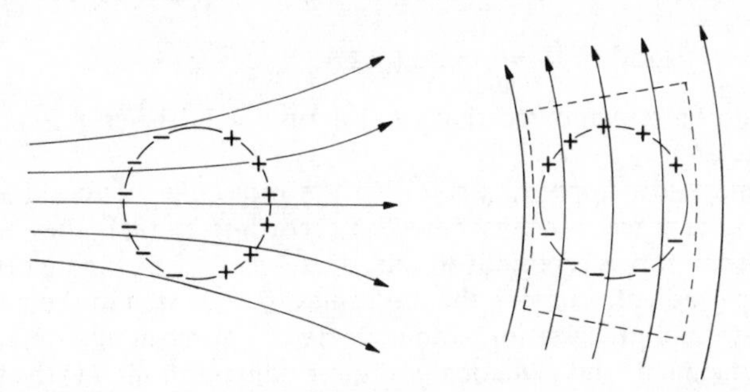

obvious, so that the negative surface charge due to polarization of the molecule is pulled more strongly to the left than the positive charge is to the right. Since the total positive and negative charges are equal, it is the pull to the left that wins. In the right-hand diagram the curvature of the lines of force implies that the field is stronger on the left. For the line integral round the dotted circuit must vanish, and the field must therefore be stronger along the shorter side. The positive surface charge now finds itself in a field which has a horizontal component pulling it to the left, and so does the negative charge, so the same attraction to stronger fields occurs.

To analyse the effect quantitatively, we need not assume any specific model, but simply consider the positive and negative charges within the molecule separately. If the electric field varies linearly with position, the total force acting on the positive charges is simply their total charge q times the strength of the electric field at the centroid of the positive charges, $\mathbf{r}_+$; similarly the total force on the negative charges is their total charge $-q$ times the electric field strength at their centroid $\mathbf{r}_-$. Consider then the x-component of the resultant force:

$$\begin{aligned} F_x &= q[\mathbf{E}_x(\mathbf{r}_+) - E_x(\mathbf{r}_-)] \\ &= q(\mathbf{r}_+ - \mathbf{r}_-).\operatorname{grad} E_x \\ &= \mathbf{p}.\operatorname{grad} E_x. \end{aligned}$$

We are concerned with situations in which $\mathbf{p}$ is proportional to the local field, and to avoid committing ourselves to any speculation about the meaning of the constant of proportionality we shall write $\mathbf{p} = \gamma\mathbf{E}$ and eliminate γ as soon as possible. Then

$$\begin{aligned} F_x &= \gamma(E_x\,\partial E_x/\partial x + E_y\,\partial E_x/\partial y + E_z\,\partial E_x/\partial z) \\ &= \gamma(E_x\,\partial E_x/\partial x + E_y\,\partial E_y/\partial x + E_z\,\partial E_z/\partial x), \end{aligned}$$

since, for example,

$$\partial E_x/\partial y = -\partial^2\phi/\partial x\,\partial y = \partial E_y/\partial x.$$

Hence

$$F_x = \tfrac{1}{2}\gamma\,\partial(E^2)/\partial x,$$

or

$$\mathbf{F} = \tfrac{1}{2}\gamma \operatorname{grad}(E^2) = \tfrac{1}{2}\operatorname{grad}(\mathbf{p}.\mathbf{E}). \tag{9.4}$$

It should be remembered that (9.4) is only valid when $\mathbf{p}$ is proportional to $\mathbf{E}$.*

We may now apply this result to the molecules of the fluid surrounding charged condenser plates; according to (9.4) they will be subjected to forces especially at the edges, where the fringing field has a strong gradient. Imagine the molecules to be frozen in their equilibrium state of polarization, and consider the force acting on a small volume element; this is made up of three contributions, (1) the forces exerted by all other molecules lying outside the element, (2) the force exerted by the charges on the condenser plates, and (3) the force resulting from any pressure gradient—and the resultant of these must be zero in equilibrium. If we were to keep the molecules frozen in their polarized state while we discharged the condenser, we should be left with the mutual forces (1), which we shall show demand no compensating pressure gradient. The pressure gradient need only be introduced to balance the force (2) between the charges on the condenser plates and the polarized molecules.

The demonstration that the mutual forces can be ignored rests on showing that they are highly local in effect, in the sense that the resultant force on any one molecule, due to all others that do not lie within a few molecular diameters, is negligibly small. This seems at first sight paradoxical, for we know that in a region where $\mathbf{E}$ changes, the charges on the condenser plates produce a force, and surely therefore the surface charges resulting from dielectric polarization produce a similar force. But the surface charges are only a convenient fiction to represent the field resulting from a discontinuity in an otherwise continuous dielectric, and when we tackle the problem from a molecular point of view we find a rather different picture emerging.

Consider, for example, two neighbouring equipotentials, A and B,

* This force of attraction towards regions of high field is the mechanism by which the Wilson cloud chamber operates. It is hard for the molecules in a supersaturated vapour, such as is produced by sudden expansion of a saturated vapour, to condense on the tiny droplets, containing perhaps 20–30 molecules, that are the best available nuclei for liquid drops to form on. If, however, these droplets are charged, as a result perhaps of an energetic fundamental particle passing through, they are surrounded by a strong radial field, and individual molecules in the vapour are attracted and condense more readily. Thus condensation proceeds along the track of ionizing particles, and renders the track visible.

whose separation z varies in such a way that $\delta\phi$, i.e., Ez, is constant. An element dS of one of the equipotential surfaces defines an element of volume $z\,dS$ which will contain on the average the centres of $nz\,dS$ molecules, if there are n per unit volume. The average value of the dipole moment in this element, which is directed normal to the surface, will then be $\kappa\varepsilon_0 Enz\,dS$, taking the same value $(\kappa\varepsilon_0 n\,\delta\phi)\,dS$ for all equal elements dS on the equipotential. The molecules whose centres lie between the two equipotentials thus create a uniformly polarized sheet. Now such a closed uniform dipole sheet generates no electric field outside it, as the reader may care to prove for himself, if he cannot wait for a proof in a later chapter. It therefore exerts no force on a molecule outside it, and the molecules whose centres lie between these equipotentials can be removed without changing the force on the particular molecule at X. The same result holds for most of the other molecules, which can be removed sheet by sheet until there are only left those whose centres lie in a very thin sheet C (of thickness a in the vicinity of X) bounded by equipotentials to either side of X. Since the condenser plates are themselves equipotentials, the molecules have been removed right up to the plates. The force on a molecule at X is therefore due to those lying in the sheet C, passing about one molecular diameter to either side of X; in all other sheets, further away, the molecules have that complete freedom to spread evenly which justifies removing them from consideration.

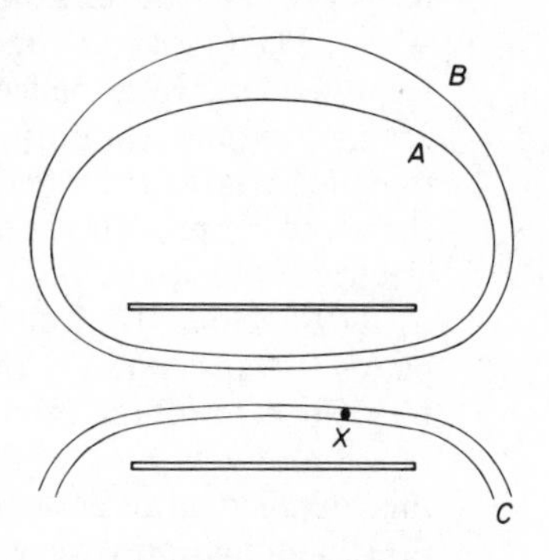

As for those that are left, only molecules lying within a few molecular diameters of X make much contribution. An annulus of the sheet between r and $r + dr$ has volume $2\pi\,ar\,dr$ and dipole moment $2\pi arP\,dr$; it therefore produces a field $aP\,dr/(2\varepsilon_0 r^2)$ at X. We may integrate over all annuli from a radius of about $\frac{1}{2}a$ out to infinity, to arrive at an approximate value P/ε_0 for the resultant field exerted by all other molecules on that situated at X. The actual magnitude is not of great consequence here (a much better way of estimating it, due to Lorentz, gives $P/(3\varepsilon_0)$), but what is important is that the contributions of different annuli fall off as $1/r^2$, so that almost the whole effect is contributed by molecules within, say, 10 molecular diameters of X. We are thus able to assert that the mutual forces between molecules can be ignored except for quite close neighbours, and moreover give rise to a resultant field which varies with position only where $\mathbf{P}$ varies with position.

If then we consider the total force acting on the very large number of molecules in a macroscopic volume element, we can ignore the short-range mutual forces, exerted by those just outside on those just inside, in comparison with the forces exerted by the charges on the condenser plates, which affect all molecules in the element, and not just those near the surface. It is this resultant force that must be balanced by the pressure gradient in the fluid. Suppose the plates to be charged and isolated, giving a field $\mathbf{E}_0(\mathbf{r})$ before the dielectric is admitted, and let the dipole moment per unit volume when the dielectric is present be $\mathbf{P}(\mathbf{r})$. All we need know, to apply (9.4), is that $\mathbf{P}$ is everywhere proportional to $\mathbf{E}_0$, and of this we have satisfied ourselves. Then (9.4), giving the force on a single molecule, may be interpreted as giving the force on unit volume to be $\frac{1}{2}$ grad $(\mathbf{P}.\mathbf{E}_0)$. By considering the equilibrium of a volume element, we see immediately that this is also the pressure gradient needed to balance the force; therefore the pressure at a point where the field was $\mathbf{E}_0$ before the dielectric was added is higher than the pressure in the field-free exterior by $\frac{1}{2}PE_0$. Now the traction per unit area of the field next to the plates is $\frac{1}{2}\varepsilon_0 E_0^2$, and the pressure difference acts in the opposite sense to yield a resultant traction of $\frac{1}{2}\varepsilon_0 E_0^2[1 - P/(\varepsilon_0 E_0)]$. Since $P/(\varepsilon_0 E_0)$ is κ/ε and $\varepsilon - \kappa = 1$, the bracketed quantity is just $1/\varepsilon$, and by this detailed argument we have achieved what energy conservation demonstrates very directly, reduction of the force by a factor ε.

Energy conservation does in fact yield the pressure difference between two regions in the field without a detailed investigation of the origin of the forces. As illustration, consider the hypothetical ex-

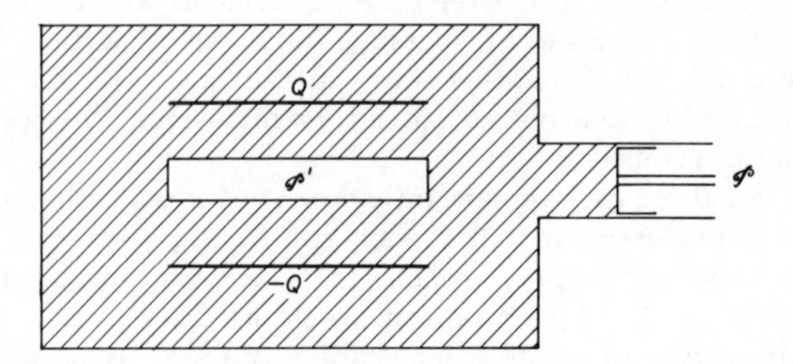

periment of creating a void (e.g., by expanding a plastic bellows) between charged and insulated condenser plates. As the void is increased in volume by ΔV, the work done by the internal pressure maintaining it is $\mathscr{P}'\Delta V$, and at the same time the fluid ΔV that is pushed out of the condenser forces back the piston maintaining the overall pressure in the fluid, and does work $\mathscr{P}\Delta V$. To apply the energy conservation law, we must also know how the stored energy in the condenser changes, and this is related to the change of capacitance when the void is expanded, since the stored energy is $\frac{1}{2}Q^2/C$. If the plate area is A and the separation is a, with a void of thickness x, the field in the void is $Q/(\varepsilon_0 A)$ just as it would be in an empty condenser, and in the dielectric $Q/(\varepsilon_0 \varepsilon A)$, so that the potential difference is $(Q/\varepsilon_0 A)[x + (a - x)/\varepsilon]$ and the stored energy

$$U = \frac{Q^2}{2\varepsilon_0 A}\left(x + \frac{a - x}{\varepsilon}\right).$$

Then

$$\Delta U = \frac{Q^2}{2\varepsilon_0 A}\left(1 - \frac{1}{\varepsilon}\right)\Delta x = \frac{\kappa}{\varepsilon}\cdot\frac{\sigma^2}{2\varepsilon_0}\Delta V.$$

To conserve energy, this must be equated to the work done, to give

$$\mathscr{P}' - \mathscr{P} = \Delta\mathscr{P} = \kappa\sigma^2/(2\varepsilon_0\varepsilon).$$

which is the same result as we obtained by direct consideration of the forces.

PROBLEM

A parallel-plate condenser has capacitance C when empty, and the plates when charged with $\pm Q$ and isolated attract one another with a force F_0. Prove the following results:

(1) a solid dielectric is inserted so as just not to touch the plates: the capacitance is now εC_0 and the force between the plates F_0.

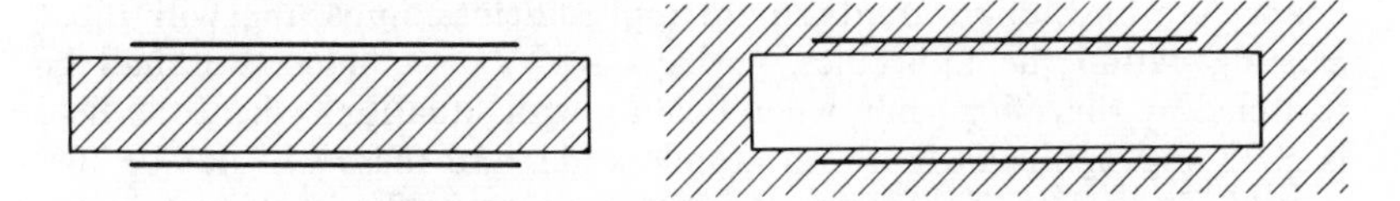

(2) a thin-walled empty box is inserted so as just not to touch the plates, and the rest of the system, including the gap between the box and the plates, is flooded with dielectric; the capacitance is now C_0 and the force between the plates F_0/ε.

Macroscopic equations for dielectrics

Our discussion so far has only treated the effect of flooding a system completely with a uniform dielectric fluid, and we have found that the field **E**, defined as the mean field averaged over the inside and outside of molecules, is simply scaled without change of pattern. We must now consider dielectric bodies only partially filling space, and these distort any field patterns that already exist before the bodies are inserted. As with conductors, our problem is to write down the differential equations and the boundary conditions that must be satisfied by the solutions at any interface between different dielectrics. There is in fact no difficulty in writing the differential equation. Since atoms consist of charges, a potential ϕ can be defined, obeying Poisson's equation $\nabla^2\phi = -\rho/\varepsilon_0$. For any given configuration of molecules the charge distribution ρ is immensely complicated and ϕ varies in a most elaborate way. But if we allow the molecules to move and take a time-average, the mean charge density at any point is zero for neutral molecules. We may therefore for macroscopic purposes interpret ϕ as a suitable average over time, or for that matter over a small region which nevertheless is large enough to contain many molecules, and assert that ϕ obeys Poisson's equation with ρ the smoothed charge density resulting only from any charges additional to those in the neutral molecules. It is ϕ defined in this way that generates, as $-\text{grad}\,\phi$, the mean internal field E.

We know, then, that a potential ϕ can be defined at all points, and that ϕ is continuous within each homogeneous piece of dielectric. It is also continuous if the properties of the dielectric vary slowly in space, but it is not obviously continuous at an interface between two media. We have indeed already discussed an example, the surface of a metal, where there is a sharp change of potential within about one atomic spacing, and we may well expect similar things to happen at a dielectric interface. These potential jumps, and associated strong electric fields are, however, present even when no external fields are applied—they are needed for the equilibrium of the structure. We can afford to ignore these fields if they are unaffected by any applied fields and if we are interested only in the boundary conditions obeyed by the extra applied field. We may assume that in most materials the discontinuity in potential is of the order of 1 V, and that to change it appreciably would require applied fields comparable with those existing within the molecules, perhaps $10^{10}\ \text{V m}^{-1}$; if so, we shall be troubled by this effect only when dealing with situations where the field is changing significantly on an atomic scale, and these will not be susceptible in any case to macroscopic averaging methods. Therefore we assume that for practical purposes the molecules, even at an interface, do not change their polarization drastically when a field is applied, and certainly do not set up a double layer of charge which alone can produce a discontinuity of ϕ. In other words, we take ϕ, describing the applied field, to be continuous at an interface. It follows immediately that the component of **E** parallel to the interface has the same value on both sides. We do not, however, assume that the gradient of ϕ normal to the interface is continuous; this requires separate examination.

For this part of the argument, the smeared-out model of uniform positive and negative charge is convenient. At the surface of a dielectric, surrounded by free space, the surface charge density due to the relative displacement of the two charge distributions is P_n, the normal component of **P**, and therefore by Gauss' theorem,

$$E_n^{(\text{out})} - E_n^{(\text{in})} = P_n/\varepsilon_0.$$

But we have seen that $P_n = \kappa\varepsilon_0 E_n^{(\text{in})} = (\varepsilon - 1)\varepsilon_0 E_n^{(\text{in})}$, from which it follows that $E_n^{(\text{out})} = \varepsilon E_n^{(\text{in})}$. More generally, at the interface between two dielectrics, the normal components of $\varepsilon\mathbf{E}$ and the parallel components of

E match on the two sides. This provides us with the rule governing the refraction of field lines at an interface; since

$$E_1 \sin \theta_1 = E_2 \sin \theta_2$$

and

$$\varepsilon_1 E_1 \cos \theta_1 = \varepsilon_2 E_2 \cos \theta_2,$$

it follows by division that

$$\tan \theta_2 / \tan \theta_1 = \varepsilon_2 / \varepsilon_1.$$

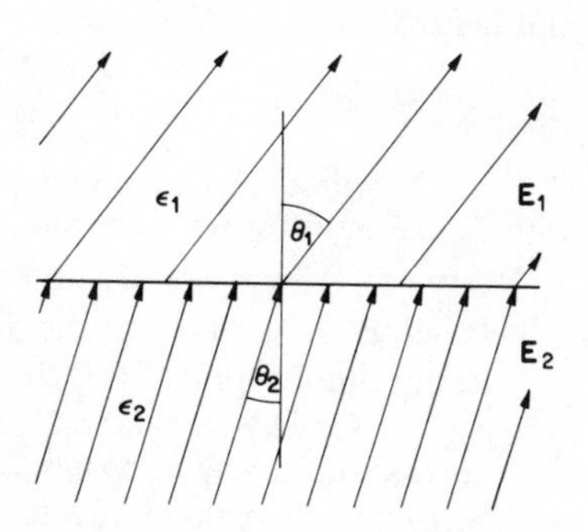

With the diagram as drawn, $\varepsilon_1 = 2{\cdot}5\,\varepsilon_2$. These boundary conditions are enough to enable the differential equations to be solved uniquely.

EXAMPLE

A dielectric sphere in a uniform field. We build on our knowledge that a uniformly polarized dielectric sphere produces a uniform field inside, and try to find a solution analogous to that for the conducting sphere in a uniform field. We try for the field outside the superposition of the original uniform field $\mathbf{E}_0$, parallel to x, and the field of a dipole $\mathbf{p}$ (also parallel to x) situated at the centre of the sphere; for the inside we try a uniform field **E**, parallel to x. Then we can write for the potential

52

$$\phi^{(\text{in})} = -Ex; \quad \phi^{(\text{out})} = -E_0 x + px/(4\pi\varepsilon_0 r^3).$$

If ϕ is to be continuous at all points on the boundary, where $r = a$, we must have

$$E = E_0 - p/(4\pi\varepsilon_0 a^3).$$

As for the normal component of field, it is enough to consider what happens in the diametral plane shown in the diagram; we put $x = r \cos \theta$ and take

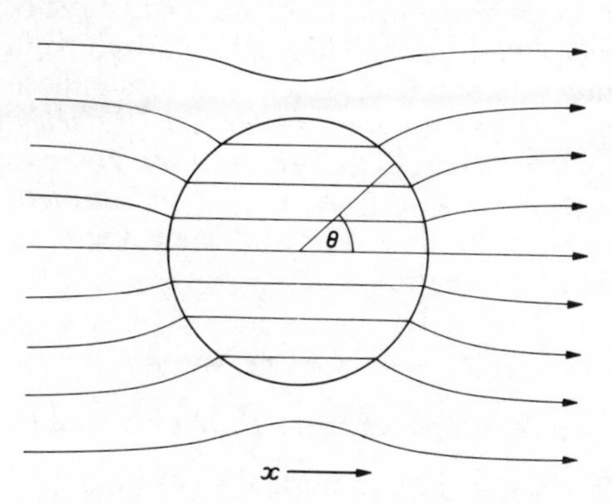

E_n as $-(\partial\phi/\partial r)_\theta$. Thus the normal field at the interface is $E \cos \theta$ inside and $E_0 \cos \theta + p \cos \theta/(2\pi\varepsilon_0 a^3)$ outside. The boundary condition can be satisfied everywhere by writing

$$\varepsilon E = E_0 + p/(2\pi\varepsilon_0 a^3).$$

Eliminating p from these two equations we find

$$E = 3E_0/(\varepsilon + 2),$$

and hence

$$p = 4\pi\varepsilon_0 a^3(E_0 - E) = \frac{\varepsilon - 1}{\varepsilon + 2}.4\pi\varepsilon_0 a^3 E_0.$$

This is, of course, just the dipole moment of a sphere uniformly polarized with dipole moment per unit volume, $\mathbf{P}$, equal to $(\varepsilon - 1)\varepsilon_0\mathbf{E}$.

It will be observed that the dielectric sphere draws the field lines into it, but that in spite of this the field inside is smaller than E_0 by a factor $3/(2 + \varepsilon)$. This is the result of the continuity of εE_n rather than E_n across the interface. When ε is very large, the internal field tends to zero, and the sphere affects the field lines in the same way as a conducting sphere.

PROBLEM

A spherical hole is cut in a dielectric to which a uniform field $\mathbf{E}_0$ is applied; show that the field in the cavity is $3\varepsilon\mathbf{E}_0/(1 + 2\varepsilon)$.

The analysis of a spherical dielectric body or hole, with its neat solution showing a uniform field within, may be generalized to an ellipsoid with arbitrary axes, for which it can also be shown, though it requires more powerful analysis to do so, that placed in a uniform field the internal field is also uniform. It is convenient to express the general statement in terms of the field produced inside an isolated, uniformly polarized ellipsoid. We have already treated the uniformly polarized sphere as a problem, and found the internal field to be uniform, directed opposite to $\mathbf{P}$, and of magnitude $\frac{1}{3}P/\varepsilon_0$; for an ellipsoid polarized parallel to one of its axes the resulting internal field, which is always opposite to $\mathbf{P}$, may be written as DP/ε_0 where D is the *depolarizing coefficient*, depending on the axial ratios of the ellipsoid. In certain cases, D is easily calculated by imagining $\mathbf{P}$ to establish a surface charge distribution of density $\mathbf{P}.\mathbf{n}$ on a surface whose normal is described by the unit vector $\mathbf{n}$, and calculating directly the field at the centre.

52

PROBLEM

Show that the depolarizing coefficient D takes the values

- 1 for a thin slab polarized normal to its face
- $\frac{1}{2}$ for a cylinder polarized transversely
- $\frac{1}{3}$ for a sphere
- 0 for a cylinder polarized longitudinally.

The more prolate the ellipsoid along its direction of polarization, the further from the centre are the main concentrations of surface charge and the smaller is the resulting field and therefore D. Values of D are tabulated* for spheroids, i.e., ellipsoids of revolution with two equal axes. It is worth noting as an occasionally useful curiosity that for any ellipsoid the values of D in the three principal directions add up to unity; thus for a sphere all three values are equal to $\frac{1}{3}$, for a cylinder we have two values of $\frac{1}{2}$ and one of 0, and for a slab one of 1 and two of 0.

We may apply this general result to calculate the field in an ellipsoidal cavity cut in a uniform dielectric carrying an otherwise uniform field $\mathbf{E}_0$. Following the previous discussion, we assume that the whole effect of the dielectric polarization may be simulated by a surface charge distribution such as would be produced by uniform polarization $\mathbf{P}'$ of the ellipsoid, and that the field $\mathbf{E}$, both inside and outside the cavity, is represented correctly by the superposition of the uniform applied field $\mathbf{E}_0$ and the field due to the surface charges, which are now imagined to be suspended in free space. All we need do is choose the value of $\mathbf{P}'$ so that the normal component of $\varepsilon\mathbf{E}$ is continuous at the boundary, and it is enough to do this at one point, say the apex A (the parallel boundary condition is automatically satisfied by our synthesis of the field from two valid solutions of Poisson's equation). At A, the field inside the ellipsoid is $E_0 - DP'/\varepsilon_0$ and the surface charge is P'. By Gauss' theorem, therefore, the field just outside A is $E_0 + (1 - D)P'/\varepsilon_0$, and P' must be chosen so that

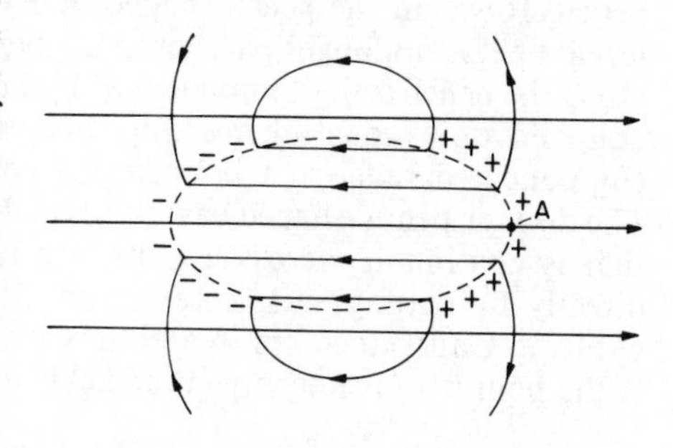

$$E_0 - DP'/\varepsilon_0 = \varepsilon[E_0 + (1 - D)P'/\varepsilon_0];$$

i.e.,

$$P' = \frac{\varepsilon_0(\varepsilon - 1)E_0}{(\varepsilon - 1)D - \varepsilon},$$

and the field inside the cavity is

$$E_0\bigg/\left(1 - \frac{\varepsilon - 1}{\varepsilon} D\right).$$

PROBLEM

Show that a dielectric ellipsoid in a uniform field, $\mathbf{E}_0$, which is not parallel to an axis, is in general polarized in a direction not parallel to $\mathbf{E}_0$, and that the torque exerted by $\mathbf{E}_0$ is such as to turn the longest axis parallel to $\mathbf{E}_0$.

* E. C. Stoner, *Philosophical Magazine*, **36**, 803 (1945).

Note that this result does *not* depend on $\varepsilon > 1$; a material of negative susceptibility would behave similarly.

There are two special cases of this last result which are of particular interest. For a long thin cavity cut parallel to $\mathbf{E}_0$, $D = 0$ and the field in the cavity is $\mathbf{E}_0$; for a thin disc-shaped cavity cut with the normal to the disc along $\mathbf{E}_0$, $D = 1$ and the field in the cavity is $\varepsilon\mathbf{E}_0$. It is not in fact necessary to go through the foregoing analysis to arrive at these special results, which the reader can easily demonstrate for himself by considering the surface charges produced when the cavities are cut. These two cavities are useful devices that enable one to define what one means by the internal field without thinking of molecular structure. Instead of defining $\mathbf{E}$ as the mean field inside and outside of molecules, we define it as the field inside a pencil-shaped cavity cut so that the field inside lies along the pencil. If we cut the pencil-shaped cavity in any other direction, the internal field is no longer parallel to the pencil, but the component of field along the pencil is the component of $\mathbf{E}$ in that direction. It is this field that obeys Poisson's equation in a uniform homogeneous dielectric, and whose component parallel to a boundary is continuous across the boundary. The disc- or penny-shaped cavity, normal to $\mathbf{E}$, measures $\varepsilon\mathbf{E}$ and it is this that is continuous across a boundary normal to $\mathbf{E}$. This can be seen directly by cutting such a cavity on either side of the boundary and applying Gauss' theorem to a pillbox whose faces lie within the cavities. If the boundary is uncharged the fields in the cavities must be equal.

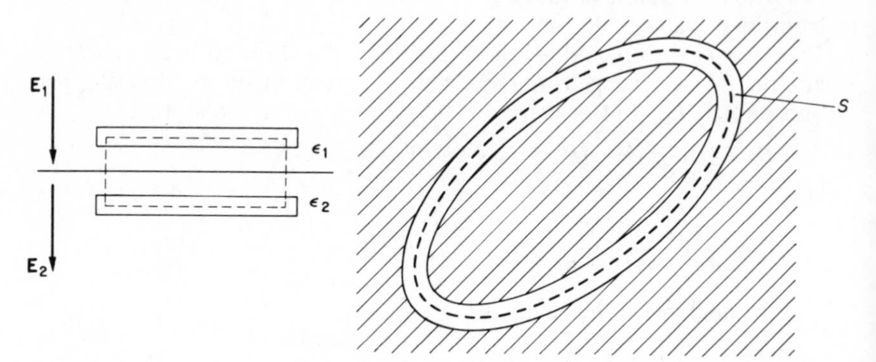

This result goes beyond what we have proved up to now, for it does not depend on the polarization in the medium being proportional to the field, but only on the absence of real surface charge. We may therefore state the boundary conditions more generally by defining fields $\mathbf{E}$ as measured in a pencil-shaped cavity and $\mathbf{D}$ as measured in a penny-shaped cavity (historically $\mathbf{D}$ came to be defined as ε_0 times the field in a penny-shaped cavity), and taking the parallel component of $\mathbf{E}$ and the normal component of $\mathbf{D}$ as continuous across an interface. In a non-linear medium $\mathbf{D}$

and **E** are not proportional to one another, and this leads in general to awkward, non-linear equations to be solved for the fields. But however complicated the relation may be between **E** and **D**, two results stand firm for static fields: **E** may be written as $-\text{grad}\,\phi$, and $\text{div}\,\mathbf{D} = \rho$ where ρ is the real charge density. The reader may satisfy himself on the latter point by application of Gauss' theorem to a thin 'eggshell' cavity cut in the dielectric, to demonstrate that $\int_S \mathbf{D}.\mathrm{d}\mathbf{S}$ equals the charge enclosed within any surface S.

We have now demonstrated how the molecular structure of matter enables us to understand the general behaviour of dielectrics; we have indicated how the theory may be built up in macroscopic terms, and we have solved a few simple examples to show the application of the theory. At this point we refer the reader to specialized textbooks for further developments both of the formal mathematical structure, with engineering and other applications, and for detailed discussion of the molecular theory of polarizability and related topics in molecular and solid state physics.

Cavendish Problems: 204, 206, 209, 210, 213, 214.

READING LIST

Molecular Theory of Dielectrics: C. Kittel, *Introduction to Solid State Physics*, Wiley.

General Theory: B. I. Bleaney and B. Bleaney, *Electricity and Magnetism*. Oxford U.P.

10

Magnetic fields due to currents

A1 A closed loop of wire carrying a current *i* produces the same field configuration as would a magnetic shell bounded by the loop and of strength m equal to *i*.

A2 If a wire carrying current *i* is placed in a magnetic field of strength *B*, the force on an element d*s* of the wire is *i* d*s* ∧ *B*.

(Note that although we shall refer to these statements as Ampère's first and second laws, they do not carry these titles by universal custom as do Newton's or Kepler's laws.)

We turn now from electric to magnetic fields, to study a phenomenon which in its normal manifestations is much stronger and more impressive. A well-rubbed stick of plastic will pick up pieces of paper—a child's horseshoe magnet will pick up nails. On the laboratory scale, a typical electromagnet may have an attractive force between its iron poles amounting to 1 ton weight, orders of magnitude larger than any attainable force between two condenser plates. Thus, a quite modest electromagnet can produce a magnetic field of 2 tesla (abbreviated to T), which exerts a tension on unit area of the pole faces equivalent to about 16 atmospheres;

to achieve the same with an electric field would require 6×10^8 V m^{-1}, several times greater than is needed to cause all but the very best dielectrics to break down, and enough to draw a significant current of electrons out of a metal into vacuum. The limitation to producing high magnetic fields in the laboratory is not the electronic breakdown that limits electric fields, but such macroscopic problems as the difficulty of extracting the heat generated by the currents in the conductors of an electromagnet, or the mechanical failure of a coil under the stresses produced by the field; at 20 T the field exerts tensions and pressures equivalent to 600 atmospheres, and at 75 T (about the largest field obtained momentarily in a coil by discharging a large condenser through it) the tension, rising quadratically with field, has reached 8500 atmospheres. If the coil does not fly to pieces during the discharge it is almost certainly so distorted as to be unfit for a second discharge. There is a good reason for magnetic effects being stronger than electric, the absence of any fundamental magnetic particle analogous to the fundamental electric particles, the electron and the proton. The 'magnetic monopole' has been conjectured to be a possible elementary particle, and it may be that one occasionally wanders into our part of the universe, but although sensitive techniques for finding one have been developed they have not so far produced any unequivocal example of a monopole. At all events they are totally absent from ordinary matter, and the magnetic fields we observe are produced not by stationary particles, but by the movement of electrons and other electrically charged particles. As we shall see when we come to discuss the forces exerted by magnetic fields on moving charges, these may be large and still not cause the steady acceleration of an electron to higher energies that is responsible for the first stages of dielectric breakdown. There is no limit to the strength of magnetic field that a material can support, before we reach a field so strong (and quite unattainable at present) that it directly interferes with the binding forces within molecules; but, as we have seen, electrical breakdown occurs at field strengths very small compared with those associated with the binding forces in molecules. The search for the magnetic monopole has indeed made use of the fact that quite a small electromagnet would exert such a force on one as would tear it out of any solid in which it was nestling; obviously if these monopoles were a constituent of matter it would not be possible to generate such fields without the materials in the vicinity falling apart.*

The nearest approach we have to static sources of magnetic fields, analogous to electrostatic field sources, are permanent magnets which are

* The ease with which strong magnetic fields can be attained makes possible the detection of very weak electric fields and currents; for simple instruments depend on the use of force, and the force between a current and a magnetic field, as described by Ampère's second law, provides a mechanism for detecting weak currents. The stronger the magnetic field available, the weaker the current that can be detected. Thus the existence of charges and absence of poles accounts in general terms for both the relative strength of magnetic fields and the relative detectability of electric fields.

known to depend on the intrinsic property that electrons possess of behaving like magnetic dipoles. The fields that iron magnets produce are in no way different in their effects from those produced by currents in wires. Magnets exert forces on other magnets and on current-carrying wires; iron filings can be used to plot out the field lines from either source. The identity of the fields is in practice taken for granted, and studies of the influence of magnetic fields on materials (their conductivity, magnetization, etc.) employ iron magnets or iron-free solenoids indifferently. We may therefore use evidence from either to build up a plausible statement of the laws obeyed by magnetic fields. In what follows, our aim is to show that Ampère's laws, quoted at the head of the chapter, are not so far removed from experiment as might seem at first. We shall not attempt to follow Ampère's reasoning (even if we knew what it was) but indicate lines of argument that are likely to prove acceptable nowadays. As always, the real test of the statements lies in the verification of their consequences rather than in their logical deduction from experiment.

48 (1) In 1750 Michell concluded from experiments performed by others as well as himself that magnets behave as if the poles at or near their ends were the sources of an inverse square law field. This is consistent with the magnets being built up of elementary magnetic dipoles producing field patterns of exactly the same form as those of an electric dipole. The field, **B**, which is a measure of the force exerted on a given magnet, may therefore be taken to be conservative and described by a magnetic potential, ϕ_B, such that $\mathbf{B} = -\text{grad}\,\phi_B$ and $\nabla^2\phi_B = 0$ —Laplace's rather than Poisson's equation since we do not believe free poles to exist.

(2) By means of a tangent galvanometer we may use a suspended magnet to study the additivity of the field due to a current in the coil and the Earth's field or that produced by a magnet or other coil in the vicinity. The magnet simply indicates the resultant field direction, but it is enough to show that if we take the field due to the coil as directed normal to the coil and of strength proportional to the current (as measured, for example, by the rate of deposition of metal in a voltameter or electrolytic cell), this field and the Earth's are always to be added vectorially.

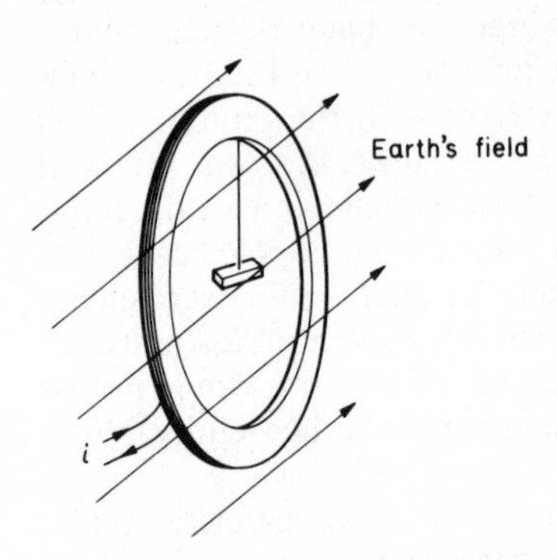

(3) A more refined instrument is the moving coil galvanometer, in which the current to be measured is passed through a coil suspended from a torsion wire in the field of a horseshoe magnet. We can now verify that the angle of twist is proportional to the current; the force exerted by a magnet on a current is, like the force of a current on a magnet, proportional to the current. This is what N3 would lead us to expect, and we may take N3 as one of the important clues to the laws of magnetic fields,

since we know it to be obeyed on the macroscopic scale by objects made of iron within which magnetic forces must be present.

(4) The moving coil galvanometer may be used ballistically to verify what we have implicitly assumed, that the current measured in a voltameter is the same thing as the current responsible for transferring charge. A capacitor C charged by the cell S can be discharged almost instantaneously through the galvanometer, which swings to a maximum deflection proportional to the charge passed through the coil, if the force on any element of the moving coil is proportional to the current. By performing the experiment with different numbers of identical cells in series at S it is easily verified that the throw is proportional to the charge passing, and that therefore the force on a current in a magnetic field is proportional to the current, defined now as the rate of transfer of charge in the wire.

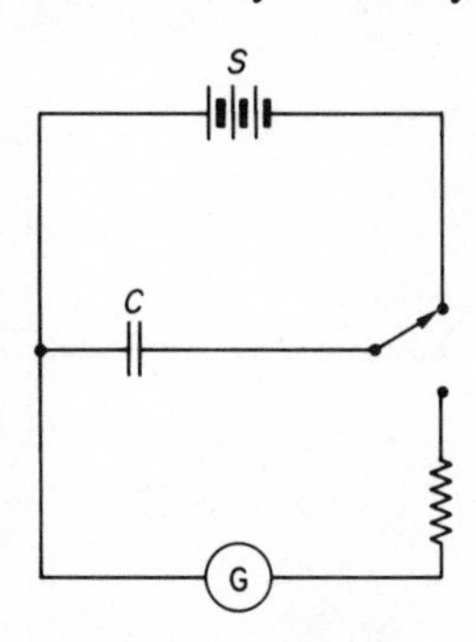

(5) By hanging a loop of wire under tension and observing the repulsion of the wires when a current is passed, it is readily verified that the force between the wires is proportional to the square of the current. By means of separate leads to provide different currents, i_1 and i_2, one may go further and show that the force $\propto i_1 i_2$ and that N3 is obeyed in this unsymmetrical situation, both wires being equally deflected. This serves to confirm that the force may be understood in terms of one current, i_1 say, producing a field B proportional to i_1, which acts on the other current with a force proportional to Bi_2, or $i_1 i_2$. This result provides an experimental procedure for comparing the strength of different magnetic fields, by hanging a current-carrying wire in the fields and determining the force on it. Needless to say, this only outlines the principle of the method, but it is enough for present purposes to recognize that a numerical measure may be ascribed to B.

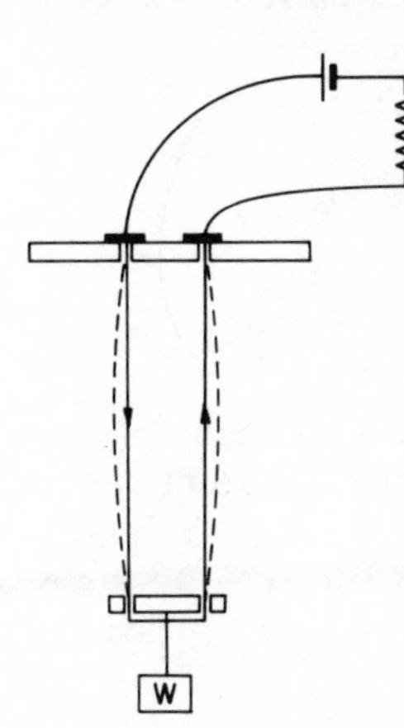

(6) The foregoing experiments serve to establish the general relationships between currents, magnets and fields, except for the actual configuration of the field near a current, which can be easily demonstrated by compass needle or iron filings. Thus with a vertical wire carrying 10–20 A and a small compass needle the circulation of field lines round the current is clearly visible. An elegant experiment originally performed by Biot and Savart shows that the field strength round a straight wire varies inversely with distance. They hung a light annular tray by threads from a point on the wire, and rested a bar magnet on it. When the current was switched on

there was no tendency for the tray to turn. Since the two poles of the magnet are at different distances from the wire the force of the field on them can only result in zero torque if the field varies inversely with distance.

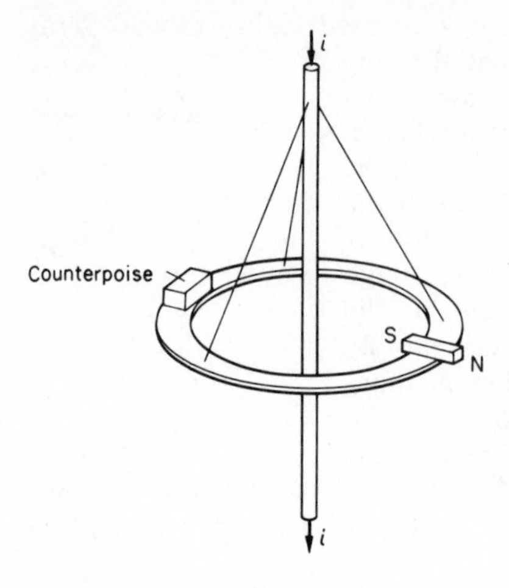

It is at this point that we may take stock of our evidence and see how Ampère's first law can be made to appear a reasonable synthesis. We have to resolve what looks distinctly paradoxical: the field has the characteristics of a conservative field when produced by a magnet, but has obvious circulatory properties when produced by a current. If we form $\oint \mathbf{B}.\mathbf{dr}$ round any circuit in the vicinity of a bar magnet the result is zero, but not if we form it round a circuit that encloses a current. It is this last statement that is significant—$\oint \mathbf{B}.\mathbf{dr}$ does not vanish if the circuit encloses a current; but even when the field is due to a current, the field is conservative so long as we do not enclose the current. This can be verified for an arbitrary circuit in the vicinity of a straight wire by noting that the two elements cut off by the same two rays from the origin (through which the current flows along a wire normal to the page) have tangential length proportional to their radial distance ρ, and if the tangential field $\propto 1/\rho$, the two elements produce cancelling contributions to $\oint \mathbf{B}.\mathbf{dr}$. We may therefore reconcile the various observations by postulating that $\mathbf{B}$ is a conservative field in the sense that $\oint \mathbf{B}.\mathbf{dr} = 0$ for any circuit not enclosing a current, but that when the circuit encloses a current i,

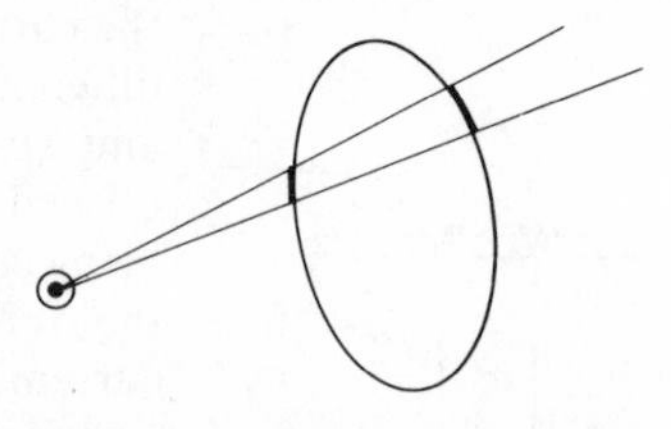

$$\oint \mathbf{B}.\mathbf{dr} = \mu_0 i, \tag{10.1}$$

where μ_0 is a constant depending on our choice of units for i and $\mathbf{B}$. It should be particularly noted that the line integral around a current has been postulated to take a value determined solely by the current and not by the configuration of the wire or the shape of the circuit round which the integral was taken. The reason for this can be seen by considering any two circuits (not necessarily in the same plane) around a given piece of current-carrying wire. By taking two paths AB and CD infinitesimally close together we can construct a new circuit that starts with AB, goes around the inner loop as far as C, along CD and around the outer loop back to A, and this circuit does not enclose the wire, since it can be shrunk steadily to a point without passing any part of it through the wire. The line integral round this must vanish, and since AB and CD contribute cancelling terms it follows that the original two circuits are equal and opposite in their contributions. Therefore all circuits round a given current in the same sense have the same value of $\oint \mathbf{B}.d\mathbf{r}$. The wire itself may be as long as we choose and be arranged in totally different patterns at various places along its length, and all these must, by the above argument, yield the same value of $\oint \mathbf{B}.d\mathbf{r}$; hence the configuration of the wire must be irrelevant, and the integral determined solely by i.

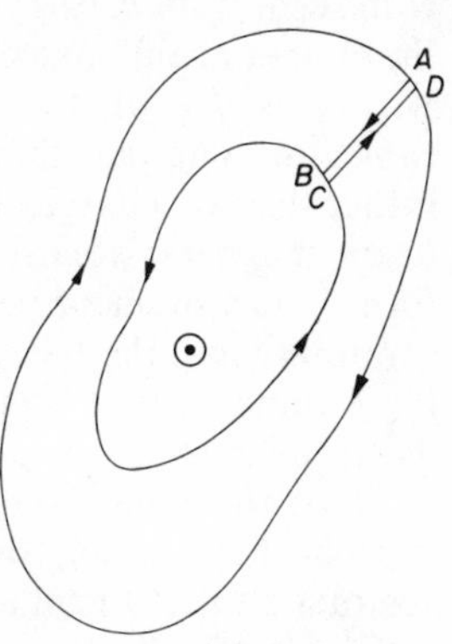

The idea of a potential function having the properties that we have to ascribe to ϕ_B is not outrageous—it is exhibited by every spiral staircase. If on the ground below we plot the height h as a function of position (r, θ) relative to the central newel post, we are forced to plot it as a many-valued function; leaving aside the discontinuities due to the steps, the height is represented by the ramp function $h = c\theta$, so that at any point h can have integral multiples of $2\pi c$ added. Any circuitous walk on the staircase that does not enclose the newel leaves the walker at the same height as he started, but any which encloses the newel n times and returns to a point vertically above or below the starting point involves a height change or $2\pi nc$. So also with the potential ϕ_B around a straight wire carrying a current i; if we write $\phi_B = -\mu_0 i\theta/(2\pi)$, we have expressed a tangential field

$$B = -|\text{grad}\,\phi_B| = \mu_0 i/(2\pi r), \tag{10.2}$$

varying as $1/r$ in accordance with experiment. And the potential at every point is well-defined apart from the possible addition of a multiple of $\mu_0 i$.

We may convert this multivalued potential into a single-valued potential by arranging that whenever we encircle the current we always add $\mu_0 i$ to ϕ_B, or (according to the sense of the circuit) subtract it, at some predetermined point. This is most easily understood when the current is flowing in a closed loop rather than a straight wire. If we draw any surface that is bounded by the loop, it is clear that any closed path encircling the current once must pass once through the surface, while any that does not encircle it either does not pass through the surface or goes through in both directions an equal number of times. The net number of passages through the surface, counting forwards and backwards as having opposite signs, measures the number of times the path encircles the current, and the sense of the circuit. If then we arbitrarily impose a discontinuity of $\mu_0 i$ on the value of ϕ_B on opposite sides of the surface, but otherwise insist that $\nabla^2\phi_B = 0$, we have constructed a single-valued potential that defines $\mathbf{B}$ correctly at every point except on the surface of discontinuity.

A dipole sheet (or *magnetic shell*) is just the thing for this purpose as we can see by analysing the electrical analogue, a parallel-plate capacitor, with separation a, carrying $\pm\sigma$ per unit area. The field within is σ/ε_0, and the potential difference between the plates is $a\sigma/\varepsilon_0$, which can also be written as p/ε_0, where $p = a\sigma$ and is the dipole moment of unit area of the capacitor. The fact that the potential drop depends on dipole moment rather than on charge allows us without hesitation to carry the idea over to the magnetic situation, and introduce a magnetic dipole sheet of uniform dipole moment per unit area, $\mathfrak{m}$, as a device for establishing a discontinuity in potential. The ideal sheet has zero thickness but $\mathfrak{m}$ is finite. If we think of it as consisting of positive and negative poles on parallel sheets, we must let the pole strength tend to infinity as the separation goes to zero; the finite potential drop is ultimately achieved by an infinite internal field acting over an infinitesimal distance—a mathematical abstraction that fortunately need not be achieved in practice.

We can now see the implication of Ampère's first law (A1) by comparing the fields due to a current loop (left) and to a magnetic sheet (right)

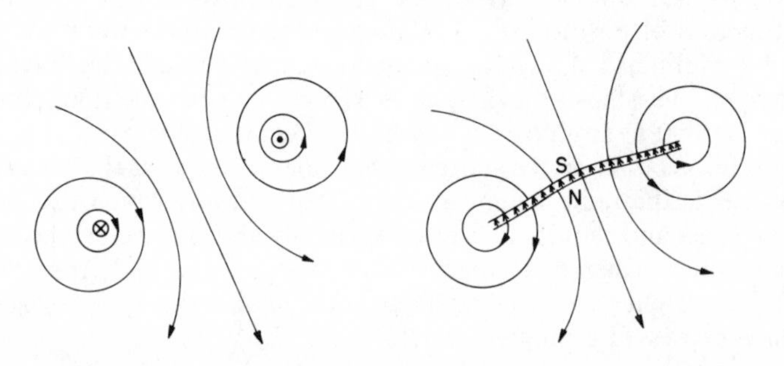

having a boundary defined by the shape of the loop. The diagrams show such a loop in section with its magnetic field, and the equivalent shell with its field. If we believe that the field of the current is described by a multi-valued potential ϕ_B obeying Laplace's equation, the arguments we have been through show that the field lines in the two diagrams are identical except within the interior of the shell, and that this will be true however we dispose the shell, provided we keep the edges fixed. This independence of the shape of the shell, though made plausible by the preceding argument, is still a little difficult to accept, but we shall give a formal demonstration presently. What we have arrived at, then, is some rationale behind Ampère's prescription for finding the field of a current loop; replace it by a magnetic shell which does not pass through the point where the field is to be calculated, and it will give the right answer. It is amusing to note that the whole field, described in these terms, is merely what is commonly enough disregarded as the 'fringing field' of a capacitor; the important part of the capacitor field, between the plates, is the very part that here we must discount as irrelevant to the description of the magnetic field of a current.

The strength of the magnetic shell is defined in terms of the discontinuity in potential, $\mu_0 i$, which it is responsible for. This is enough to define quantitatively the field produced by any element of the shell, as we can see from the electrical analogue. An electrical dipole layer of moment p per unit area produces a potential discontinuity p/ε_0, and an element of area dS, with dipole moment p dS, makes a contribution to the potential outside the shell equal to $p\ \mathrm{d}\mathbf{S}.\mathbf{r}/(4\pi\varepsilon_0 r^3)$, which is $\mathrm{d}\mathbf{S}.\mathbf{r}/(4\pi r^3)$ times the discontinuity. Carrying this result over to the magnetic case, we see that the potential at any point X due to an element dS may be written

$$\mathrm{d}\phi_B = \mu_0 i\,\mathrm{d}\mathbf{S}.\mathbf{r}/(4\pi r^3), \tag{10.3}$$

where $\mathbf{r}$ is the vector joining X to dS. To determine the potential due to the complete shell, all that is needed is to integrate over the surface S of the shell, as is easily done since the element dS subtends at X a solid angle $\mathrm{d}\Omega$ equal to $\mathbf{r}.\mathrm{d}\mathbf{S}/r^3$. Therefore

$$\mathrm{d}\phi_B = \mu_0 i\,\mathrm{d}\Omega/(4\pi),$$

and

$$\phi_B = \mu_0 i\Omega/(4\pi), \tag{10.4}$$

where Ω is the solid angle subtended by the whole shell, i.e., by the current loop, at X. This result demonstrates that it is not the shape of the shell, but only the shape of its boundary, that matters, and it provides a very neat formula from which the field can be calculated.

The form of (10.4) also confirms that ϕ_B is a multivalued potential which changes by $\mu_0 i$ in the course of one circuit round the current, since in such a circuit Ω changes by 4π. The geometry involved is easier to appreciate in a two-dimensional example. Suppose we have two fixed points A and B in a plane, and define the angle they subtend at a point O by the angle turned through in going from OA to OB clockwise. As the point moves on a circuit of B, successive stations 1, 2, 3 etc., have the angle steadily increasing until at 4 it becomes greater than 2π if we are not to impose an arbitrary discontinuity in our measure of the angle. By the time the point returns to O the angle has increased by 2π. Similarly if we take X in the diagram above through the current loop and round the outside back to its original position, the cone formed by joining X to all points of the loop expands steadily and

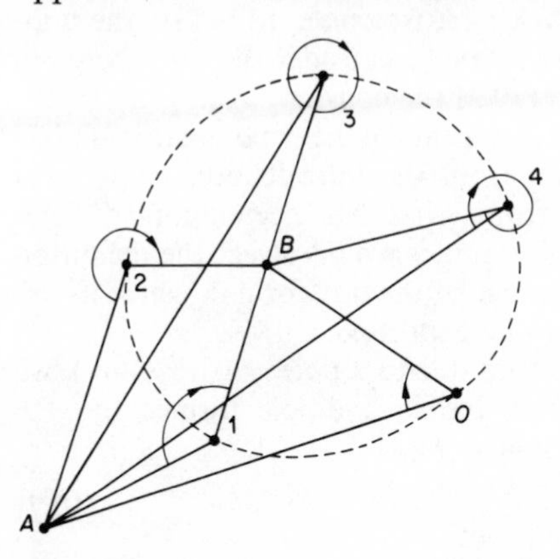

performs a feat, impossible with umbrellas, of being turned inside-out so thoroughly that it returns to its first shape having described a complete solid angle of 4π.

Before proceeding further, it is convenient to define some of the field variables quantitatively. Up to this point, we have passed over the definition of the ampere, the unit of current, which carries with it the definition of the coulomb, the unit of charge. The reason for not defining them earlier is that according to present international convention it is the electromagnetic properties of a current, rather than its electrolytic properties or the electrostatic properties of charge, that are agreed to be the most useful for the purpose of definition. And the way in which the definition is achieved involves assuming the truth of Ampère's laws, or equivalent statements, and assigning the value $4\pi \times 10^{-7}$ to μ_0. The 217 reason for this seemingly eccentric choice will appear later. It is not, however, sufficient just to define μ_0, since μ_0, i and $\mathbf{B}$ are all related by the single expression $\oint \mathbf{B}.\mathrm{d}\mathbf{r} = \mu_0 i$, and so far neither $\mathbf{B}$ nor i has units assigned to it. We therefore need to define one more unit, from which the other will be derived by application of this expression. A convenient way of achieving this is to define what we mean by the strength of a magnetic dipole; this is something we have avoided in the preceding discussion of A1; we were concerned only with the field produced by the dipole sheet and not with assigning a measure to its dipole moment per unit area. Simpler still, and an equivalent procedure, is to define the unit pole by writing the force between two poles P_1 and P_2 in the form

$$\mathbf{F} = \mu_0 P_1 P_2 \mathbf{r}/(4\pi r^3), \tag{10.5}$$

and the force acting on a pole P in a field $\mathbf{B}$,

$$\mathbf{F} = P\mathbf{B}. \tag{10.6}$$

Once μ_0 is fixed, the unit of $\mathbf{B}$ is uniquely determined by these definitions, even though it may not be easy, lacking a free pole, to realize the unit. But this will sort itself out when we come to consider the force between currents. It may be noticed that in (10.5) the constant μ_0 appears in the numerator, while in Coulomb's law ε_0 appears in the denominator. This is largely a historical accident arising from the introduction of μ_0 by a different sequence of arguments from what has been adopted here. Obviously there is no basic physical significance involved, the definition (10.5) having been chosen solely because of the tidy form in which its experimental consequences turn out to be expressed.

With (10.5) and (10.6) giving the field due to a pole as $\mu_0 P\mathbf{r}/(4\pi r^3)$, we can immediately write the potential due to a dipole formed by $\pm P$ separated by $\mathbf{a}$, and having moment $\mathrm{m} = P\mathbf{a}$,

$$\phi_B = \mu_0 \mathrm{m}.\mathbf{r}/(4\pi r^3). \tag{10.7}$$

Comparing this with (10.3), we have that the moment per unit area of the shell equivalent to a current i is just i, as stated in the version of A1 quoted at the head of the chapter.

Returning now to the physical content of A1, we note that it is an integral law; it tells how a complete physical system, a current loop, will behave. We may ask whether it is possible to break it down into differential form, to allow each element of the circuit to make its own contribution to the field, all contributions to be added vectorially at the end. The following argument shows how this can be achieved. Suppose we attempt

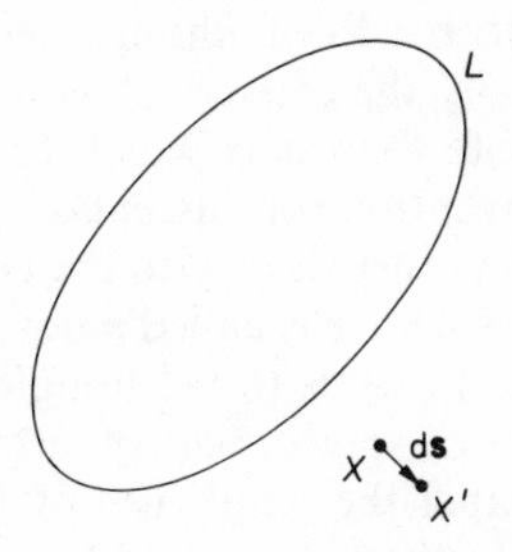

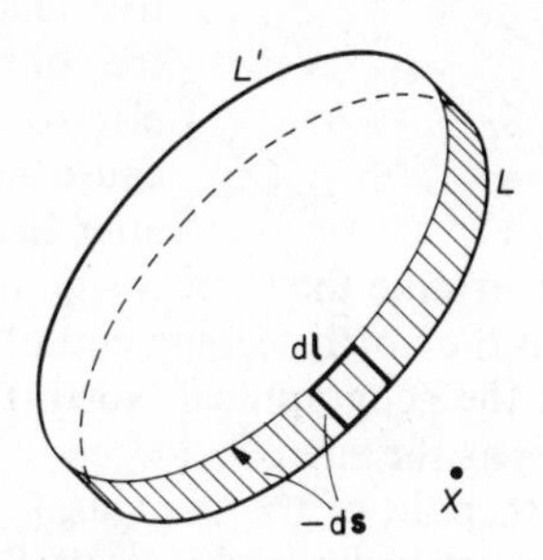

to calculate the component of $\mathbf{B}$ in the direction XX' at a point X in the vicinity of a closed loop L carrying current i. The change in potential from X to X' is $-\mathbf{B}.\mathrm{d}\mathbf{s}$, and this is related, through (10.4), to the difference in solid angle subtended by L at X and X'. We can therefore calculate the component of $\mathbf{B}$ along XX' simply by determining this change in solid angle, and this is just as well achieved by remaining at the point X while the circuit is shifted parallel to itself, to L', through a distance $-\mathrm{d}\mathbf{s}$. Now the solid angle subtended by L' is the same as that subtended by L together with the side of the pillbox formed by L and L'; therefore the change in solid angle is just the solid angle subtended by the side shown shaded.* Here we see how the field due to the whole circuit may be dissected into the contribution of each element, for the solid angle subtended by the side may be considered as made up of elements of which a typical example is shown in the diagram. The element $\mathrm{d}\mathbf{l}$ of L in being moved through $-\mathrm{d}\mathbf{s}$ sweeps out an area $\mathrm{d}\mathbf{l} \wedge \mathrm{d}\mathbf{s}$ which subtends at X the solid angle $\mathrm{d}\Omega = (\mathrm{d}\mathbf{l} \wedge \mathrm{d}\mathbf{s}).\mathbf{r}/r^3$. If then we imagine $\mathrm{d}\mathbf{l}$ being responsible for a contribution $\mathrm{d}\mathbf{B}$ to the field at X, we may write its contribution to the potential difference between X and X' as $\mathrm{d}\mathbf{B}.\mathrm{d}\mathbf{s}$ and equate it to $\mu_0 i\, \mathrm{d}\Omega/(4\pi)$. Therefore

$$\begin{aligned}\mathrm{d}\mathbf{B}.\mathrm{d}\mathbf{s} &= \mu_0 i(\mathrm{d}\mathbf{l} \wedge \mathrm{d}\mathbf{s}).\mathbf{r}/(4\pi r^3)\\ &= \mu_0 i(\mathrm{d}\mathbf{l} \wedge \mathbf{r}).\mathrm{d}\mathbf{s}/(4\pi r^3), \text{ by VI.}\end{aligned}$$

Since we could have chosen $\mathrm{d}\mathbf{s}$ in any direction we must interpret this result as implying

$$\mathrm{d}\mathbf{B} = \mu_0 i(\mathrm{d}\mathbf{l} \wedge \mathbf{r})/(4\pi r^3). \tag{10.8}$$

* At this point we abandon any attempt to keep track of signs; we shall put the sign of the answer right by inspection afterwards. If the reader regards this as slovenly, he is welcome to carry through the calculation properly himself, but he may find he is devoting too much attention to signs and too little to the real argument and his understanding may thereby suffer.

Here we have anticipated the following analysis of the signs by stating the correct answer. The result (10.8) is the law of Biot and Savart, an alternative to A1 as a basic formulation of the magnetic field due to a current. The vector **dl** is to be taken in the sense in which the current flows, which by convention is the direction that positive charges would have to move to produce the observed effects. Even though negative electrons are usually the carriers, it would now cause more trouble than it is worth to try to alter the sign convention for currents.

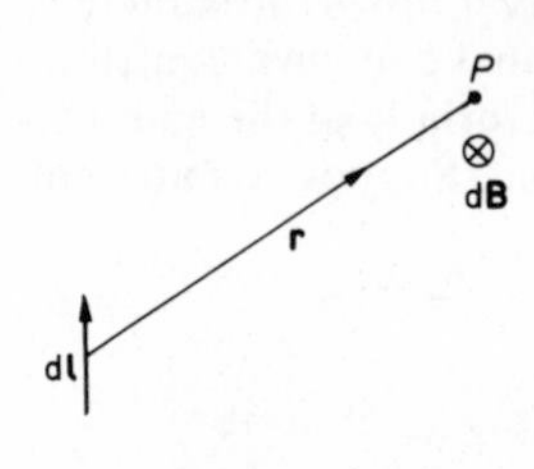

To determine the correct sign in (10.8) we must start with the convention that the north-seeking end of a compass needle is called a north pole (so that the geographical North Pole of the Earth is to be thought of, if you like, as the site of a large magnetic pole of south polarity, attracting the north pole of the compass), and we label the field lines of **B** with arrows pointing in the direction (South to North) towards which they push a north (or positive) pole. Thus the north-seeking compass needle points along the line of **B**. If we pass a current through a vertical wire, determining its sense (if we do not already know it from the battery connections) by putting a copper voltameter in the circuit to see which electrode has metal deposited on it,* and place a compass needle in the vicinity, we find that the field circulates round the current in a right-handed corkscrew sense—a corkscrew turned to move along the direction of i turns in the same sense as **B**. Since the field at P is made up of contributions from such elements as **dl**, it is clear that $\mathbf{dl} \wedge \mathbf{r}$ correctly describes the sign of **B**, as in (10.8).

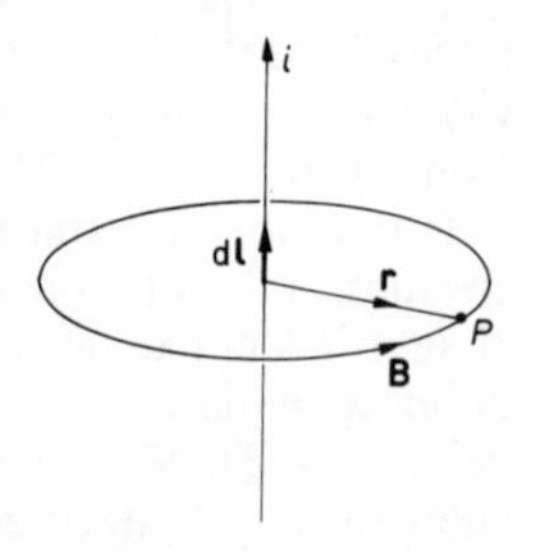

The formulation by Biot and Savart is not a unique dissection of the complete circuit into the elements; anything could be added to (10.8) that cancelled automatically when the integration round the complete circuit was performed. There is no point in enquiring into the most general form of (10.8), still less in asking which is the right form, since the dissection does not correspond to any physically possible operation. An element of current cannot exist on its own, and therefore one should look on (10.8) as a convenient aid to calculating the effects of complete circuits and not as something with significance in its own right. This point of view may save us from becoming worried when we find (10.8) to be on occasion in conflict with N3.

* The current flows in the solution towards this electrode.

EXAMPLES

Calculation of magnetic field due to a current-carrying circuit

(1) *Single circular loop of current; to calculate the field on the axis.* First we use the Biot and Savart law. Any element $\mathbf{dl}$ produces $\mathbf{dB}$ at P of magnitude

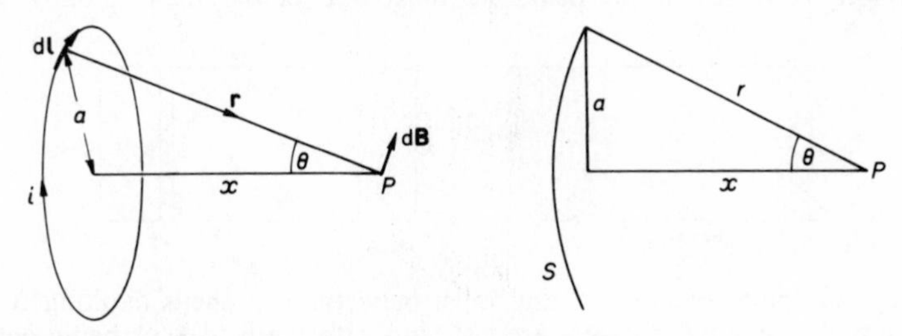

$\mu_0 i\,\mathrm{d}l/(4\pi r^2)$. When all the contributions are summed, only the axial component, $\mathbf{dB}\sin\theta$, i.e., $\mu_0 a i\,\mathrm{d}l/(4\pi r^3)$, is left. Hence

$$B(x) = \tfrac{1}{2}\mu_0 a^2 i/r^3 = \tfrac{1}{2}\mu_0 a^2 i/(a^2 + x^2)^{3/2}.$$

To obtain the same result by use of a magnetic shell, we note that the area of the spherical cap S is $2\pi r(r - x)$, so that the circuit subtends a solid angle $2\pi(1 - x/r)$, i.e., $2\pi[1 - x/(a^2 + x^2)^{1/2}]$, at P. Hence

$$\phi_B(x) = \tfrac{1}{2}\mu_0 i[1 - x/(a^2 + x^2)^{1/2}],$$

and

$$B(x) = -\partial\phi_B/\partial x = \tfrac{1}{2}\mu_0 a^2 i/(a^2 + x^2)^{3/2}.$$

(2) *Finite solenoid.* Any one layer of the solenoid consists of a helix of wire, usually wound closely so that the pitch ($1/n$ for n turns per unit length) is much smaller than the radius of the solenoid. Unless we are interested in the

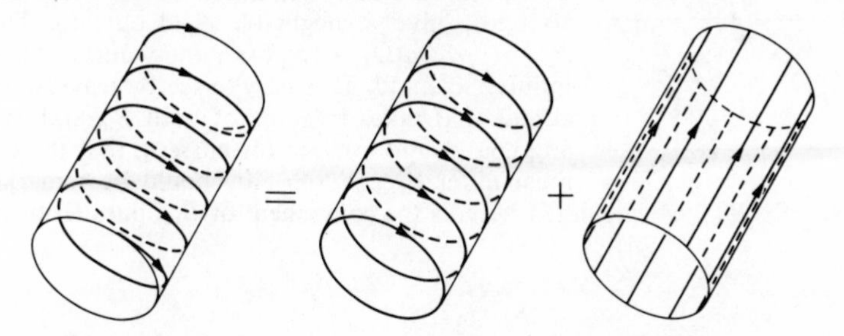

field close to the windings, we may disregard them except in so far as they provide a circulating current ni per unit length of solenoid. At the same time there is a longitudinal current i distributed evenly round the solenoid, the two current components combining to give a resultant uniform current sheet whose flow direction is parallel to the windings. The longitudinal current can be neglected if we are interested in the field inside the solenoid, for it makes no contribution here.

To calculate the field at any point on the axis, we may integrate the result obtained for the single turn. Alternatively we may apply A1 directly. Any length $\mathrm{d}x$ of the solenoid, carrying $ni\,\mathrm{d}x$, may be replaced by a shell of strength

ni dx. If we give it thickness dx, it will carry pole strength $\pm ni$ per unit area of its faces. The shells stack together with their faces virtually touching, and therefore neutralizing each other so far as producing an external field is concerned. However, the end pole faces remain uncompensated, and so do those on either side of the point X where the field is to be measured; for in accordance with Ampère's prescription we must not let the shell actually pass

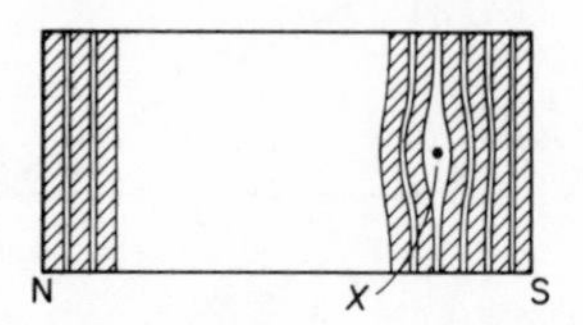

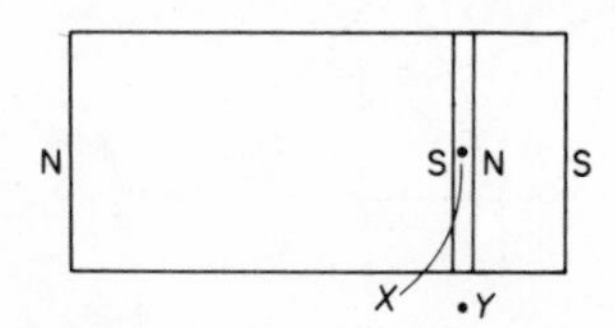

through the point of interest, and must bend the two shells flanking X infinitesimally away from it. We are left then with the field at X being determined by four sheets of polarity $\pm ni$. The reader may easily satisfy himself that the field on the axis of a circular sheet of magnetic poles is just the pole strength per unit area times $\mu_0\Omega/(4\pi)$, if the sheet subtends a solid angle Ω at the point considered. Adding together the contributions of the four sheets that matter here, we see that on the axis

$$B = \mu_0 ni\left(1 - \frac{\Omega_1 + \Omega_2}{4\pi}\right),$$

where Ω_1 and Ω_2 are the (positive) solid angles subtended by the ends. If the solenoid is long, Ω_1 and Ω_2 are small and $B \approx \mu_0 ni$, the standard result for an infinite solenoid.

The field at Y, just outside the solenoid, is produced by the pole sheets at the ends of the solenoid, with no contribution from the two flanking X which, being infinitesimally separated but of finite pole strength, give a negligible field outside. Thus $B_{\text{out}} = -\mu_0 ni(\Omega_1 + \Omega_2)/(4\pi)$, and vanishes for an infinite solenoid. It is easy to see by traversing a circuit that takes in some of both B_{in} and B_{out}, and goes round some of the current, that the two must differ by $\mu_0 ni/(4\pi)$. It should be remarked that strictly B_{out}, as calculated here, is the component of B_{out} parallel to the axis.

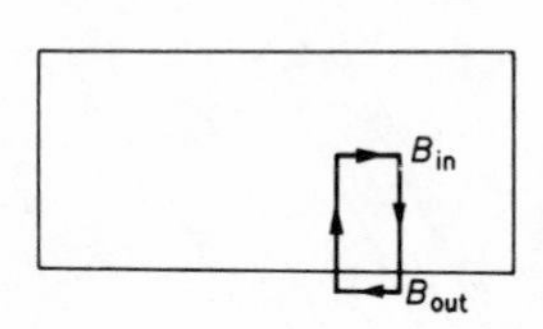

Comments. (1) The similarity of the expressions for the field due to a sheet of poles and the potential due to a sheet of dipoles is not accidental. If we know the potential ϕ_B at X due to a sheet of poles, we may calculate the field by taking the difference between ϕ_B at X and at a neighbouring point. Alternatively we may keep X fixed and move the sheet, taking the difference by reversing the polarity and adding the two contributions; but this simply determines the potential due to a dipole sheet.

(2) The examples show how A1 may be used directly in the calculation of fields, but there are very few cases where it is really helpful.

When the field of an actual coil has to be determined, the formulation of Biot and Savart is a more generally convenient starting point, since it allows computation to proceed directly. The accurate calculation of the field due to a coil, given its precise dimensions, is a real necessity when electrical standards are being established, and it is a very tedious business indeed.

It might be imagined from this that it would be more sensible to start with Biot and Savart's law rather than Ampère's, and indeed many authors do take this point of view. We prefer, however, to take as a basic principle something that does at least refer to real systems, i.e., complete current loops, especially as it is so simple to demonstrate Biot and Savart's law from A1. It is not nearly so simple to derive A1 from Biot and Savart, though of course it can be done but only after a certain amount of mathematical manipulation.

Distributed currents

Up to this point, we have considered only the magnetic fields produced by currents flowing in wires, and have found them to be described by a many-valued potential ϕ_B. As soon as we consider currents distributed in space, as in the interior of conductors, we have to abandon the potential. For the change in ϕ_B in the traversing of a closed curve is μ_0 times the current enclosed, and must now depend on the actual form of the curve; no definite value or set of values can thus be assigned to ϕ_B, which becomes useless as a description of **B**. We must fall back on the basic statement (10.1) which for a distributed current of density $\mathbf{J}(\mathbf{r})$ takes the form

$$\oint \mathbf{B}.\mathrm{d}\mathbf{r} = \mu_0 \int \mathbf{J}.\mathrm{d}\mathbf{S}, \qquad (10.9)$$

the line-integral being taken round the boundary of the not necessarily plane surface, S, over which the surface integral is evaluated.

Let the surface be plane, with vector area **S** (normal to the surface), and be made very small, so that in the limit **J** is constant all over it; then the right-hand side of (10.9) becomes $\mu_0\mathbf{J}.\mathbf{S}$ and has magnitude proportional to S if the orientation of the surface is kept fixed. Hence $\oint \mathbf{B}.\mathrm{d}\mathbf{r}$ becomes in the limit proportional to the area enclosed, and $1/S \oint \mathbf{B}.\mathrm{d}\mathbf{r}$ tends to a definite limit. We now define a vector, **C**, parallel to **J** and of magnitude

$$\mathbf{C} = \lim \left[\frac{1}{S}\oint \mathbf{B}.\mathrm{d}\mathbf{r}\right], \qquad (10.10)$$

the limit being evaluated for an area **S** oriented parallel to **J**. This vector has the property, from (10.9), that $\mathbf{C} = \mu_0\mathbf{J}$ at every point; and if we draw an arbitrarily oriented surface element **S**, the flux of current through this area is just $1/\mu_0$ times the component of **C** parallel to **S**.

The vector defined by (10.10) is conventionally written as curl **B** (or rot **B**). Since the component in any direction is to be found by applying

(10.10) to an element of area pointing in that direction, we can write curl **B** in Cartesian coordinates by choosing suitable rectangular elements. To find the z-component, for example, we choose the element δx–δy and traverse it clockwise, looking along the z-axis, as indicated by the curved arrow. If the mean field strength along AB has a component B_x parallel to AB, along CD it will have a component $B_x + (\partial B_x/\partial y)\,\delta y$, and the contribution of these two sides to the line-integral is $-(\partial B_x/\partial y)\,\delta y\,\delta x$. Similarly BC and DA contribute $(\partial B_y/\partial x)\,\delta y\,\delta x$. Adding these contributions and dividing by the area, $\delta y\,\delta x$, we have for the z-component of curl **B**, $(\text{curl}\,\mathbf{B})_z = \partial B_y/\partial x - \partial B_x/\partial y$. The Cartesian form of the curl of a vector is fully displayed in XII. The curl, as its name is intended to imply, measures the rotational properties of a vector field. Any irrotational field, such as may be derived from a potential, necessarily has vanishing curl; for both $\partial B_y/\partial x$ and $\partial B_x/\partial y$ may be written as $-\partial^2\phi_B/\partial x\,\partial y$, so that their difference vanishes identically. The electrostatic field is a typical irrotational field, whose basic properties in free space may be written:

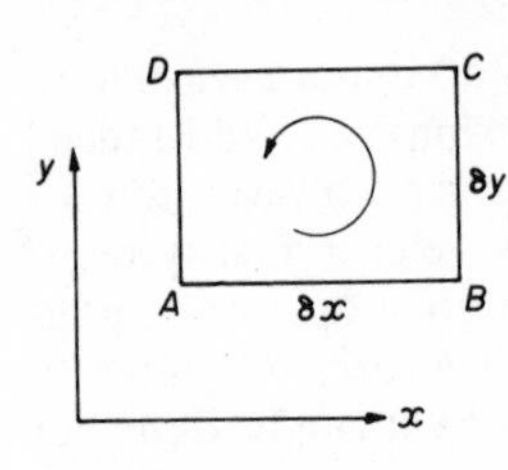

155

$$\text{div}\,\mathbf{E} = \rho/\varepsilon_0 \quad \text{and} \quad \text{curl}\,\mathbf{E} = 0.$$

The magnetostatic field, however, which is the subject of this chapter, obeys the equations

$$\text{div}\,\mathbf{B} = 0 \quad \text{and} \quad \text{curl}\,\mathbf{B} = \mu_0\mathbf{J}. \tag{10.11}$$

These equations represent in very compact form the essential properties of the magnetostatic field, its non-divergence and its relation to the currents that generate it. They form the starting-point for developments of the mathematical formulation of electromagnetism, some of which we shall touch on later. For most developments, however, the reader will do better to consult specialized treatises on electromagnetic theory.

For the rest of this chapter we are concerned once more only with currents in wires. It should be appreciated that the field patterns produced by localized currents may look as if they had rotational properties, in that $\oint \mathbf{B}.d\mathbf{r}$ does not vanish if the path of integration encloses a current, but in fact curl $\mathbf{B} = 0$ almost everywhere. The 'curliness' of the field is confined to the wire, within which curl **B** takes a very large value, tending to infinity as the wire diameter is imagined shrunk to zero while the current

39 is maintained constant. A similar situation occurs in the bath-tub vortex, where conservation of angular momentum causes the tangential velocity to vary as $1/r$, just like the field round a current-carrying wire; although the flow obviously possesses a strong rotational character, the occurrence of circulation while at the same time curl $\mathbf{v} = 0$, is achieved through the catastrophe at the very centre, where the simple flow pattern is abandoned. Circulation in a curl-free field always involves one or more *singularities* where the field locally takes an extraordinary character.

Forces in a magnetic field

Let us consider now the force exerted on a current-carrying wire by a magnetic field, as stated in Ampère's second law (A2). We may seek to make this plausible by applying N3 to the interaction of a current loop with a magnetic pole, P. From (10.8) we write down the field at P and hence the force on the pole in the form

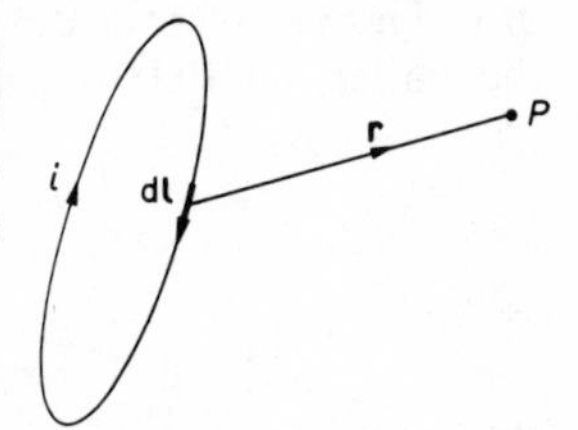

$$\mathbf{F}' = \frac{\mu_0 iP}{4\pi}\oint\frac{\mathbf{dl}\wedge\mathbf{r}}{r^3}.$$

If we believe that the opposite of this is the force $\mathbf{F}$ exerted by the pole on the current loop, we write

$$\mathbf{F} = -\frac{\mu_0 iP}{4\pi}\oint\frac{\mathbf{dl}\wedge\mathbf{r}}{r^3} = i\oint\mathbf{dl}\wedge\mathbf{B},$$

in which $\mathbf{B}$ is $-\mu_0 P\mathbf{r}/(4\pi r^3)$, the field due to the pole P at the element $\mathbf{dl}$, distant $-\mathbf{r}$ from it. The form of this result encourages us to believe that it may hold in detail and not only in integral form, and this is the statement of A2:

$$\mathbf{dF} = i\,\mathbf{dl}\wedge\mathbf{B}. \tag{10.12}$$

This is in no sense a proof of A2, but only an argument to make it seem plausible, so as to encourage experimental tests. It should be noted that this is not a statement about an incomplete physical system in the way that (10.8) is, for the current element in (10.12) may be part of a complete circuit. It is not possible to isolate $i\,\mathbf{dl}$ to ask what field it produces; it is perfectly sensible to ask what force is acting on $\mathbf{dl}$ when a field is applied to a current-carrying wire, for if the wire is flexible, the shape it will take up is determined not by the total force only but by its detailed distribution along the length of the wire.

Nevertheless there is implicit in our argument a potential source of difficulty, in that the force between a current element and a pole, even though it may obey N3, does not obey the more stringent requirement of conserving angular momentum. For the force is not radial, but normal to the line joining the pole and the current element, and the action and re-action, though equal, form a couple. So long as this is a property solely of the isolated current element we need not be troubled, but it is desirable to show that the difficulty is absent when a complete loop is considered. It is easy to see what we have to prove. Since each element of the loop is supposed to act on the pole directly, the resultant force on the pole obviously passes through it; we must therefore show that the resultant of all forces exerted by the pole on elements of the loop is a force passing through the pole. In other words, we must show that the resultant moment of these forces about the pole vanishes. Since the force on any element is

29

proportional to $\mathbf{r} \wedge \mathrm{d}\mathbf{l}/r^3$, the moment about P is proportional to $(\mathbf{r} \wedge \mathrm{d}\mathbf{l}) \wedge \mathbf{r}/r^3$, and we must show that $\oint (\mathbf{r} \wedge \mathrm{d}\mathbf{l}) \wedge \mathbf{r}/r^3$ vanishes.

Consider an element $\mathrm{d}\mathbf{l}$ of the loop, making an angle ϕ with the direction of $\mathbf{r}$. The vector product $\mathbf{r} \wedge \mathrm{d}\mathbf{l}$ has magnitude $r\,\mathrm{d}l \sin\phi$ and is normal to the paper, so that $(\mathbf{r} \wedge \mathrm{d}\mathbf{l}) \wedge \mathbf{r}$ lies in the paper, normal to $\mathbf{r}$, and has magnitude $r^2\,\mathrm{d}l \sin\phi$. Thus the element contributes to the integral a vector normal to $\mathbf{r}$, lying in the plane of motion of $\mathbf{r}$, and of magnitude $\mathrm{d}l \sin\phi/r$. Now a point U on the radius vector, unit distance from P, traces out just such a vector, and as the loop is followed, the subsequent vector contributions to the integral are all laid end to end by the movement of this point. The closing vector after one circuit of the loop vanishes identically, and the required result is proved. There is thus no conflict between A2 and the conservation of angular momentum in the interaction of a complete current loop and an isolated pole.

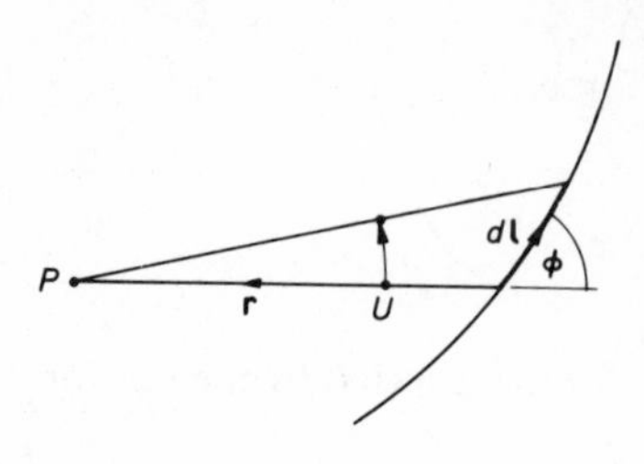

Having shown that the interaction between a complete loop and an isolated pole obeys the stringent form of N3, we may conclude without any further calculation that if A2 is true the action of a pole on a magnetic shell constructed according to A1 is exactly the same as on a current loop. For the loop and the shell produce the same field, and we know, all forces between poles being central, that the interaction between the shell and a pole must obey the stringent form of N3; this is enough to prove the exact equivalence. We may go further; all magnetic fields can be thought of as being produced by shells, i.e., by distributions of poles, and we may therefore be confident that A2 leads to no violation of the law of conservation of angular momentum in the interaction of current loops with each other, since there can be no violation when two shells interact. We shall see this in detail presently.

Ampère's two laws enable us, if we wish, to replace current loops by magnetic shells not only to calculate the field due to a loop, but to calculate the force and torque experienced by a current loop in any configuration of magnetic field. We may, for example, assert without further analysis that the torque on a coil of area $\mathbf{A}$ carrying current i in a uniform field $\mathbf{B}$ is $i\mathbf{B} \wedge \mathbf{A}$, since this is the vector sum of the torques on every element of the shell considered as an infinitesimal dipole. This result is of course not confined to plane coils, since any closed loop has a vector area of which any component can be found by casting a shadow of the loop on a plane normal to the direction of the desired component.

If the wires carrying the current are not rigid, we cannot readily use the shell method to calculate their distortions in a field, and direct application of (10.8) and (10.12) is the best approach.

PROBLEM

Two straight wires 3 m long hang vertically with a 200 g mass attached to each, and are held 10 mm apart at the top and bottom but are otherwise free. Show that when the same current is passed through both in the same direction they attract one another, and estimate the current needed for them to become unstable; for this purpose you may find it convenient to assume that they form circular arcs. [*Answer:* a little over 10 A]

It is not too hard to solve this problem completely, provided you are prepared for some numerical work at the end. One way of expressing the result of a more complete analysis is that the critical current at which the wires become unstable is $(8\pi T/\mu_0)^{1/2}a/L$ times the maximum value of $[\exp(-z^2)\int_0^z \exp(z^2)\,\mathrm{d}z]$; here T is the tension in each wire, a the separation and L the length. The maximum value of the quantity in square brackets is 0·57, giving a value for the critical current only 14% greater than that estimated by assuming circular arcs.

What happens when the critical current is exceeded? Assume the wires to be insulated. If the current is then reduced, describe the behaviour. See if you can set this up in the laboratory, and check your analysis. The wires need to be very straight initially, but if after setting everything in place you stretch them a bit, this should be achieved.

The results of simple experiments on the behaviour of coils and the forces and torques they exert on one another and on permanent magnets have apparently provided adequate experimental backing for Ampère's laws, since they have not been subjected to the degree of critical attention that Coulomb's law has received. This is not very surprising, for once the inverse-square law has been established with high precision for electrical charges it is most unlikely that any experiment with lesser sensitivity to departures from this law will reveal anything interesting about magnetic forces, in the light of what is now known about the intimate connection between the two types of field. The fact that no simple test comparable to Cavendish's experiment has been invented for magnetic forces is sufficient reason for the lack of critical study. Nevertheless Ampère's laws are accepted as completely as the others, and indeed form the basis for the definition and determination of unit current in terms of the force between coils, and the agreement between different standards laboratories, using different designs of current balance, argues strongly for the correctness of the laws within the limited precision of the measurements, perhaps a few parts in a million.

53

The current balance, and the values of μ_0 and ε_0

The current balance is well worth discussing on its own merits as a precision instrument, and we may begin by deriving an explicit formula for the force between two coils, starting from (10.8) and (10.12). The element $\mathbf{dl}_1$ carrying i_1 produces a field $\mathbf{dB}$ at the position of the element $\mathbf{dl}_2$, where

$$\mathbf{dB} = \mu_0 i_1(\mathbf{dl}_1 \wedge \mathbf{r}_{12})/(4\pi r_{12}^3),$$

and a contribution $\mathbf{dF}_{12}$ to the force acting on $\mathbf{dl}_2$:

$$\mathbf{dF}_{12} = \mu_0 i_1 i_2\, \mathbf{dl}_2 \wedge (\mathbf{dl}_1 \wedge \mathbf{r}_{12})/(4\pi r_{12}^3).$$

The total force is then obtained by integrating round both circuits:

$$\mathbf{F}_{12} = \frac{\mu_0 i_1 i_2}{4\pi} \oint_1 \oint_2 \mathbf{dl}_2 \wedge (\mathbf{dl}_1 \wedge \mathbf{r}_{12})/r_{12}^3,$$

$$= \frac{\mu_0 i_1 i_2}{4\pi} \oint_1 \oint_2 \left[\frac{(\mathbf{r}_{12}\,.\,\mathbf{dl}_2)}{r_{12}^3}\,\mathbf{dl}_1 - \frac{(\mathbf{dl}_1\,.\,\mathbf{dl}_2)}{r_{12}^3}\,\mathbf{r}_{12} \right], \quad \text{by VII.} \qquad (10.13)$$

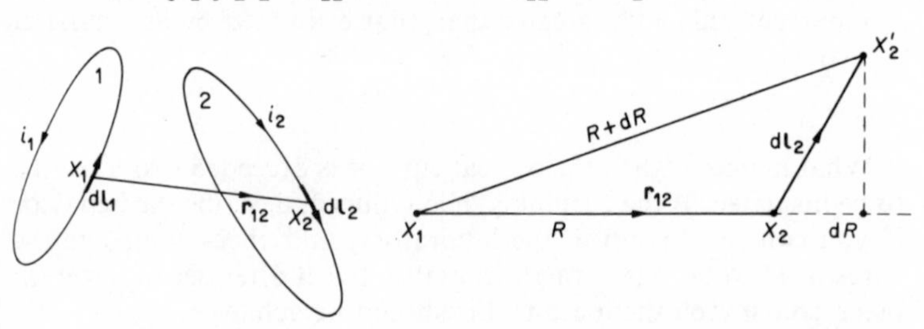

We may now show that the first term in the integral vanishes on integration. If we first carry out the integration round the second circuit, i.e., calculate the force exerted by a given element $\mathbf{dl}_1$ on the whole of the second circuit, we need to evaluate $\oint_2(\mathbf{r}_{12}\,.\,\mathbf{dl}_2)/r_{12}^3$, I say, and then as a second operation evaluate $\oint_1 I\,\mathbf{dl}_1$. As it turns out, however, I is identically zero and the second operation is unnecessary. A diagram shows this clearly. If X_1 is the position of the element $\mathbf{dl}_1$, we choose it as origin of polar coordinates, writing the scalar R for $|\mathbf{r}_{12}|$, the distance from X_1 to a point X_2 on the second circuit. Then $R + \mathrm{d}R$, the polar coordinate of X_2', is just $|\mathbf{r}_{12}| + \mathbf{r}_{12}\,.\,\mathbf{dl}_2/|\mathbf{r}_{12}|$, the increment $\mathrm{d}R$ being the projection of $\mathbf{dl}_2$ on to the line X_1X_2. Hence we may write $\oint_2(\mathbf{r}_{12}\,.\,\mathbf{dl}_2)/r_{12}^3$ as $\oint_2 \mathrm{d}R/R^2$. But this integral, considered as an indefinite integral, is just $-1/R$, and considered as a definite integral has identical limits, since the beginning and end are the same point of the circuit. The integral therefore vanishes. As a consequence we may write for the force between two circuits:

$$\mathbf{F}_{12} = -\frac{\mu_0 i_1 i_2}{4\pi} \oint_1 \oint_2 \mathbf{r}_{12} \frac{(\mathbf{dl}_1\,.\,\mathbf{dl}_2)}{r_{12}^3}. \qquad (10.14)$$

This has exactly the form it would take if two current elements attracted one another with a central inverse-square force whose magnitude depended on their relative orientation. The fact that the force between two circuits can be cast into such a form confirms our earlier conclusion that there is no violation of the strict form of N3, but it is important to remember that (10.14) applies only to the total force between two coils. If one wishes to find the local force acting on one section of a coil it is in general incorrect to use the central inverse-square force law that seems to be implied by (10.14). For the total force on $d\mathbf{l}_2$, for example, it is necessary to go back to (10.13) and integrate round the first, not the second, circuit; the theorem just proved, that the first term in the integral vanishes identically, is no longer true, so that both terms have to be evaluated to obtain the right local force.

212

In a current balance, however, we are concerned to measure the force on a coil, attached to a balance, exerted by another fixed coil close to it and carrying the same current i. From the force and the dimensions of the apparatus, i is then found, (10.14) being the most convenient expression and valid for this purpose. The form of balance devised by Lord Rayleigh, and used at the National Bureau of Standards, has two fixed coils with current circulating in opposite directions, and a smaller coil hung from a balance arm midway between them. All carry, in series, the same current whose magnitude is to be determined by weighing the force on the moving coil, which is proportional to i^2. There are many points of interest in the detailed design, such as the problems of taking current to the moving coil without exerting any irreproducible forces, and of cooling the coils and minimizing convection currents in the air which can be quite disastrous to an accurate measurement. The maximum force due to the currents is only about 1% of the weight of the suspended coil, so that weighing to an accuracy of 1 part in 10^8 is needed if the current is to be determined to one part in a million (1 p.p.m.). These matters are discussed at length in specialist texts, and we shall confine our attention here to general matters that affect the design.

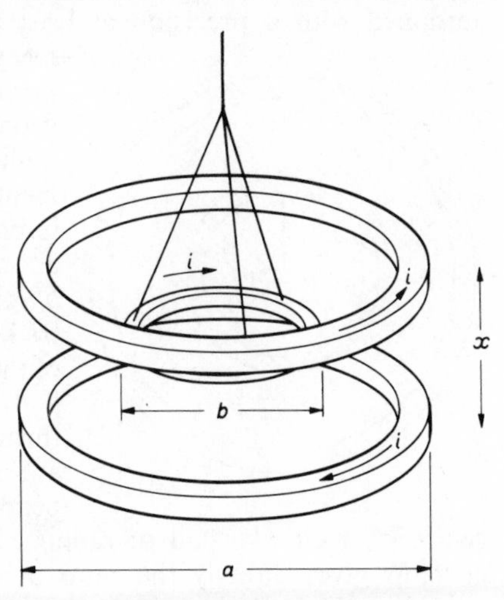

The form of (10.14) shows that the double integral is dimensionless, so that the force depends on $\mu_0 i^2$ and on a function of the dimensionless ratios x/a and b/a, but not on their absolute values. The accuracy of the final determination of i therefore depends, among other things, on the accuracy with which these ratios can be measured. The dependence on x/a can however be minimized by choosing the value at which the force is maximal, since at this point the force is unaffected to first order by changes in x/a. It is clear that a maximum exists in the variation of force with x/a, since there can be no force when $x = 0$ (apart from the currents cancelling each other, coplanar coils exert no force on one another normal to their plane), and the force must also vanish when x becomes large. When $b = \frac{1}{2}a$, the maximum occurs with x/a roughly equal to 0·4. By choosing this value of x/a, it is unnecessary to measure it to the very high precision that otherwise would be needed. The same is not so of b/a. There is no value which maximizes the force, if one excludes the impracticable case, $b/a = 1$ and x as small as possible. The choice of b/a is conditioned by other factors; if it is too small, the force is also too small, but as it approaches unity the precision with which the moving coil must be set midway between the fixed coils increases rapidly, since the maximum of force in the midway position becomes very sharp. A compromise value of about $\frac{1}{2}$ seems most convenient. It is this ratio, b/a, which must be determined with a precision at least as great as that demanded in the measurement of i, and it should be borne in mind that the coils are not geometrical circles but rather bulky, multi-layer windings. Direct measurement of the dimensions, though not impossible, is unlikely to be so accurate as an indirect electrical determination of b/a; this is achieved by the ingenious trick of mounting the moving and one of the fixed coils vertically and coplanar, and finding what currents passed through them produce no magnetic field at the centre, as revealed by a short suspended magnet. Since the field at the centre of a circular coil of radius r is, from (10.8), $\frac{1}{2}\mu_0 i/r$, the ratio of currents gives directly the ratio of radii. The corrections required in extending the formula for a single turn to a multi-layer coil are very much the same here as in the current balance, so that the ratio determined in this way may be used directly with relatively minor corrections that do not demand an impossible degree of precision in the knowledge of the form of the windings. Moreover, small irregularities of shape and in the evenness of the windings are to a great extent allowed for automatically. The precision measurement thus is reduced to the ratio of two currents, which is found by means of a potentiometer.

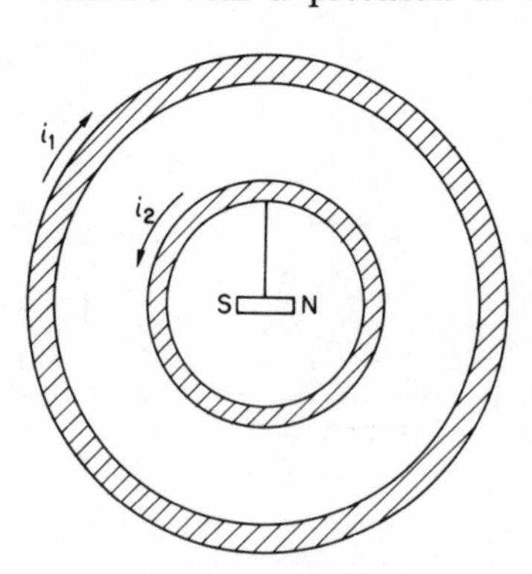

Finally it is worth observing that the force is to be expressed in terms of the weight of a certain mass, and this involves an accurate knowledge of g. Until recently Kater's pendulum, or a variant, was the standard instrument for determining g, and accuracy of 1 p.p.m. was a very difficult matter. But Cook has developed an alternative method that involves timing the flight of a glass sphere projected vertically upwards by a catapult, and has achieved a considerable enhancement of accuracy. The value of g is no longer a limitation to the accuracy of a current balance, which is something like 4 p.p.m. at the present time. This still does not compare well with the precision available for comparing masses (0·001 p.p.m.) lengths (0·01 p.p.m.) or times (0·0002 p.p.m.).

Although we have devoted some space to the current balance, because of its importance as a means of relating currents to the fundamental unit and because its design presents points of interest, it is indeed a very delicate instrument that is not to be found except in a standards laboratory. Moreover, a standard or sub-standard ampere is not something that can be stored in a cupboard like a kilogramme or a metre. It is more convenient to keep a calibrated resistance and a cell (e.g., the Weston cadmium cell) which when made to the correct specification reproduces well a known voltage. To measure a current with a precision greater than a few parts per thousand, to which only the best ammeters are reliable, is most simply performed by passing it through a calibrated resistance and comparing potentiometrically the voltage developed with that of a standard cell. We shall return in Chapter 14 to the question of calibrating 299 a resistance.

It is now convenient to clear up a point raised earlier—why is μ_0 defined 204 to be $4\pi \times 10^{-7}$? In the old c.g.s. electromagnetic system of units there was no independent electrical unit, the abampere being defined as the unit current that would allow the force law (10.11) to be written with $i_1 i_2$ rather than $\mu_0 i_1 i_2/(4\pi)$ outside the integral. Thus once the choice of mechanical units had fixed unit force, there was no choice left for current. In terms of the notation adopted here, μ_0 was defined to be 4π. In the middle of the nineteenth century this unit, the abampere, was thought to be too large for convenience and the ampere was chosen as one-tenth of an abampere; after some vicissitudes of definition, it has reverted to this by international agreement. In the transition from c.g.s. to m.k.s. the unit of force ($[F] = [M][L][T]^{-2}$) increases by a factor of 10^5, and μ_0, having dimensions $[F][I]^{-2}$, is to be measured in terms of a unit which increases by 10^7. Hence the measure of μ_0 alters from 4π to $4\pi \times 10^{-7}$. It may be left to the reader to discuss, if he has a taste for futile argument, whether the definition of μ_0 as $4\pi \times 10^{-7}$ makes it dimensionless, in which case the ampere is not an independent unit, or whether the choice of the ampere as 0·1 abampere implies that any other fraction of an abampere could have been chosen, in which case it is independent.

Once μ_0 has been given a numerical measure, the unit of current is fixed and therefore the unit of charge. It follows then that the constant ε_0 that appears in Coulomb's law (4.2) is not adjustable, but is to be determined experimentally either by measuring the force between known charges or by some more convenient but equivalent method. We shall now discuss a rather primitive method of estimating ε_0, similar to that used by Weber and Kohlrausch (1856) and by 291 Rowland (1889). A condenser of accurately known dimensions (Rowland's was a concentric spherical condenser) is charged to a 300 measured potential difference (p.d.) V, so that the stored charge CV is known in terms of the dimensions, ε_0 and V. The condenser is then discharged through a moving-coil galvanometer whose sensitivity to charge can be related to its sensitivity to steady currents, the latter being directly measurable in terms of the current standard established by means of a current balance. Two points here need a little elucidation—the measurement of p.d. and the relation between current and quantity sensitivities.

The p.d. can be determined by direct application of Coulomb's law in an attracted disc electrometer, which is essentially a parallel-plate condenser of which one plate is hung from a balance. If the plates have area A and separation d, a p.d. of V produces a field V/d 176 between them, which exerts a tractive force $\frac{1}{2}\varepsilon_0 V^2/d^2$ on unit area. Thus the change in apparent mass of the plate is given by

$$\delta m = \varepsilon_0 V^2 A/(2\,d^2 g). \qquad (10.15)$$

For plates $10^{-2}\ \text{m}^2$ in area, separated by 5 mm and charged to 1000 V, this amounts to 160 mg which is capable of being weighed at least within 0·1%, and much better given care. Now the capacitance of a condenser can be written as $4\pi\varepsilon_0 af$, where a is some dimension of the condenser and f a dimensionless factor determined by its shape (e.g., for a concentric spherical condenser, a might be the radius of the 159 inner sphere and b that of the outer, f being then $[1 - a/b]^{-1}$). If this condenser is charged to the same p.d. as is measured by the electrometer, the charge acquired may be written:

$$Q = 4\pi\varepsilon_0 af V = C(g\varepsilon_0\,\delta m)^{1/2}, \qquad (10.16)$$

where C is a quantity determined by the dimensions of the apparatus, having magnitude $4\pi adf(2/A)^{1/2}$, by use of (10.15). It is clear from (10.16) that if Q can be measured by discharging the condenser through a ballistic galvanometer, the value of ε_0 can then be found.

A suspended-coil galvanometer consists of a light coil held on a torsion wire in the field of a permanent magnet, so that when current i is passed through it, it experiences a torque proportional to i, Gi say, and finds a new position of equilibrium at an angle θ from its original position, where this torque is balanced by the torque $\gamma\theta$ exerted by the torsion wire. The angle θ is measured by means of a mirror, lamp and scale, and the *current sensitivity*, S_i, of the instrument is defined as θ/i which is clearly G/γ. To determine the quantity sensitivity, S_q, we must write down the equation of motion of the coil,

$$I\ddot{\theta} = Gi - \gamma\theta,$$

where I is the moment of inertia of the coil. We have neglected any viscous or other damping, but this can be allowed for by a systematic procedure that need not concern us so long as we are seeking only the basic principles and not enquiring into all the details of the actual measurement. In the ballistic use of the galvanometer, the condenser is discharged through the coil while the latter is at rest and undeflected; moreover, it is usually not difficult to ensure that the discharge takes place in less than one hundredth of the natural period of swing of the galvanometer, so that it is a good assumption that θ is sensibly zero while the current i is flowing, and that thereafter the galvanometer swings freely without current. Integrating the equation of motion through the short time when the current flows and θ is zero, we see that immediately after the discharge the angular velocity is given by the equation

$$I\dot{\theta} = G\int i\,\mathrm{d}t = GQ.$$

The galvanometer then swings freely according to the equation $I\ddot{\theta} = -\gamma\theta$, whose relevant solution has the form, $\theta = \theta_0 \sin \omega t$, where $\omega^2 = \gamma/I$. Since the value of $\dot{\theta}$ must be GQ/I as θ goes through zero, we see immediately that $\omega\theta_0 = GQ/I$, and that the maximum deflection, θ_0, is $GQ/(I\omega)$. The *quantity sensitivity*, S_q, is defined as the ratio of ballistic swing to charge, θ_0/Q, i.e.,

$$S_q = G/(I\omega),$$

and

$$S_q/S_i = \gamma/(I\omega) = \omega = 2\pi/T,$$

if T is the natural period of the galvanometer. Since S_i may be determined by passing a known current, S_q is also readily found and the galvanometer may be used to determine Q and hence ε_0.

If we were to perform this experiment with easily accessible modern equipment, we might procure a galvanometer of 2-second period with a sensitivity such as would give a deflexion of 1 mm, on a scale 1 m from the galvanometer, when a current of 1 nA is passed. The corresponding quantity sensitivity is π times as great, say 3×10^6m/C. A spherical condenser of outer radius 125 mm and inner radius 100 mm has a capacitance of $2\pi\varepsilon_0$, i.e., 56 pF, and charged to 1000 V carries a charge of 56 nC, so that the ballistic deflection would be about 170 mm. There should be no great difficulty in measuring this with an accuracy of a few tenths of one per cent, so that a determination of ε_0 to rather better than 1% seems feasible. It is easy, however, to overlook some of the practical difficulties such as arise from the proximity of a 1000 V battery and a sensitive galvanometer. On paper one may assume that they are insulated from each other; in the laboratory it may be hard to ensure that the insulation resistance is higher than 10^6 MΩ (which still can cause a deflection of 1 mm); again, when the switch is thrown to discharge the condenser, it is quite possible for changes in potential of different parts of the circuit to cause charge movements in the vicinity that contribute to the ballistic kick. Experience with precision electrical measurements involving very high sensitivity shows that paper estimates of the accuracy attainable must be interpreted with great caution.

In fact, the early measurements of ε_0 by means such as these managed to achieve about 1% accuracy. A better method, devised by Maxwell and carried out by Rosa and Dorsey, will be described in Chapter 14.

298

Electric motors

The forces between current-carrying conductors, and between currents and magnets, are used in many different ways to convert electrical energy into mechanical work, and the study of the design and performance of motors is an important branch of electrical engineering. We shall describe one example, and that the simplest in principle, which until recently might have been regarded as of academic interest only but which now seems to promise the possibility of building stronger and more compact motors than ever before. In the *homopolar motor* current is carried radially by a conducting disc situated in the field of a stationary magnet; the resulting tangential force on the disc drives the motor. The theory is simple: if the total radial current is I, the tangential force on the annulus of the disc lying between radii r and $r + \mathrm{d}r$ is $BI\,\mathrm{d}r$, and its contribution to the driving torque is $BIr\,\mathrm{d}r$; since this is $I/(2\pi)$ times the magnetic flux passing through the annulus, it is clear that the total torque G is $I\Phi/(2\pi)$, where Φ is the total flux passing through the disc.

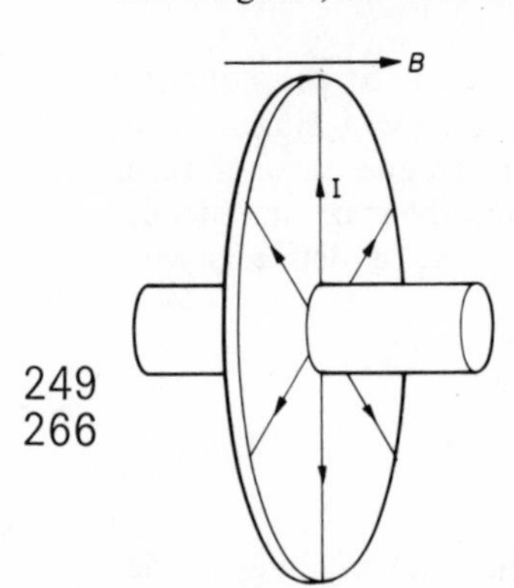

249
266

The torque is developed even when the motor is at rest, in which respect it differs from most a.c. motors whose starting torque is weak; there are many applications where this is a very considerable advantage. When running at an angular velocity ω, the rate at which the motor performs work is $G\omega$, and this is W, the power developed; therefore

$$W = \frac{I\Phi\omega}{2\pi} = \frac{1}{60}\,I\Phi R.$$

if R is the speed in revolutions per minute. To quite a real example, an experimental motor operated at Fawley (Hampshire, England) had a rotor diameter of about 2 m and a mean field strength of about 2 T, giving a flux Φ of 6·45 Wb. At a speed of 200 rev/min, with 120,000 A radially through the rotor, the power developed was nearly 2·6 MW, or 3400 horsepower. It should be made clear that the enormous current was not simply fed in at the centre and taken out at the rim; the rotor in fact was constructed of radial spokes which, by an arrangement of slip rings, carried 6000 A through twenty spokes connected in series.

The technical development that has made large homopolar motors worth while is the superconducting magnet that provides the extended strong magnetic field. Wires of niobium-titanium alloy, embedded in copper for ease of handling, lose all resistance when cooled below about 14 K, and at the temperature of liquid helium, 4 K, can carry current densities of 4×10^7 A m^{-2} in a strong magnetic field without any dissipa-

The 3400 h.p. homopolar motor (on right) installed at Fawley power station to drive the water pump on the left. (Photograph supplied by International Research & Development Co. Ltd.)

tion. A magnet coil made of such wire consumes only the power required to run its helium refrigerator, which is quite negligible compared with the power of the motor. A similar coil made of copper and running at room temperature would generate as heat several times more power than the motor produces. Let us estimate how much current is needed in a coil 1·4 m in radius, i.e., rather larger than the rotor, to develop a field of 2·5 T at the centre. If it is thought of as a single turn of radius a, the field is $\frac{1}{2}\mu_0 i/a$ at the centre, and i must be $5{\cdot}5 \times 10^6$ A; field windings of 0·14 m^2 total cross-sectional area, carrying 4×10^7 A m^{-2}, will provide enough current. It does not matter very much how many turns there are in the coil, provided the total circulating current has the required value. It is usually convenient to use many turns of comparatively small cross-sectional wire. The photograph of the Fawley motor, which is probably only a precursor of very much more powerful motors, shows how the rotor is mounted, at room temperature, inside the annular, vacuum-insulated container for the superconducting coil.

Cavendish Problems: 19, 193–195, 198, 223–227, 230, 232–234, 250.

READING LIST

Current Balance: A. GRAY, *Absolute Measurements in Electricity and Magnetism*, Dover.

Determination of g: A. H. COOK, *Phil. Trans. Roy. Soc.* **261**, 211 (1967).

Electric Motors: J. HINDMARSH, *Electrical Machines*, Pergamon; A. D. APPLETON, *Cryogenics*, **9**, 147 (1969).

11

Charged particles in electric and magnetic fields

Lorentz's law **A particle carrying charge e and moving in a magnetic field $\mathbf{B}$ experiences a force $e\mathbf{v} \wedge \mathbf{B}$.**

The last chapter was concerned with the phenomenology of magnetic fields, without any thought of the atomic processes involved. Believing, as we do, that electric currents are due to the motion of charged particles, we naturally enquire what magnetic field is produced by a single moving particle, and what forces it experiences when it moves in a magnetic field. To take the second question first, let us postulate a simple model of the current in a wire, imagining it to result from the motion of n charged particles per unit length of wire, each carrying charge e and each moving with the same velocity $\mathbf{v}$. Then in unit time the particles contained in a length v pass a given point, so that $i = nev$. The force on a length $\mathrm{d}\mathbf{l}$ of wire when a field $\mathbf{B}$ is applied is $i\,\mathrm{d}\mathbf{l} \wedge \mathbf{B}$, which may be written as $(n\,\mathrm{d}l)e\mathbf{v} \wedge \mathbf{B}$ as if each of the $n\,\mathrm{d}l$ particles experienced a force $e\mathbf{v} \wedge \mathbf{B}$ and, being unable to escape from the wire, transmitted this force to the lattice structure of the metal by collisions either within the wire or at the boundary.

The idea that the electrons in a metal are at rest until a field is applied, when they all acquire the same drift velocity, is far from the truth; the electrons in a typical metal have speeds up to 10^6 m s^{-1} or more, even at the absolute zero, but the velocities being randomly distributed in all directions have zero resultant in equilibrium. When a field is applied, there is very slight tendency for more to move in one direction than in the reverse direction; how slight may be seen by calculating what drift velocity would be needed to produce a large current density, say 10^8 A m^{-2}, corresponding to 100 A in a wire of 1 mm^2 cross-section. In copper there are about 10^{29} mobile conduction electrons per cubic metre (i.e., one per atom), or 10^{23} per metre length of the wire, each carrying $1{\cdot}6 \times 10^{-19}$C so that they have to move at only 7 mm per second to produce 100 A, and burn up the wire unless it is actively cooled. The large random motions do not, however, affect the present argument. For if we consider all the electrons in a straight length of wire, the current they carry is determined by the average velocity, since this is the rate at which the centroid of the charge is moving down the wire. Thus $i = \sum e\mathbf{v}$, the sum being taken over all electrons in unit length. And if we suppose each to be subject to a force $e\mathbf{v} \wedge \mathbf{B}$, the total force on unit length of wire is $\sum e\mathbf{v} \wedge \mathbf{B}$, or $\mathbf{i} \wedge \mathbf{B}$ if $\mathbf{i}$ is a vector pointing along the direction of current flow. This picture is therefore consistent with A2, and we shall postulate that the force on a charged particle moving in fields $\mathbf{E}$ and $\mathbf{B}$ takes the form

$$\mathbf{F} = e(\mathbf{E} + \mathbf{v} \wedge \mathbf{B}), \tag{11.1}$$

and results in an equation of motion of the particle

$$m\dot{\mathbf{v}} = e(\mathbf{E} + \mathbf{v} \wedge \mathbf{B}). \tag{11.2}$$

The magnetic force, usually called the *Lorentz force*, is always normal to the direction of motion of the particle and does no work on it. If $\mathbf{E}$ is absent, the particle moves at constant speed, only its direction changing in response to the Lorentz force, and this is true no matter how $\mathbf{B}$ varies from point to point.

The general truth of (11.2) is readily verified by simple experiments with a cathode-ray tube. In principle this consists of a source of electrons which are accelerated to a high voltage by means of a system of electrodes (*electron gun*) which at the same time forms them into a well-collimated beam. The electrons in the beam travel through the evacuated tube until they hit a fluorescent screen and produce there a spot of light. The beam may be deflected by electrodes near its path, arranged of course in a commercial tube to fit the particular application, but nonetheless suitable for simple tests of the equation of motion. It is also possible to apply magnetic fields and study the resulting deflection. Before discussing this, however, it is worth examining briefly the mode of operation of so important a device, especially as it affords an example of what is now the highly de-

169

veloped art of electron-beam handling. The electrons are produced by thermionic emission from a filament or, more usually, a heated button coated with barium oxide or some other material with a low work function. An example of the electrode configuration that focuses the emitted electrons to a fine point, while accelerating them to a high speed, is shown in the diagram. It will be appreciated that a detailed analysis of this sys-

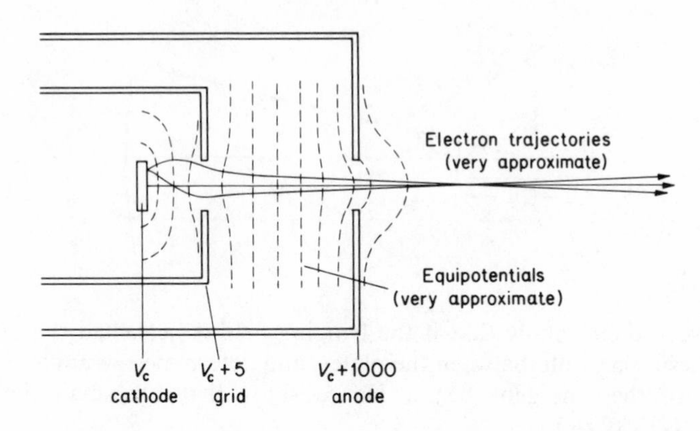

tem, involving determination of the field configuration by one of the methods described earlier, such as an electrolytic tank, followed by solution of the equation of motion for an electron, is a major undertaking, and we shall do no more than indicate a few electron trajectories that may make the focusing action seem plausible. This gun behaves as a divergent small source of electrons, and needs to be followed by a further electrode whose function is like that of a converging lens, to convert the divergent pencil of rays into a narrow parallel pencil. It is not difficult to see that this is possible with a diaphragm at a slightly lower positive potential than the main accelerating electrode, repelling the electrons that try to diverge from the point source. If this last electrode is at Earth potential there is a substantially field-free region to the right, in which the narrow beam keeps its pencil qualities. The cathode, which emits the thermionic electrons, clearly must be maintained at a negative potential, V, so that the electrons can be accelerated to earth potential. The kinetic energy of the electrons in the beam is eV, or V electron volts, and may vary in different tubes from a few hundred to many thousands of volts. Apart from small variations arising from the energy they possess on emission (of the order

157

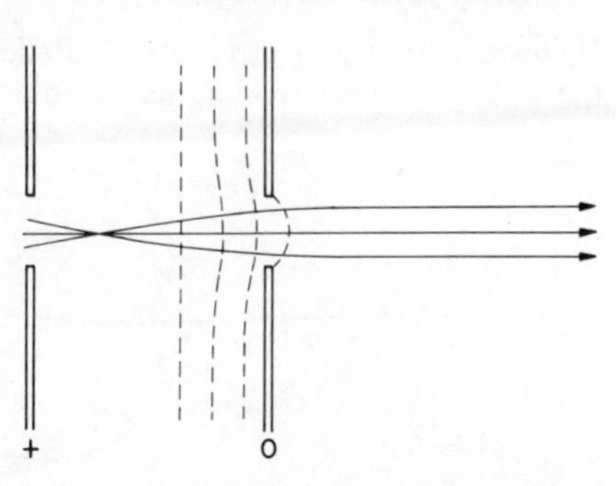

of 1 electron volt), they all have the same energy and respond to deflecting fields in the same way without fanning out into a broader beam.

PROBLEM

Electrostatic deflection. The beam of electrons of V electron volts is passed between condenser plates with a potential difference of V_1

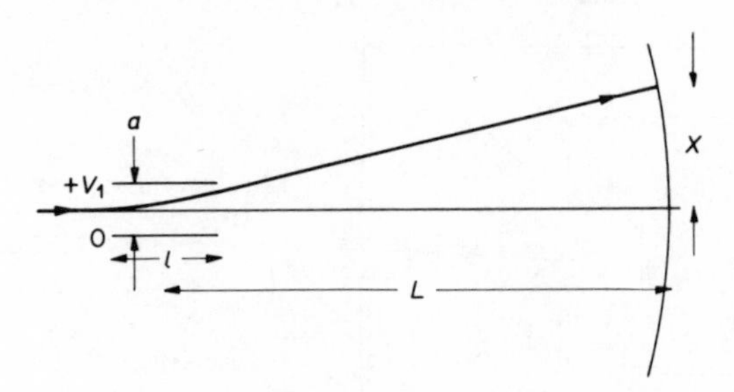

between them. Show that if the fringing field is neglected the beam describes a parabola between the plates, and determine the angle of deflection of the emergent beam. Hence show that for small deflections $X = LlV_1/(2aV)$.

The deflection depends only on the geometry of the tube and the applied voltages, and tells one nothing of the nature of the particles forming the beam (except that the sign of the electrode voltages shows them to be negatively charged).

Magnetic deflection

If, instead of using electrostatic deflection, we pass the beam between the poles of an electromagnet, so that it experiences effectively a uniform

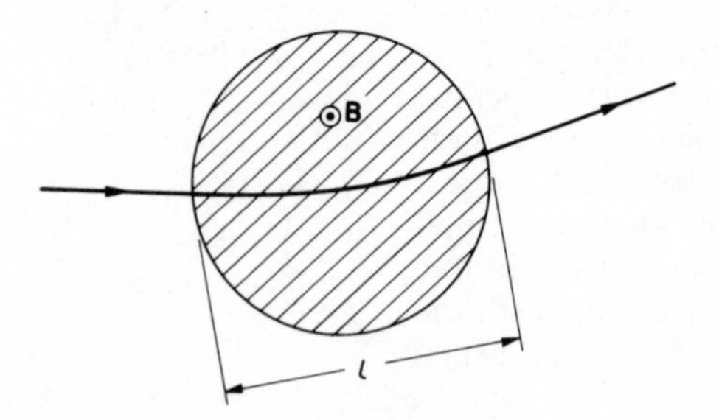

transverse field $\mathbf{B}$ over a length l, we may use (11.2) to calculate the resulting deflection. The electrons move in a circle under the constant transverse Lorentz force evB, the radius being such that

$$mv^2/r = evB;$$

i.e.,

$$r = mv/(eB). \tag{11.3}$$

The angle of deflection is therefore l/r or $eBl/(mv)$, and the deflection on the screen is given by

$$X = eBlL/(mv).$$

It will be seen that magnetic deflection is conditioned by the momentum of the electrons rather than their energy, which is the significant property for electric deflection. The difference lies in the form of (11.1) which shows how the Lorentz force involves $\mathbf{v}$ while the electric force does not. In fact, by comparing the two methods of deflection one may deduce the velocity of the electrons. The simplest way is to apply $\mathbf{E}$ and $\mathbf{B}$ simultaneously so that together they cause no deflection. Then from (11.1) the velocity of the electrons must be E/B.

An electron accelerated to 1000 V moves at a speed of about 2×10^7 m s^{-1}. The magnetic deflection by a field of 10 mT, such as is easily produced by a pair of coils flanking the tube, is thus neutralized by an electric field of 2×10^5 V m^{-1}, or 2000 V across condenser plates 10 mm apart.*

From such a measurement of the velocity, and knowledge of the accelerating potential V, one can deduce something important about the electrons. For their kinetic energy $\frac{1}{2}mv^2$ is equal to eV, and therefore the ratio of their charge to their mass, e/m, is $v^2/(2V)$ and experimentally determined. Note that e and m are not separable in this experiment, nor in any that depends on electric and magnetic deflections and can only give the velocity. This particular method of finding e/m is simple but not very accurate unless one is prepared to take great pains in plotting out the exact distribution of deflecting fields along the beam. More precise methods give a value of $1{\cdot}75888 \times 10^{11}$ C kg^{-1}, with an accuracy of about 20 p.p.m. This value should be compared with what is obtained for hydrogen ions by measuring the mass liberated at the cathode by a given charge in electrolysis; the ratio of charge to mass here is $9{\cdot}57 \times 10^7$ C kg^{-1}, some 1840 times less than for electrons. This is quantitatively explained by the fundamental assumption of Bohr's atomic theory that the hydrogen atom consists of a proton and an electron, the latter about 1840 times lighter than the proton and carrying an equal charge of opposite sign. In electrolysis, it is the proton which carries the current to the cathode. An independent estimate of the mass ratio is obtained by comparing the corresponding spectral lines of

* In the Earth's magnetic field, whose magnitude is about 0·1 mT, these electrons move in an arc of about 1 m radius, so that even this weak field produces a not quite negligible effect.

hydrogen and of ionized helium; the latter, with a nucleus four times as massive, has a correspondingly smaller correction for the motion of the nucleus about the common centre of mass (see Problem on p. 77). Although the effect on the spectral wavelengths only amounts to about one part in two thousand, this is large enough for accurate measurement since spectroscopes are especially sensitive to small wavelength shifts; in fact, the proton/electron mass ratio obtained in this way, 1838 ± 1, agrees well enough with the accepted value of 1836·1 to leave little doubt about the model's acceptability. It is also something of a test of (11.1), since electric and magnetic deflections are involved in the best methods of measuring the mass ratio but not in the spectroscopic method.

If a charged particle moves in a uniform magnetic field it experiences no force parallel to **B**, and its component of velocity in this direction is therefore constant. Its motion projected on to a plane normal to **B** is governed by the component of **v** in this plane, whose magnitude is also constant though its direction changes. The particle executes a regular helical trajectory whose radius r is given by (11.3) if v is now interpreted as v', the velocity component normal to **B**. The time taken to execute one revolution of the helix is $2\pi r/v'$ or $2\pi m/(eB)$; otherwise expressed, the *cyclotron frequency*, ω_c, which is the angular velocity of the particle round the axis of the helix is v'/r, i.e.,

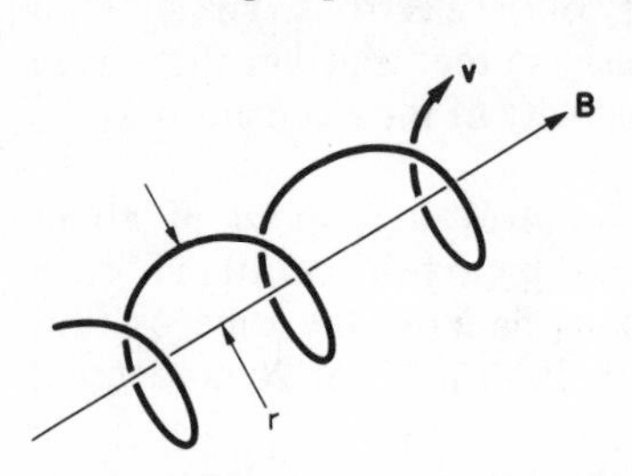

$$\omega_c = eB/m. \tag{11.4}$$

If the particle is positively charged it is seen, by an observer looking along the direction of **B**, to be rotating anticlockwise in its orbit, as may be verified by considering the sign of the Lorentz force $e\mathbf{v} \wedge \mathbf{B}$, which must be directed inwards to provide the centripetal acceleration.

PROBLEMS

(1) A large iron-yoked magnet can provide a field of 2 T without too much expenditure of energy in the field-windings. What is the orbit radius of a proton of 100 MeV kinetic energy moving in the plane normal to **B**? What is its cyclotron frequency?

(2) A typical electron in a metal has a kinetic energy of 5 eV. What are its cyclotron frequency and orbit radius in a field of 10 T such as can be supplied by a small superconducting solenoid?

The origin of the term *cyclotron frequency* is the application of the constancy of ω_c by Lawrence in his invention of the cyclotron for accelerating protons and other nuclear particles. Problem (1) above illustrates the size of magnet needed to contain the orbit of an energetic proton; it also shows that at about this energy the particle velocity approaches that of light, and then (11.4) ceases to predict constant ω_c because the mass of the particle is increased by relativistic corrections. This sets a limit to the operation of the simplest form of cyclotron, but does not preclude development of the idea so as to make possible acceleration to energies 1000 times greater. We shall not discuss cyclotrons here, since many good accounts are accessible, and in any case an account which neglects to consider, at greater length than is here justifiable, the difficult problems of beam focusing (i.e., how to keep the particles from straying out of their proper orbits and hitting the walls of the vacuum vessel) gives a quite misleading picture of what is really involved in the design, not to speak of the construction, of a real accelerator.

We shall illustrate instead the motion of particles in a uniform magnetic field by examples taken from the behaviour of pure metals at low temperatures, under conditions where the electrons are able to travel quite considerable distances, perhaps 1 mm or more, between collisions with impurities and other structural defects. Since, as problem (2) above shows, the electron orbits can be bent into a very small radius by a magnetic field, the electrons are able to accomplish many revolutions in their orbits between collisions, and with suitable methods of observation reveal the characteristic features of their orbits. Consider for example a transverse acoustic wave travelling through the metal, at a speed so much lower (~ 1000 times) than the electron speeds that it may be treated almost as if it were at rest. If we think of the oscillating quartz plate that generates this wave as pushing the ionic lattice backwards and forwards, but not acting directly on the conduction electrons, it will be necessary for a transverse electric field to travel with the wave so as to pull the electrons into synchronous motion with the lattice. Since an electric field **E** acting on a current density **J** gives rise to energy dissipation in the form of heat at a rate **E**.**J** per unit volume, we can appreciate that the heating rate, and therefore the loss of energy from the wave, will be determined by the magnitude of **E** required to establish that electronic current **J** which keeps ions and electrons moving together. The higher the conductivity, the smaller the loss.* If now we imagine a

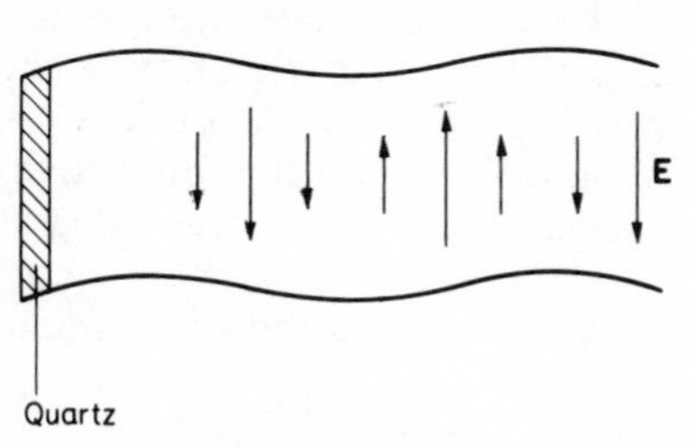

* The argument here is deliberately over-simplified; there is nothing wrong with it in essence but, as the reader will easily appreciate, the detailed working-out is far more elaborate than would be proper to illustrate the point at issue—the direct observation of an effect due to electron orbits.

transverse magnetic field to bend the electron orbits into circles, the size of the circle in relation to the wavelength determines how strongly the electron is able to respond to the electric field. In the left-hand diagram, where the diameter is about half a wavelength, the electron is steadily accelerated (1) or decelerated (2) and is in a state

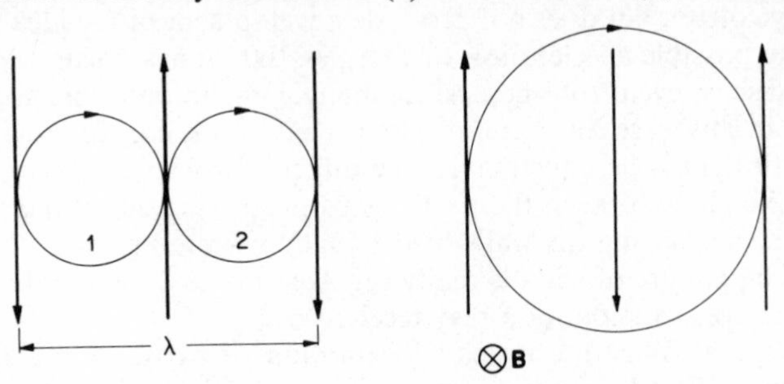

to respond to the field—the conductivity is high. But in the right-hand diagram, where the diameter equals the wavelength, what acceleration it gets on the right it loses on the left, and the net effect of going round the orbit is just about nothing—the conductivity is low. We expect the conductivity to show minima whenever the orbit diameter is close to an integral multiple of a wavelength, i.e., from (11.3) when $2mv/(eB) = n\lambda$, n being an integer. So we expect the attenuation to show maxima under the same conditions, when

$$1/B = n\left(\frac{e\lambda}{2mv}\right),$$

i.e., at regular intervals of $1/B$. The experimental curve shows how this expectation is realized in practice. From the period in $1/B$ one

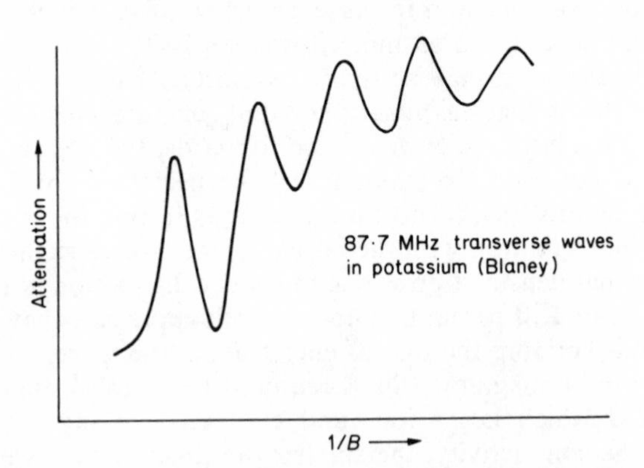

can deduce the momentum mv of the electrons responsible. The argument has been presented here as if the electrons in a metal were perfectly free, which is not true. But it can be generalized to apply to the more complicated real situation, and measurements of this sort give direct information on how far the motion of electrons in a metal differs from the motion of electrons in empty space.

Another experiment with pure metals at low temperatures enables the cyclotron frequency ω_c to be measured. This depends on the *skin effect* which will be discussed in a later chapter. 262 All that is needed for the present purpose is to know that when a high-frequency alternating field is applied to the surface of a highly conducting metal, it can only penetrate a very short distance, e.g., with a frequency of 10^{10} Hz perhaps only 10^{-7} m, which may be 100 times less than the orbit radius of an electron in that magnetic field ($\sim\frac{1}{3}$ T) which causes it to rotate in its orbit 10^{10} times a second.

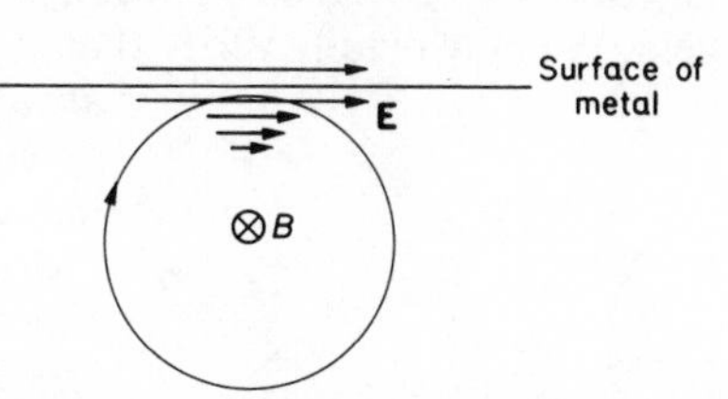

With a steady magnetic field applied parallel to the surface of the metal, some of the electrons travel in orbits that skim close to the surface without touching it, and if they have long free paths between collisions they pass through the region containing the electric field time and time again. Only near the surface do they experience this field, and their net response to it depends on whether in successive traverses of the surface layer they find the field pointing in the same direction or whether, as may happen, the time between traverses is such as to allow field reversal. If the oscillating field frequency ω is an integral multiple of the cyclotron frequency, successive traverses find the field unchanged, and the electron is influenced strongly. The surface layer then behaves as if the metal were an especially good conductor, and the energy dissipation (as measured possibly by heating of the sample) shows a minimum. The condition for a minimum, that $\omega = n\omega_c$, may be written, from (11.4), in the form

$$\frac{1}{B} = n\left(\frac{e}{m\omega}\right).$$

Thus when the surface resistance is plotted against $1/B$, the frequency ω being held constant, the observed oscillations should be evenly spaced, and their spacing should be determined by e/m (or what in a real metal serves instead of this parameter). For technical reasons, it is usual to measure the derivative of the surface resistance with respect to magnetic field, $\mathrm{d}R/\mathrm{d}B$ rather than R itself; this enhances the visibility of the oscillations, which are very clear in this

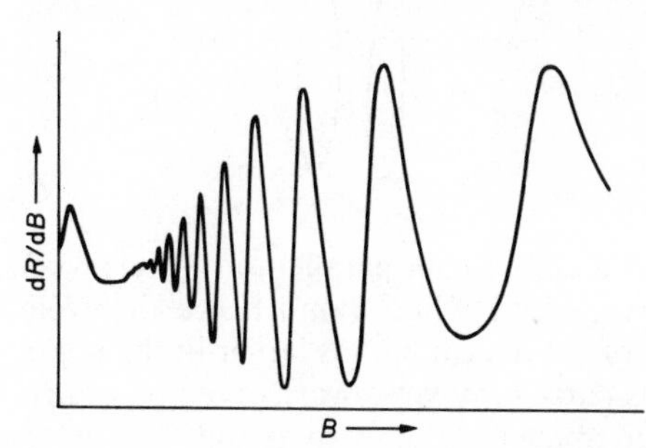

experimental curve of dR/dB *versus* B in a crystal of very pure copper. The crowding of the oscillations at low values of B is a consequence of their regular periodicity as a function of $1/B$.

Spatially varying magnetic field

Suppose that the lines of magnetic field spread out gently and symmetrically about a central straight line, and let a charged particle move in this field. We may readily verify that as it travels its orbit changes radius so as always to enclose the same flux. Imagine a conical 'tube of force' formed from field lines, having as axis the symmetry axis of the field configuration, and suppose the trajectory to lie on the surface of this tube. Since the Lorentz force is normal both to the trajectory and to $\mathbf{B}$, it is normal to the surface, and therefore is directed towards the axis; hence the angular momentum of the particle must stay constant. But this is precisely the condition for its staying on the tube; for if its component of velocity in the plane normal to the axis is v', its angular momentum L is $mv'r$, and from (11.3)

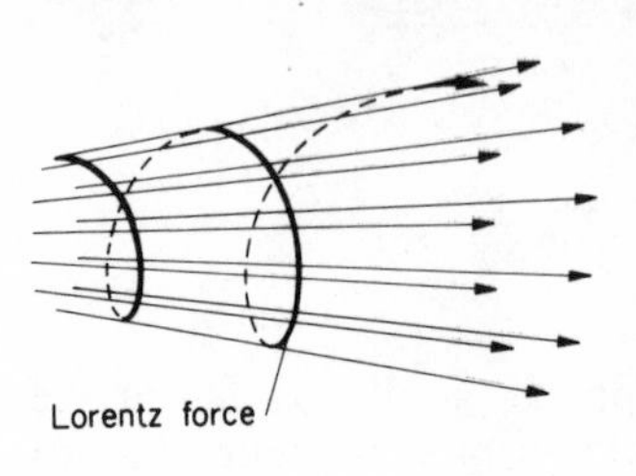

$$L = mv'r = eBr^2 = e\Phi/\pi, \tag{11.5}$$

where Φ is the flux, $\pi r^2 B$, enclosed in the orbit. Constancy of L implies constancy of Φ, and Φ is in any case constant on all cross-sections of a tube of force. Thus the postulated solution of the problem is verified.

The kinetic energy can be written as $\frac{1}{2}m(v'^2 + v''^2)$, v'' being the axial component of the velocity, and as the orbit radius changes so do v' and v'' in such a way as to maintain constant energy. Constancy of L demands that v' shall vary inversely as r, so that v'' increases as r increases. The diagram above shows how this is brought about by the Lorentz force

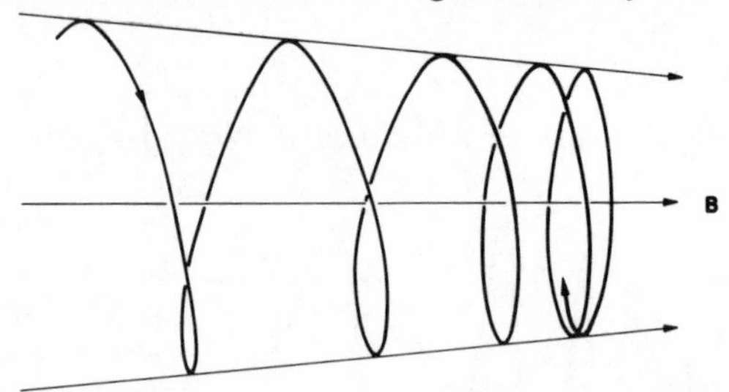

which, being normal to $\mathbf{B}$, has a component parallel to the axis when the lines diverge. Since v'' increases with r while B and hence the cyclotron frequency decrease, the pitch of the helical path is larger in the regions of weaker field. Conversely, if the particle moves towards a region of stronger field, the pitch decreases and may become zero if the field is strong enough to ensure that $\frac{1}{2}mv'^2$ accounts for the whole of the kinetic energy. At such a point there is still a component of the Lorentz force from right to left, which causes the particle to accelerate towards the left and retrace its path. This configuration of magnetic field therefore acts as a mirror for the particles.

In experiments aimed at heating ionized gas (*plasma*) to very high temperatures, in the hope of creating conditions in which nuclear reactions can occur and generate useful amounts of power, magnetic fields are used to contain the plasma without the necessity of solid walls which would cool it immediately. The solenoid shown in the diagram, with extra field-enhancing windings at the ends, is an

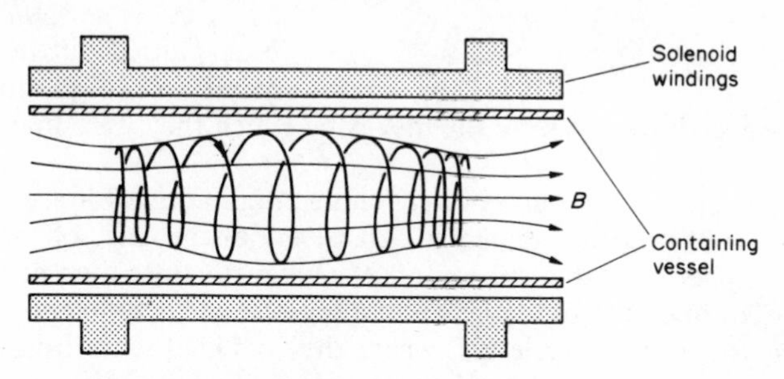

example of such a 'magnetic bottle', which can trap individual charged particles in orbits which are reflected to and fro by the stronger fields at the ends. The actual vessel in which the low-pressure gas is contained, whose heating creates the ionized plasma, may be just inside the solenoid windings, so that many particles never get a chance of touching the walls. Unfortunately the particles collide with one another and get thrown out of the bottle, and for this and other more subtle reasons plasma containment has not yet been achieved at such high temperatures ($\sim 10^8$ K) as allow nuclear reactions at a useful rate.

On the large scale, however, effects such as this are observed, as for example in the stable belts of protons that surround the Earth and are kept in being by the containing action of the Earth's magnetic field. It should be realized that the arguments applied above to axially symmetrical varying fields also apply when the field lines are gently bent. At any stage in its motion a particle executes a helix whose axis is parallel to the direction of **B** at the centre of the orbit. The path is therefore bent to follow the direction of **B**, and the constancy of the contained flux Φ, exactly as before, determines the variations of size and pitch of the bent helix. The Earth's field lines define magnetic bottles because of the convergence of field lines toward the poles, and protons trapped in contained orbits may run in helical paths between the poles for a very long time indeed if the mirror points lie at a sufficient height ($\gtrsim 2500$ km) for the rarefied atmosphere to permit no more than very infrequent collisions of the protons with other particles. The protons are produced, among other mechanisms, by the spontaneous decay of neutrons emitted from the Sun; the neutrons, being neutral, enter the region of the Earth's field without being deflected, but if one then decays into a proton and an electron, the proton may happen to move in such a direction as not to escape from the magnetic bottle. The exact form of the various belts of trapped particles (*Van Allen belts*) has been plotted out in many experiments with high-altitude rockets, which indeed were the means by which they were first discovered in 1958.

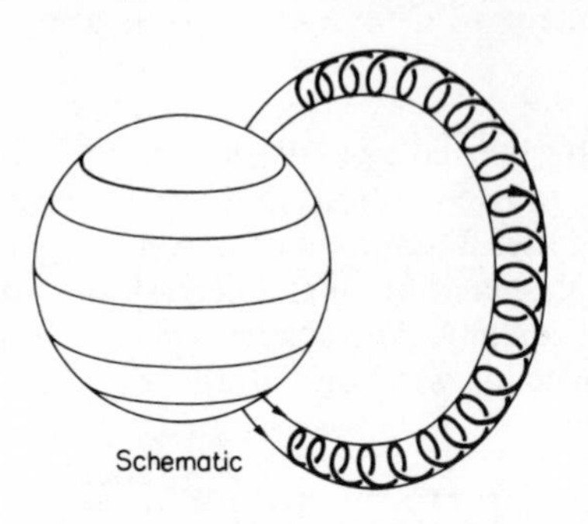

In 1962, a nuclear weapon test high above Johnson Island in the South Pacific created so extensive a disturbance of the upper atmosphere and ionosphere that the magnetic lines were distorted enough to open the bottleneck and allow the particles in the belt concerned to reach lower levels, where they collided with atmospheric atoms and were lost to the belt. Six years after the event, the belt had still not recovered its original form. This is enough to indicate how long an individual particle may survive in the belt; it also provides a salutary warning of how easily the Earth's environment may be grossly disturbed through inadequate prevision of the consequences of experiments.

Quantum effects in magnetic orbits

We have seen that when a charged particle moves in a spatially varying field the flux through its orbits is invariant. This suggests that we might try to apply quantum rules to the motion by permitting only those orbits for which the contained flux takes certain values. From (11.5) it is seen that Φ and L are uniquely related for a given charge e, so that quantization of Φ is closely analogous to quantization of L as in the Bohr atom. Alternatively we might note that the cyclotron frequency is directly proportional to B, so that ω_c/B is also an adiabatic invariant which might be quantized. In fact a full quantum-mechanical analysis shows that all these quantization schemes (which are only different ways of looking at the same thing) are to be carried out in a way similar to Planck's original quantization of the harmonic oscillator; the kinetic energy permitted to a particle in a plane orbit, with no motion along $\mathbf{B}$, is a half-integral multiple of $\hbar\omega_c$,

78

$$E = (n + \tfrac{1}{2})\hbar\omega_c. \qquad (11.6)$$

PROBLEM

Show that this quantum rule allows such orbits as enclose flux in half-integral multiples of the 'flux quantum' $2\pi\hbar/e$, and that the permitted values of L are $(2n + 1)\hbar$.

If in addition to its orbital motion the particle has a velocity component parallel to **B**, this is not subject to any quantal restriction; (11.6) still applies to the motion normal to **B**.

Quantization of the orbits leads to oscillatory behaviour of the magnetic properties of metals at very low temperatures; these variations are worth outlining briefly for their intrinsic interest, and because they allow searching investigation of the details of electron motion in metals. As before, we shall ignore the real complexities and pretend that the electrons move freely in a box of the same size as the metal sample. The first point to note is that a metal shows only very weak magnetic properties. This is perhaps surprising, since each electron executes an orbit around the lines of **B**, and therefore contributes a magnetic moment which can be calculated from Ampère's first law. If the electron velocity is v, (11.3) gives the radius and hence the area of the orbit, while (11.4) gives the cyclotron frequency and hence the current, which is the charge passing any point in unit time, i.e., $e\omega_c/(2\pi)$. In this way we find that an electron contributes a magnetic moment equal to its kinetic energy divided by B, and since all electrons have orbits in the same sense, the total moment is the sum of all individual contributions. Now this is not only very large, but it also has the remarkable property of varying inversely as B, and it is rather comforting to find that there is another effect which cancels the moment. This is illustrated in the diagram, which shows how a few electrons bounce round the perimeter of the metal so as to execute a large orbit in an anticlockwise sense. It can be shown, by a very general statistical argument that does not depend on the simple bouncing model used here, that the electrons near the surface always cancel the moment due to the electrons that do not touch the surface. If the orbits were not quantized there would be strictly no magnetic effect due to the electrons as a whole. But quantization of the orbits slightly upsets the balance of interior and surface electrons, and the degree of upset is determined by whether the most energetic electrons present are permitted to move with the Fermi velocity entirely lying in a plane normal to

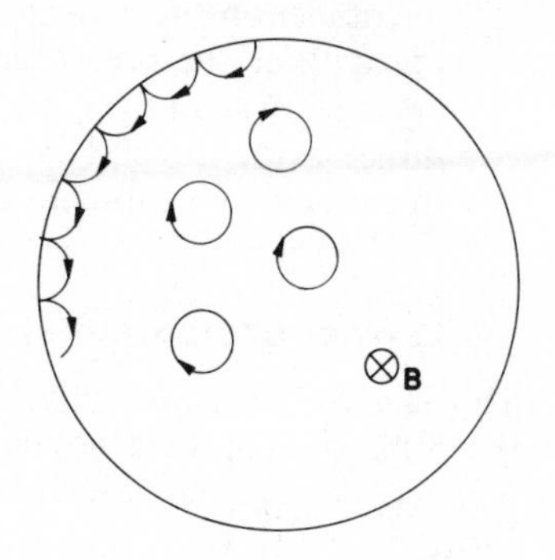

B, or whether the quantum condition imposes a component of **v** parallel to **B**, so that their moment is less than it might be in the unquantized metal. If we write the kinetic energy of the most energetic electrons as E_F, (11.4) and (11.6) show that wholly transverse motion satisfies the quantum condition at such values of B as are given by

$$\frac{1}{B} = (n + \tfrac{1}{2})\frac{e\hbar}{mE_F}. \tag{11.7}$$

As B is altered, the moments of interior and surface electrons swing in and out of balance periodically, so that measurements of the magnetic moment of a sample (at a low temperature, not more than a few degrees above the absolute zero, for the maximum energy E_F

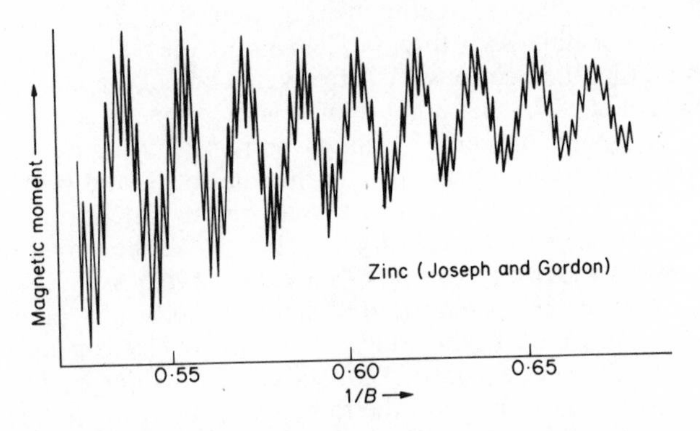

to be sufficiently well defined) reveal the effects of orbit quantization as an oscillation that has constant period in $1/B$. In practice the phenomenon (*de Haas-van Alphen effect*) is often rather more complicated, as the presence of several distinct periodicities in the diagram shows, and it is the analysis of these complications that gives direct evidence of how the electrons in the metal differ in their dynamical behaviour from free electrons.

Larmor's theorem

This is a very elegant analysis of the effect of a steady magnetic field on the orbit of a particle bound by a central force. In the absence of the magnetic field the equation of motion, as discussed in Chapter 5, takes the form

$$m\ddot{\mathbf{r}} = \mathbf{r}f(r), \tag{11.8}$$

and the central force on the right must now be supplemented by the Lorentz force to give the equation of motion

$$m\ddot{\mathbf{r}} = \mathbf{r}f(r) + e\dot{\mathbf{r}}\wedge\mathbf{B}.$$

Let us examine the motion from the point of view of an observer in a rotating frame of reference whose axis passes through the centre of force and is parallel to **B**. If the angular velocity is $\boldsymbol{\omega}$, the equation of motion in this frame, according to (2.3) and (2.4), takes the form

$$m\ddot{\mathbf{r}}' + 2m\boldsymbol{\omega}\wedge\mathbf{r}' + m\boldsymbol{\omega}\wedge(\boldsymbol{\omega}\wedge\mathbf{r}') = \mathbf{r}'f(r') + e(\dot{\mathbf{r}}' + \boldsymbol{\omega}\wedge\mathbf{r}')\wedge\mathbf{B}.$$

If we now set $\boldsymbol{\omega}$ equal to $-\frac{1}{2}e\mathbf{B}/m$ (i.e., half the cyclotron frequency), the Lorentz and Coriolis forces cancel each other, leaving

$$m\ddot{\mathbf{r}}' = \mathbf{r}'f(r') + m\boldsymbol{\omega}\wedge(\boldsymbol{\omega}\wedge\mathbf{r}'). \tag{11.9}$$

Apart from the last term, (11.9) has the same form as (11.8). The last term is quadratic in ω, and therefore in B, and provided the orbit is finite so that r' has a maximum value, we can always choose B small enough for this term to be negligible compared to the central force. This demonstrates that the effect of **B**, if not too large, on a finite central orbit can be annulled by viewing the orbit from a rotating framework, or alternatively that the effect of **B** is to cause the orbit to precess around an axis defined by **B** at one-half the cyclotron frequency. Thus a charged particle in a Coulomb field, which without **B** executes an elliptical orbit, is caused by **B** to move in a rosette.

As a consequence of this Larmor precession, the magnetic moment of the orbit is changed, and it was this that Langevin suggested as the origin of the diamagnetism exhibited by many atoms. The easiest way to calculate the size of the diamagnetism is to work out the average effect produced by **B** on a large number of identical atoms whose orbits lie in different planes with their normals evenly distributed in all directions. Before **B** is applied the mean velocity of the electrons at a given point **r**, each relative to its own centre of attraction, will be zero since for every electron moving with **v** there will be another traversing the orbit in the opposite sense with velocity $-\mathbf{v}$. However, **B** causes each orbit to precess with the same angular velocity $-\frac{1}{2}e\mathbf{B}/m$, so that the average behaviour involves currents circulating round an axis parallel to **B**. If the electron spends a fraction $f(\rho)\,\mathrm{d}\rho$ of its time between ρ and $\rho + \mathrm{d}\rho$ away from the axis through the centre of the atom, we may calculate the contribution of this ring of radius ρ to the moment by supposing a charge $fe\,\mathrm{d}\rho$ to be moving round it with angular velocity $-\frac{1}{2}e\mathbf{B}/m$, and acting as a current $fe^2B\,\mathrm{d}\rho/(4\pi m)$ in an orbit of area $\pi\rho^2$. Then by A1 the moment, integrated over all ρ, is given by

$$\mathfrak{m} = \frac{e^2B}{4m}\int f\rho^2\,\mathrm{d}\rho = \frac{e^2B}{4m}\,\overline{\rho^2},$$

where $\overline{\rho^2}$ is the average value of ρ^2 for the electrons in the atom. Since we are concerned with all orientations of the orbits, we may relate $\overline{\rho^2}$ to $\overline{r^2}$, the mean square distance of electrons from the nucleus. For if **B** is directed along the z-axis,

$$\rho^2 = x^2 + y^2 \qquad \text{and} \qquad \overline{\rho^2} = \overline{x^2} + \overline{y^2},$$

while

$$r^2 = x^2 + y^2 + z^2 \qquad \text{and} \qquad \overline{r^2} = \overline{x^2} + \overline{y^2} + \overline{z^2}.$$

Now for random orientation x, y and z must be equivalent, so that

$$\overline{x^2} = \overline{y^2} = \overline{z^2} \qquad \text{and} \qquad \overline{\rho^2} = \tfrac{2}{3}\overline{r^2}.$$

Hence

$$\mathfrak{m} = -\frac{e^2\mathbf{B}}{6m}\overline{r^2}. \tag{11.10}$$

The vector form of $\mathfrak{m}$ and the negative sign have been inserted here by inspection. The Larmor precession is in such a direction that the extra current sets up a field inside the orbit in opposition to **B**. This direction of magnetization, which is in the same sense as that of an orbiting free electron, is called *diamagnetism*.* By contrast a magnetic dipole tends to turn so that $\mathfrak{m}$ is parallel to **B**, and this is called *paramagnetism*. An atom containing an orbiting electron whose moment is not cancelled by another in the reverse orbit exhibits both effects, paramagnetism due to the turning of the atom to align the moment nearer **B** and diamagnetism due to the Larmor precession modifying the orbits of all the electrons. Under these circumstances the paramagnetism usually dominates, but of course if the resultant moment of all orbits in an atom is zero, there still remains the diamagnetism.

273 These ideas will be developed a little further at a later point.

Crossed electric and magnetic fields

Consider the motion of a charged particle in uniform fields **E** and **B** which are normal to each other. There is no force acting along the direction of **B** and this component of velocity is therefore constant. We shall neglect such motion and concentrate on motion in the plane transverse to **B**. It is convenient to use complex notation and to take **E** along the real axis. Then the equation of motion has the form

$$m\ddot{r} = eE + ieB\dot{r}.$$

Substituting $y + iE/B$ for $\dot{r}$, we find y to be governed by the equation:

$$m\dot{y} = ieBy,$$

* Langevin's expression (11.10) for the diamagnetic moment is in conflict with a very general theorem, due to van Leeuwen, which states that the magnetic moment of a system of classical particles in equilibrium is identically zero under all circumstances. It should be noted, however, that the treatment given here applies Larmor's theorem to an orbiting electron, a system which in classical mechanics is unstable. In effect we have tacitly fed in a non-classical assumption, along the lines of Bohr's theory of the hydrogen atom, and need not be worried at getting an answer that is also non-classical. The answer expressed by (11.10) is indeed exactly what comes out of a proper quantum-mechanical analysis.

with solution $y = y_0 \exp(i\omega_c t)$, ω_c being the cyclotron frequency eB/m and y_0 a constant of integration. Integration with respect to t gives the general solution:

$$r = r_0 + iEt/B - iy_0/\omega_c \exp(i\omega_c t). \qquad (11.11)$$

This describes uniform motion in a circle of radius y_0/ω_c (y_0 is arbitrary until fixed by the energy of the particle) superposed on a steady drift, upwards in the diagram, with velocity E/B. This drift, whose direction is the same for both positive and negative particles and whose magnitude is independent of charge, arises through variations of the radius of the otherwise circular trajectory as movement along **E** changes the kinetic energy of the particle. The positive particle in the diagram has greater energy and therefore greater radius on the right side, the negative particle on the left side, and the diagram shows how this results in the same direction of drift for both.

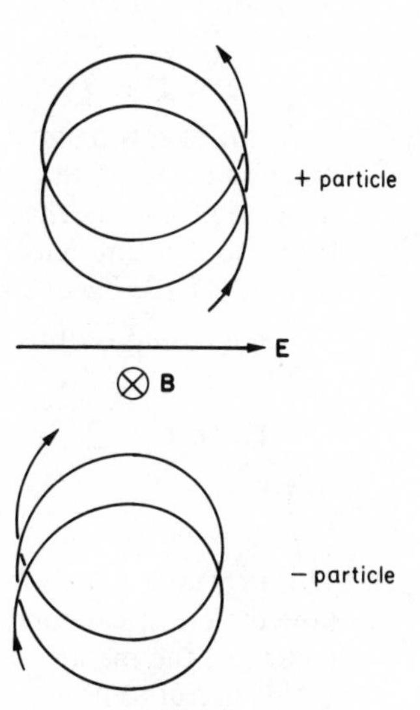

It is interesting to note that there is no mean velocity along the direction of the electric field; in principle, any magnetic field, however weak, should be able to prevent **E** from accelerating charged particles. Leaving aside the breakdown of this theory, for relativistic reasons, when the drift velocity E/B exceeds the velocity of light, in practice the limitation of its application is set by collisions of the particles with other particles or obstacles, or with the boundaries of their container. Between collisions, a particle executes a drifting helical path, and the effect of collisions is to permit a gradual shift of the helix in the direction of **E**, and thus to set up a current component parallel to **E**. With certain simplifying assumptions, the effect is readily calculated for the electrons in a metal.

First we examine the situation when $\mathbf{B} = 0$, and derive the steady current by writing down the factors tending to change the momentum per unit volume of the electron assembly. On the n electrons contained in unit volume, the field **E** exerts a force $ne\mathbf{E}$, which therefore represents the rate of change of momentum density. This is countered by collisions, which may be supposed to transfer the electronic momentum to the ionic lattice at a rate proportional to the momentum present:

$$\partial \mathbf{P}/\partial t = -\mathbf{P}/\tau,$$

where τ is a constant for a given metal at a given temperature and is called the *relaxation time*. If **E** was used to give the electrons momentum $\mathbf{P}_0$ and was then switched off, the momentum would, according to this equation, be dissipated exponentially, $\mathbf{P}(t) = \mathbf{P}_0\, \mathrm{e}^{-t/\tau}$, and after an interval of a few times τ would be almost entirely lost. In the continued presence of **E**, however, a stationary situation is achieved when the momentum loss by collisions matches the gain from the field,

$$ne\mathbf{E} = \mathbf{P}/\tau. \tag{11.12}$$

Now an electron moving with velocity **v** has momentum $m\mathbf{v}$ and makes a corresponding contribution $e\mathbf{v}$ to the current carried by the metal; no matter what the distribution of velocities may be, therefore, momentum density **P** is associated with current density $e\mathbf{P}/m$, so that (11.12) becomes

$$ne\mathbf{E} = m\mathbf{J}/(e\tau),$$

or

$$\mathbf{J} = \sigma_0\mathbf{E},$$

where

$$\sigma_0 = ne^2\tau/m. \tag{11.13}$$

This expresses *Ohm's law*, the proportionality of **J** and **E**, and shows how the electrical conductivity, σ_0, is related to the electronic parameters of the metal.

It is useful to note, when we come to introduce the magnetic field, that the loss of momentum to the lattice by collisions is equivalent to a force per unit volume of magnitude $-\mathbf{P}/\tau$, acting on the electrons, a force which can also be written as $-ne\mathbf{J}/\sigma_0$. The force exerted by **E** and **B** on a single electron is given by (11.1), and on summing over all electrons in unit volume we have a resultant force density of $ne\mathbf{E} + \mathbf{J}\wedge\mathbf{B}$, which in the stationary state is balanced by the collision force $ne\mathbf{J}/\sigma_0$, i.e.,

$$\mathbf{E} = \mathbf{J}/\sigma_0 - \mathbf{J}\wedge\mathbf{B}/(ne).$$

This equation has a simple geometrical interpretation; because of the Lorentz force, $\mathbf{J}\wedge\mathbf{B}$, the current and electric field are no longer parallel, as when $\mathbf{B} = 0$. The angle, θ, between them is clearly $\tan^{-1}[\mathbf{B}\sigma_0/(ne)]$ i.e., $\tan^{-1}(\omega_c\tau)$ from (11.4) and (11.13). When a current is passed through a conducting strip in the presence of a transverse magnetic field, the component of **E** parallel to **J** can be detected (and measured with the aid of a potentiometer) as a potential difference between the electrodes *A* and *B*, while the component normal to **J** and **B** is detected by means of the electrodes *C* and *D*. The parallel field associated with the current **J** may be written $E_{\parallel} = \rho J$, and ρ is the *resistivity*. In this simple free-electron model, ρ does not change with **B** and takes its zero field value ρ_0, or $1/\sigma_0$. The normal field $E_{\perp}$ is $\rho_{\mathrm{H}}J$, where ρ_{H} is the *Hall resistivity*; in this simple case, and not uncommonly in practice, ρ_{H} is proportional to

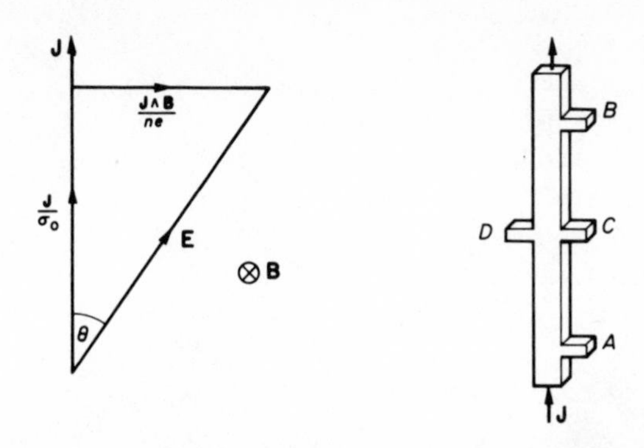

B; the quotient ρ_H/B is the *Hall constant*, and is here given by $1/(ne)$. The Hall effect thus measures the density of electrons taking part in conduction, and has proved to be a very valuable aid in the study of the technically important semiconductors (e.g., Ge, Si, InSb) in which n is small compared with its value in metals, and the Hall effect is correspondingly large and easily studied.

Metals in which the behaviour of the conduction electrons is closest to motion in free space do indeed approximate fairly well to the model presented here, in that the Hall effect agrees quantitatively with the number of electrons calculated from the number of atoms per unit volume, and the resistivity does not change much when a transverse magnetic field is applied. A sample of potassium at 4·2 K (with conductivity 560 times higher than at room temperature) increased in resistance by only 2% when a field of 5 T was applied, though $\omega_c\tau$ was then 18 and the transverse Hall field was 18 times as strong as the field along the wire. Other metals, however, show a pronounced *magneto-resistance* which depends very strongly on the orientation of **B** with respect to the crystal axes; increases of ρ by a factor of 10^6 have been observed in bismuth, and by many thousands in a number of other metals. These dramatic deviations from the simple free-electron theory have proved extremely useful in determining how the motion of electrons is modified when they are contained in a metal and have to move through the potential variations of the ionic lattice.

Cavendish Problems: 32, 192, 231.

READING LIST

Electron Optics: V. E. Cosslett, *Electron Optics*, Oxford U.P.
Cyclotron: E. Segré, *Nuclei and Particles*, Benjamin.
Electrons in Metals: J. M. Ziman, *Electrons in Metals*, Taylor and Francis.
Plasmas: L. Spitzer, *Physics of Fully Ionized Gases*, Interscience.
Van Allen Belts: S. F. Singer and A. M. Lenchek, 'Geomagnetically Trapped Radiation', in *Progress in Elementary Particle and Cosmic-ray Physics*, Vol. 6, North-Holland.

12

Electromagnetic induction

***Faraday's law.* When the flux of magnetic field through a circuit is changing, an electromotive force is set up in the circuit, of magnitude proportional to the rate of change of flux.**

We have discussed the motion of a charged particle in various configurations of electric and magnetic fields, and shall now examine some of these cases from the point of view of another observer moving at a uniform velocity $\mathbf{u}$. In Chapter 3 we applied this method of examination to the dynamical behaviour of particles under the influence of such recognizable forces as are exerted by strings etc., and we now extend the argument to charged particles influenced by electric and magnetic fields. For the sake of definiteness we shall suppose that the fields are produced by charged conductors and permanent magnets at rest relative to the observer referred to in the last chapter. The moving observer therefore observes events resulting from the movement of particles in fields which are themselves in motion, in the sense that he is aware of the sources of the fields moving at velocity $-\mathbf{u}$ in his frame of reference. If u is not comparable to the velocity of light we may accept, in the light of the discussion in Chapter 3, that both observers agree on the measurement of mass and acceleration, but we shall quickly find that they do not agree on the origin of the force responsible for the acceleration.

Consider for example a particle executing a helical path in a magnetic field which varies slowly with position (e.g., radiating from a pole P), and let us look at the motion from the point of view of an observer moving to the right at constant speed u, so that he sees the particle when it reaches A with no component of velocity momentarily in the direction of **B**. Like 232

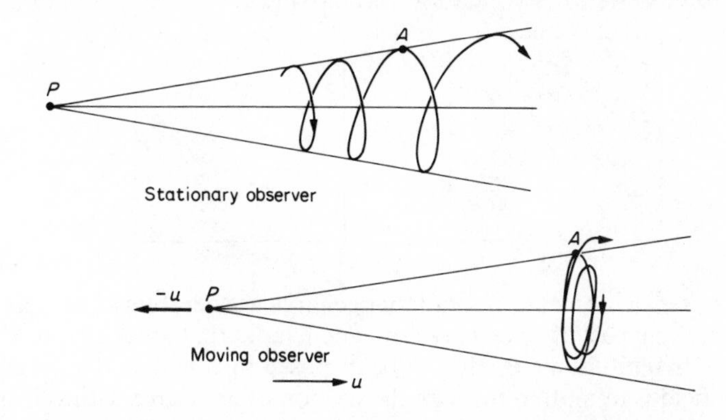

the fixed observer, he sees that the orbiting particle has constant angular momentum and that the orbit expands to include constant magnetic flux; but whereas the fixed observer sees the decreasing orbital contribution to the kinetic energy compensated by the increasing energy associated with motion along **B**, the moving observer sees no such compensation and will be liable to conclude that there is an extra tangential force decelerating the particle.* A circulating electric field would meet the need.

Consider again the motion of a charged particle in crossed **E** and **B**, which according to (11.11) consists of a helical orbit superposed on a steady drift at speed E/B. If an observer moves at this speed he will see the particle executing a simple helical orbit without drift and in the absence of any other information will conclude that it is acted upon by a magnetic field **B** but no electric field. In both these examples, we see how different observers, seeing a particle acted upon by the same force and suffering the same acceleration, nevertheless may disagree on the origin of the force, attributing different importance to the two terms that make up the right-hand side of (11.1). If we wish to retain our belief in Galilean invariance (invariance of Newton's laws of motion with respect to uniform relative motion of different observers) in systems involving electric 238

* The reader may well wonder how angular momentum is conserved while the kinetic energy changes in response to an apparent tangential force. The answer is that because of the distortion of the trajectory, the Lorentz force as seen by the moving observer does not pass through the axis of the trajectory as it did for the stationary observer. The moving observer might just as well have inferred the existence of an extra tangential force from the fact that something was needed to balance the tangential component of the Lorentz force and keep angular momentum constant.

and magnetic effects, we must, while preserving the general concepts of electric and magnetic fields, abandon any thought that they are absolute, and recognize instead that different observers need not agree about the magnitudes of **E** and **B**.

One final example, simpler than those already discussed, should make the position clear. An observer watching a charge e moving with velocity $\mathbf{v}$ past a stationary pole P attributes its acceleration to the action of the Lorentz force $e\mathbf{v} \wedge \mathbf{B}$. A second observer, moving at velocity $\mathbf{v}$ with respect

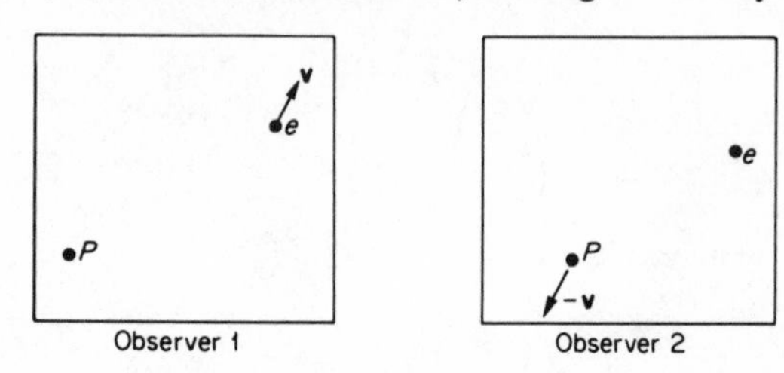

to the first, sees a momentarily stationary charge accelerating as a result of the pole moving past at velocity $-\mathbf{v}$, and concludes that an electric field is present, of magnitude $\mathbf{v} \wedge \mathbf{B}$. He will be disposed to conclude that when a magnetic field is in motion through the motion of its source with velocity $\mathbf{u}$ (in this case $\mathbf{u} = -\mathbf{v}$), it generates locally an electric field given by the equation:

$$\mathbf{E} = \mathbf{B} \wedge \mathbf{u}. \qquad (12.1)$$

The argument may be put into more general terms by enquiring what relations must exist between the field strengths seen by different observers if they are all to agree that (11.1) is the correct force law. If one observer sees **E** and **B**, and a second moving at velocity $\mathbf{u}$ relative to the first sees $\mathbf{E}'$ and $\mathbf{B}'$, the force on a charged particle is $e(\mathbf{E} + \mathbf{v} \wedge \mathbf{B})$ in the first frame and $e[\mathbf{E}' + (\mathbf{v} - \mathbf{u}) \wedge \mathbf{B}']$ in the second. If these are to be the same for all $\mathbf{v}$, the terms independent of $\mathbf{v}$ and those proportional to $\mathbf{v}$ may be equated separately, to give $\mathbf{B}' = \mathbf{B}$ and $\mathbf{E}' = \mathbf{E} + \mathbf{u} \wedge \mathbf{B}$. It should be noted that in this argument we do not introduce any concept of motion of the field lines; we simply record the relations between what is seen by different observers without stating why they see certain field strengths. It is only when one relates the observed fields to their sources that one may find it convenient to think in terms of moving field lines and similar mechanical analogies. So long as the sources—charges, poles, dipoles, etc.—are at rest relative to the observer we regard them as pure sources of either **E** or **B** in accordance with Coulomb's or Michell's laws. When they are moving, we may imagine them to carry their lines with them and generate extra fields thereby.

There is, however, a curious asymmetry about the rule we have developed. The movement of **B** with velocity $\mathbf{u}$ apparently generates an electric field $\mathbf{B} \wedge \mathbf{u}$, but if we are to accept the transformation $\mathbf{B}' = \mathbf{B}$ which we have just derived we must conclude that the movement of **E** does not reciprocally generate a magnetic field. This is not a very satisfactory situation, since we know that the movement of charges in a conductor when a current flows generates a magnetic field, and we should be prepared to regard this as a process entirely analogous to the generation of an electric field by a moving pole. For the present, we note the existence of an inconsistency, which will turn out to be more than a minor affair; indeed it hints at the inadequacy of the Newtonian assumptions when high speeds are involved, and leads us to the threshold of relativity theory. 305

Leaving this aside, then, we concentrate on the way in which a moving magnetic field generates an electric field, which is the phenomenon of *electromagnetic induction*, discovered by Faraday and expressed by him in terms similar to the statement at the head of the chapter. The discovery was made many years before the Lorentz force was enunciated, and our treatment departs entirely from the historical sequence. We shall proceed to show how the ideas of the last few pages enable us to derive Faraday's law, but it should not be supposed that the law is thereby made redundant and to be regarded as a straightforward corollary of the Lorentz force law. It is only after one has discovered the essential simplicity of the structure of the fundamental laws that one can afford to neglect the possibility that they might have been something quite different.

In the light of this discussion let us exemplify Faraday's law by seeing what effect is produced when the magnetic field threading a circuit is changed, either by moving the sources of the field or by deforming the circuit. Suppose for simplicity that the field through the circuit is provided by a single magnetic pole which is then moved with velocity $\mathbf{u}$. As a result each electron in the wire experiences an additional field, $\mathbf{E}_1 = \mathbf{B}\wedge\mathbf{u}$, by (12.1), and under its influence drifts round the circuit. But, if the circuit is not closed, this drift causes an accumulation of charge which produces further electric fields opposing the motion, and soon an equilibrium arrangement results in which the charge density at all points is such that there is no resultant force on the electrons to impel them round the circuit. This charge distribution therefore produces an electric field $\mathbf{E}_2$, of magnitude $-\mathbf{B}\wedge\mathbf{u}$, at every point within the metal of the circuit. What then is the situation in the vicinity of the terminals AB which we suppose to be well away from the region where $\mathbf{B}$ is strong? There is no significant field resulting from the movement of the magnetic pole, but there is a field from the charge distribution; for the line integral of $\mathbf{E}_2$ round a closed circuit must vanish if $\mathbf{E}_2$ is caused by real charges. It follows, by taking a line integral from A round the circuit to B and then across the gap back to A that

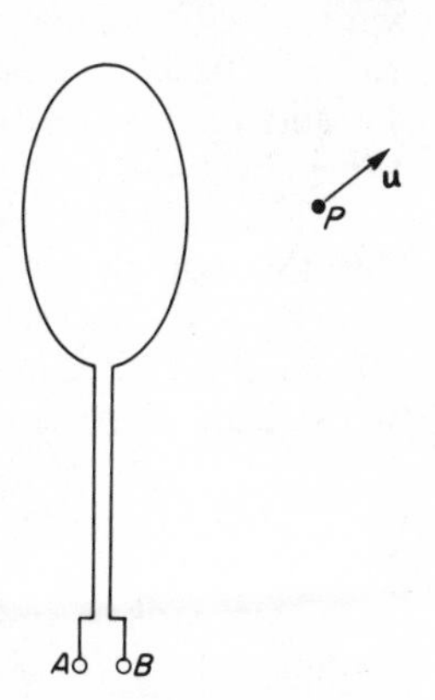

$$0 = \oint \mathbf{E}_2 . \mathrm{d}\mathbf{r} = \int_A^B -(\mathbf{B}\wedge\mathbf{u}) . \mathrm{d}\mathbf{r} + \int_B^A \mathbf{E} . \mathrm{d}\mathbf{r},$$

the first integral being round the circuit and the second across the gap. Writing the second as V, the *electromotive force* (e.m.f.) that would be recorded by a voltmeter attached to AB, we have

$$V = \pm \int_A^B (\mathbf{B} \wedge \mathbf{u}).\mathrm{d}\mathbf{r}. \tag{12.2}$$

For the present we shall not worry about which sign to use in (12.2).

As we have described the experiment, movement of the magnetic field generates an electric field $\mathbf{E}_1$ at all points, irrespective of whether or not the wire circuit is there to respond; the response of the circuit is to set up another balancing electric field $\mathbf{E}_2$ which is detected at the terminals. The total field present at any point, $\mathbf{E} = \mathbf{E}_1 + \mathbf{E}_2$, but since $\mathbf{E}_2$ is irrotational it follows that any evaluation of $\oint \mathbf{E}.\mathrm{d}\mathbf{r}$ round a closed loop will take the same value, $\oint \mathbf{E}_1.\mathrm{d}\mathbf{r}$, whether or not there are wires present. Thus (12.2) expresses the value of $\oint \mathbf{E}.\mathrm{d}\mathbf{r}$ round any closed loop in terms of the movement of magnetic field in the vicinity. It should be particularly noted that $\mathbf{E}$ is no longer a conservative (irrotational) field of force, and that a potential ϕ cannot now be defined such that $-\mathrm{grad}\,\phi = \mathbf{E}$ everywhere.

If we were to examine the experiment from the point of view of an observer moving with the magnetic pole, we should see a circuit moving with velocity $-\mathbf{u}$ in a fixed magnetic field, and should expect the electrons to drift until they set up an irrotational electric field balancing the Lorentz force at all points in the wire (but not necessarily outside). The same effect will be produced at the terminals, as expressed by (12.2). In this frame it is easier to see how the e.m.f. may be written in terms of the flux through the circuit. As any element $\mathrm{d}\mathbf{l}$ of the circuit moves with velocity $-\mathbf{u}$, it sweeps out area at a rate $\mathbf{u} \wedge \mathrm{d}\mathbf{l}$, and increases the flux of magnetic field through the circuit at a rate $\mathbf{B}.(\mathbf{u} \wedge \mathrm{d}\mathbf{l})$ (again we ignore signs). If then we write the total flux through the circuit as Φ, we have that

$$\mathrm{d}\Phi/\mathrm{d}t = \pm \oint \mathbf{B}.(\mathbf{u} \wedge \mathrm{d}\mathbf{l}) = \pm \int (\mathbf{B} \wedge \mathbf{u}).\mathrm{d}\mathbf{l}, \qquad \text{by VI.}$$

By comparison with (12.1) we write the e.m.f. in the form given by Faraday's law:

$$\mathbf{V} = \pm\, \mathrm{d}\Phi/\mathrm{d}t. \tag{12.3}$$

It will be seen that we expect the e.m.f. to be exactly equal to $\mathrm{d}\Phi/\mathrm{d}t$ and that there is no undetermined constant of proportionality.

PROBLEM

Extend the preceding analysis to the case where a circuit is arbitrarily deformed, by moving different parts at different rates, and show that (12.3) still holds.

We passed lightly over the question of signs in (12.2) and (12.3) although it does not in fact present any difficulties, the direction of the field between A and B being clear by inspection and our knowledge of the direction of the Lorentz force. If the circuit be imagined to lie in a horizontal plane and a positive pole to be above it, when the circuit is raised towards the pole with velocity $\mathbf{v}$ the flux is increased. The field $\mathbf{E}_2$ that annuls the effect of the Lorentz force is $-\mathbf{v}\wedge\mathbf{B}$ which since $\mathbf{B}$ has an outward-pointing component in the plane of the circuit, circulates in a clockwise sense, as seen from above. For the line integral of $\mathbf{E}_2$ to vanish when taken round the circuit, the field must run from A to B as shown, and a voltmeter will record A as the positive terminal. Whether this means a positive or negative sign in (12.3) is entirely a matter of convention, and convention has decreed that it should be negative. To remember the sign of the e.m.f. it is helpful to think about what would happen if A and B were joined by a conductor. There would now be nowhere for space charge to build up and create the electric field needed to annul the Lorentz force, and a current of positive charge would be pushed round by the Lorentz force in an anticlockwise sense. Such a current generates a magnetic field that rises through the circuit, in opposition to the increment of flux caused by motion of the circuit. This exemplifies Lenz's law, that any current set up by electromagnetic induction flows in such a direction as to oppose the change of flux responsible for the induction. If the circuit is now broken, the current will continue for a very short space of time until it has established $\mathbf{E}_2$, and the positive terminal is clearly that from which current would flow into a wire completing the circuit.

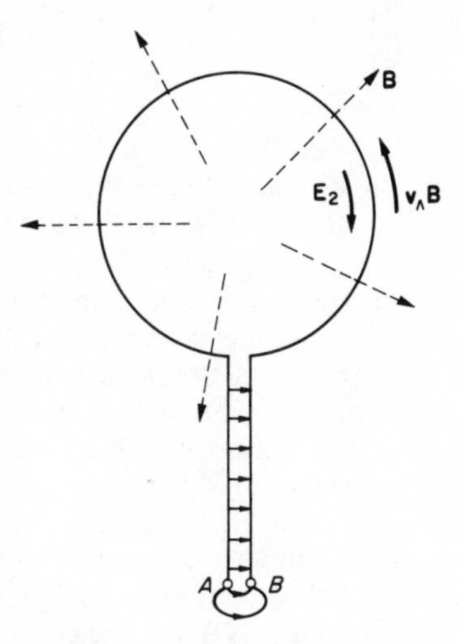

255

We have developed (12.3) for the special case where a single magnetic pole is moved with respect to a circuit, but the form of the result in terms of the total flux linkage shows that in a more complex situation, where the field is due to one or more permanent dipoles moving arbitrarily, we may expect the superposition of the effects of individual poles to yield the correct answer in exactly the same form—it is the change of Φ, no matter how it is produced, that is responsible for the induced e.m.f. It would be dangerous to assert that (12.3) must still apply when the flux change is brought about by changing the current flowing in neighbouring circuits, for we have deliberately avoided interpreting the magnetic field of a current in terms of the movement of electric fields, and cannot with any assurance apply the methods of Galilean transformation that work so well with poles. Nevertheless, experiments on the e.m.f. induced by changing currents leave one in no doubt about the validity of Faraday's law here also. The law, applied to current-induced e.m.f., is indeed the basis of the discussion of self- and mutual inductance that soon follows, and the use of such inductances in precision measurements confirms beyond doubt that the application is valid.

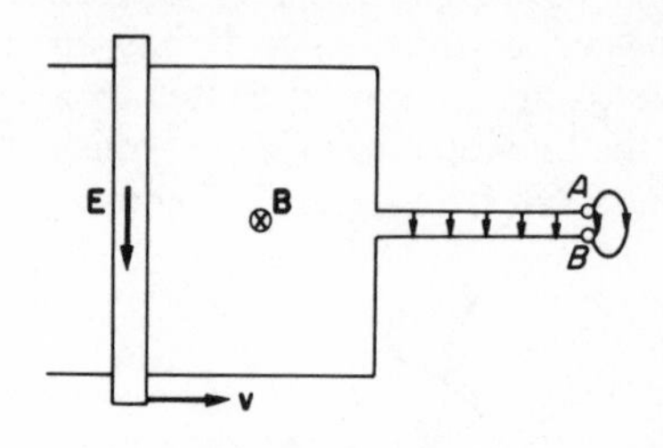

So far our examples have extended to deformable circuits, and we now proceed one step further to circuits containing sliding contacts. The first example presents no problems. The metal slider moving in the field **B** experiences the Lorentz field $\mathbf{v} \wedge \mathbf{B}$ and becomes polarized just as did the moving circuit we have already discussed. The field due to the resulting space charges acts along the slider and between the terminals in the sense shown, and the reader will easily satisfy himself that (12.3) holds in this case.

It is interesting to extend this argument to a case not covered by the formulation expressed in (12.3). Suppose the slider to be an insulator, not a metal; then each molecule experiences the Lorentz force and acquires a dipole moment just as if it were subject to an electric field $\mathbf{v} \wedge \mathbf{B}$. Unless the slider is a very long, thin rod, the surface charge distribution due to this polarization sets up a real electric field in the material. If, for simplicity, we suppose the slider to be an ellipsoid with depolarizing coefficient D, this real field has magnitude $-D\mathbf{P}/\varepsilon_0$, so that the effective polarizing field acting on the molecules is $\mathbf{v} \wedge \mathbf{B} - D\mathbf{P}/\varepsilon_0$. Hence

$$\mathbf{P} = \kappa\varepsilon_0(\mathbf{v} \wedge \mathbf{B} - D\mathbf{P}/\varepsilon_0)$$

and

$$D\mathbf{P}/\varepsilon_0 = \frac{\kappa D}{1 + \kappa D}\,\mathbf{v} \wedge \mathbf{B}.$$

It is this depolarizing field, $D\mathbf{P}/\varepsilon_0$, that is picked up by the stationary wires and should be recorded by a suitable voltmeter at the terminals; and if $\kappa D > 1$ as is easily achieved, the potential difference is almost as great as if the slider were metallic. It is important to note, however, that a 'suitable' voltmeter must be used. The dielectric rod cannot pass a current, and as soon as current goes through the voltmeter the wires become charged and acquire a potential difference determined by their mutual capacitance. The only hope of observing the effect would be to use an electrostatic voltmeter or an electronic circuit of very high resistance; better still, to use the latter and also to oscillate the dielectric rod so as to generate an alternating signal that can be picked up and amplified. It would be inappropriate to enter into further details of how to observe the effect and how the capacitance of the instruments would further alter the result, but the the example should suffice to show that inductive effects are in principle possible even when Faraday's flux law (12.3) cannot be applied.

For a second example we consider the *homopolar generator*, a spinning metal disc with **B** parallel to its axis, and sliding contacts at rim and hub. This is clearly the same machine in principle as the homopolar motor already described, but worked in reverse; instead of converting electrical into mechanical energy, the generator uses mechanical energy to produce electrical energy. If we were to think of the circuit as made up of the wires AP and BQ, together with a radius of the disc joining P and Q, we might be misled into believing that the flux linkage was constant and therefore no e.m.f. should appear across AB. But clearly this is wrong, since the Lorentz force on the electrons moving with the disc must be balanced by a radial electric field. In fact, at radius r where the speed is ωr, the electric field must be $\omega r B$, and the e.m.f. between centre and rim must be $\omega B \int_0^a r dr$, i.e., $\frac{1}{2}\omega B a^2$, or $\omega \Phi / 2\pi$ if Φ is the flux through the whole disc. This will be recorded by a voltmeter at AB. To apply (12.3), one should consider the circuit as completed, not by the radius PQ, but by any radius fixed with respect to the disc, such as PR, and the variable arc RQ, which does not cut flux as the wheel spins. The rate of change of flux in this changing circuit gives the correct result.

220
266

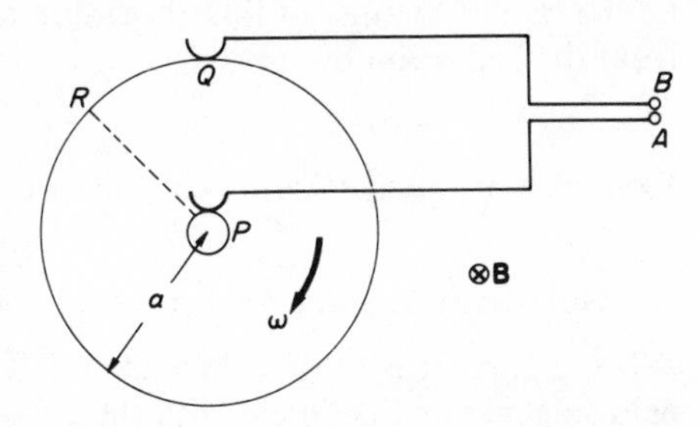

The different examples given should be enough to indicate that various points of view are useful in discussing electromagnetic induction. In spite of its freedom from what many would regard as undesirable mechanistic models, like field lines, the approach by way of (12.3) is not always the safest—indeed we have had one example where it simply does not apply. These awkward cases tend to be rare and should not be given undue prominence; but when they arise, it seems to be a good rule to think in terms of Lorentz forces if the magnetic field is at rest, or moving lines if it is not. When the source of the magnetic field is itself composite, with different parts moving at different speeds, it is always desirable to imagine the field built up from the contributions of individual dipoles, or even poles, each of which carries its field with it as it moves. Doing this, we are not tempted to suppose that when a bar magnet rotates about its axis, the magnetic lines spin round with it; the resulting inductive effects are correctly predicted only by supposing each elementary dipole to carry its own lines without rotation, only translation.

Finally, persuasive though the field line concept is, one must never be lulled into believing that the lines are in any sense real. After all, even the fields which they represent have magnitudes about which different observers must disagree. Nevertheless as fictions, as representations of the technique that gives the right answer to a calculation, field lines have great value simply because the fact that they can be visualized enables intuitive powers to be brought to bear to guess solutions without elaborate analysis.

If, however, exact analysis is necessary, it is convenient to express Faraday's law in differential form, and this presents no difficulty; if $\oint \mathbf{E}.\mathrm{d}\mathbf{r}$ is formed round an infinitesimal closed path, its magnitude must be the rate of change of flux threading the circuit, from which it follows from the definition of curl that

209

$$\operatorname{curl} \mathbf{E} = -\dot{\mathbf{B}}. \tag{12.4}$$

Examples of the application of this equation will be given later.

Self-inductance and mutual inductance

When a current is passed through a coil of wire, in general the magnetic field so generated is linked with the coil and with others in the vicinity so that changing the current generates an e.m.f. across the terminals of each coil. If current i in coil A produces flux $M_{AB}i$ through coil B, according to (12.3) the e.m.f. across the terminals of B is $M_{AB}\, \mathrm{d}i/\mathrm{d}t$; M_{AB} is the *mutual inductance* between A and B. Similarly if the current i in A produces flux $L_A i$ through A itself, the e.m.f. across the terminals of A when i is changed is $L_A\, \mathrm{d}i/\mathrm{d}t$; L_A is the *self-inductance* of A. The phenomenon of mutual inductance is readily demonstrated by winding two separate layers of wire on the same former, one on top of the other so that the field produced by a current in either is intimately linked with the other. If one coil is connected to a galvanometer and a current through the other is suddenly reversed or switched off, the galvanometer will give a kick to indicate the induced e.m.f. The size of the kick is found to be proportional to the current in the first coil, as expected from (12.3). The phenomenon of self-inductance is not so simply demonstrated because the same terminals must be used to inject the current and to measure the e.m.f., and the resistance of the coils leads to the presence of an obscuring steady e.m.f. even for a steady current. Nevertheless if the flux linkage is large, and L correspondingly large, as is most easily achieved with iron-cored coils and the windings of electromagnets, dramatic effects of self-inductance may be produced by switching off a current too quickly. If the switch of the circuit shown is opened while a current i is flowing, so as to cause the current to disappear in a short time τ, the rate of change of flux linkage is of the order of Li/τ, and with short enough τ the e.m.f. produced may be extremely high. This e.m.f. appears across the air-gap, and may be enough to cause breakdown of the air and arcing between the poles of the switch. It is a convincing demonstration of self-inductance to switch off a magnet current and find the switch

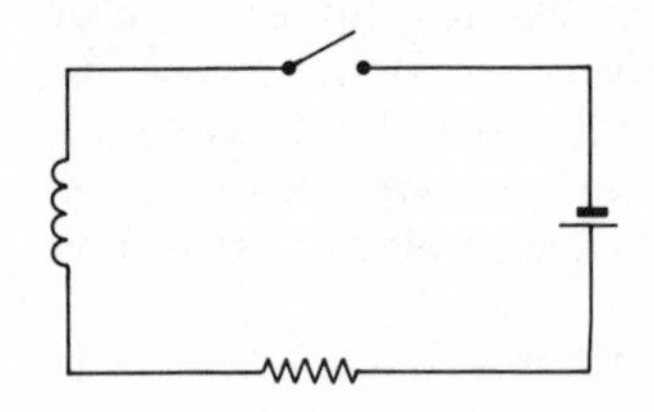

bursting into flames. The design of switches for controlling very heavy currents in coils is a specialized branch of electrical engineering, and the resulting device may be massive and complicated, even including air blasts for blowing out the arc before it can build up into something dangerous.

To calculate L or M it is necessary to find the flux linking a circuit when a current flows in it or in a coil close at hand. In general this involves an extended numerical computation, starting for example from the formula due to Neumann (which we shall not prove, as it involves a number of steps in vector analysis that need careful explanation, and in any case is well treated in standard reference works):

$$M_{AB} = \frac{\mu_0}{4\pi} \oint_1 \oint_2 \frac{\mathbf{dl}_1 . \mathbf{dl}_2}{r}, \tag{12.5}$$

in which r is the distance between elements $\mathbf{dl}_1$ and $\mathbf{dl}_2$ of the two circuits, and the integrals extend around both circuits. In simple cases one may estimate L or M directly, as for example if a short secondary coil is wound on top of a long primary coil. The field through the secondary windings due to current i in the primary is then uniform, $B = \mu_0 n_1 i$, if the primary has n_1 turns per unit length. This field extends over the cross-section A of the primary, and links flux $\mu_0 n_1 i A$ with each turn of the secondary. For n_2 turns of secondary, then, the total flux linkage is $\mu_0 n_1 n_2 i A$, so that

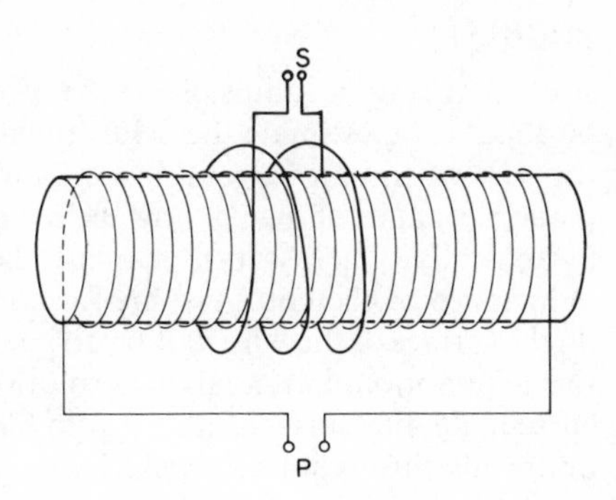

$$M_{12} = \mu_0 n_1 n_2 A.$$

By the same argument, if we ignore the diminution of field towards the ends of the solenoid, we may estimate the self-inductance of the long primary coil by supposing the same flux linkage $\mu_0 n_1 i A$ to occur for each of the $n_1 l$ turns, l being the length of the primary. Then

$$L_1 = \mu_0 n_1^2 A l = \mu_0 n_1^2 V,$$

where V is the volume of the coil. It should be noted that the inductance of a coil of given dimensions varies as the square of the number of turns, since the field produced is proportional to the number, and this field is itself linked to each of the turns.

The meticulous reader may be worried by the precise definition of the flux linkage for a solenoid or any other coil of complicated shape. Strictly one should construct a surface that is bounded by the circuit, and calculate Φ as $\int \mathbf{B} . \mathbf{dS}$ taken over this surface. However, it is hard to visualize the surface, even for a simple helical coil such as the solenoid just treated; if one pictures a spiral staircase, smoothed out to a spiral ramp, with its newel post reduced to zero radius, one will have an example of the required surface—the singular behaviour along the axis is unavoidable. There is no doubt, however, that the projected area of each turn of the ramp on to a plane normal to the axis is exactly the area of the coil, so that the flux linkage of a uniform field with this surface is correctly calculated by ignoring the awkward shape and, in effect, treating each turn as a separate closed coil. In this connection, it may be worth pointing out that the value of Φ does not depend on the choice of surface, provided the boundary is fixed by the coil. Thus, if a simple coil, as shown in section by CC′, serves as boundary for two different surfaces S_1 and S_2, the flux of **B** through each must be the same since div $\mathbf{B} = 0$ and the field lines are continuous.

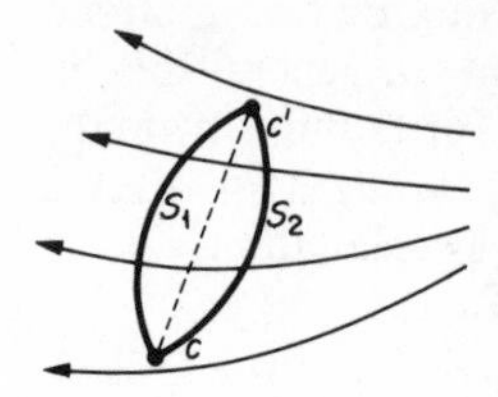

PROBLEM

A circular ring of radius a is made of wire whose radius is b. Taking b to be about $0{\cdot}1a$, estimate the field strength at the centre of the ring and near the wire when unit current flows. By joining these two extremes to give a plausible curve showing how field varies with radius, and integrating (graphically if necessary), estimate the flux linkage and hence the self-inductance of the coil. Assume for convenience that the current all flows on the surface of the wire, but try to guess what error is likely to result from this assumption. [An analytical solution gives L as $\mu_0 a\{\ln(8a/b) - 2\}$ for current on the surface, and $\mu_0 a\{\ln(8a/b) - 7/4\}$ for current distributed uniformly throughout the wire.]

An interesting example of mutual inductance is presented by two coplanar coils, one large (radius a) and the other small (radius b). If there is unit current round the larger, the field at the centre is $\mu_0/(2a)$ from (10.8), and the flux linked with the smaller coil is $\mu_0\pi b^2/(2a)$. This is one estimate of the mutual inductance, which can be compared with what is found by working the problem in reverse, passing unit current through the small coil and calculating the flux linkage with the larger. For this purpose it may be noted that any field line that comes up through the small coil must go down through the plane again somewhere, and if that somewhere is inside the larger coil, the flux linkage is cancelled out, but if it is outside the original flux linkage is maintained. We may therefore measure the flux linkage as that flux which links with the whole of the plane *outside* the larger coil. Since the small coil carrying unit current behaves, according to Ampère's law, as a dipole of strength πb^2, and by (8.3) produces a field $\mu_0\pi b^2/(4\pi r^3)$ at a distance r in the plane, the required flux linkage is

$$\int_a^\infty \frac{\mu_0\pi b^2}{4\pi r^3}\cdot 2\pi r\,dr,$$

or

$$\mu_0\pi b^2/(2a)$$

as before. In this case $M_{12} = M_{21}$, and it does not matter which coil is regarded as primary and which as secondary when M is calculated. Neumann's formula (12.5) shows by its symmetry that this is a general property, and it is one which we shall now prove directly, without the somewhat tricky analysis leading to (12.5).

The proof depends on demonstrating this reciprocity for two elementary circuits, for the following reason. If we construct surfaces S_1 and S_2 bounded by the two coils, we may dissect each of S_1 and S_2 into elements of area; then a current i_1 round S_1 produces the same field as the sum of the effects of i_1 flowing round each element $\mathrm{d}S_1$, since at all edges common to two elements the currents cancel; correspondingly the flux through S_2 is the sum of the fluxes through each element $\mathrm{d}S_2$.

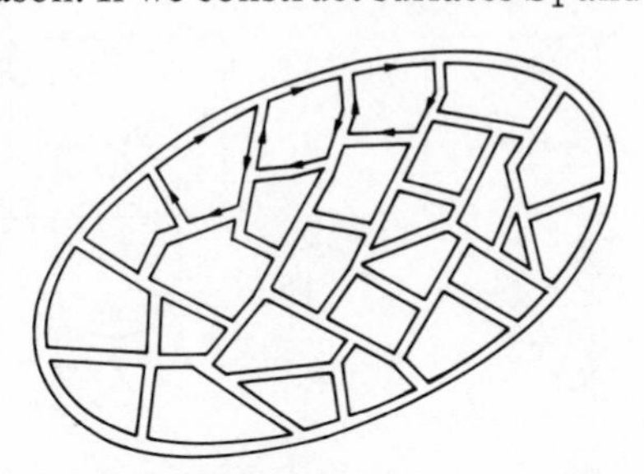

If then a typical element $\mathrm{d}S_1$ produces field $m_{12}i_1\,\mathrm{d}S_1$ at the position of a typical element $\mathrm{d}S_2$, the flux linkage of these two elements is $m_{12}i_1\,\mathrm{d}S_1\,\mathrm{d}S_2$, and the linkage for the complete circuits is $i_1 \iint m_{12}\,\mathrm{d}S_1\,\mathrm{d}S_2$. Clearly if we can show that $m_{12} = m_{21}$, the reciprocity of M_{12} and M_{21} for the complete circuits follows. The proof that $m_{12} = m_{21}$ is easy. Unit current round the element $\mathrm{d}\mathbf{S}_1$ causes it to generate the field of a dipole of strength $\mathrm{d}\mathbf{S}_1$, so that if $\mathrm{d}\mathbf{S}_2$ is distant $\mathbf{r}_{12}$ from $\mathrm{d}\mathbf{S}_1$ the field at $\mathrm{d}\mathbf{S}_2$ is given by (8.3):

$$\mathbf{B} = \frac{\mu_0}{4\pi}\left\{\frac{3(\mathbf{r}_{12}\,.\,\mathrm{d}\mathbf{S}_1)\mathbf{r}_{12}}{r_{12}^5} - \frac{\mathrm{d}\mathbf{S}_1}{r_{12}^3}\right\}.$$

The flux linkage is $\mathbf{B}\,.\,\mathrm{d}\mathbf{S}_2$; i.e.,

$$m_{12}\,\mathrm{d}S_1\,\mathrm{d}S_2 = \frac{\mu_0}{4\pi}\left\{\frac{3(\mathbf{r}_{12}\,.\,\mathrm{d}\mathbf{S}_1)(\mathbf{r}_{12}\,.\,\mathrm{d}\mathbf{S}_2)}{r_{12}^5} - \frac{\mathrm{d}\mathbf{S}_1\,.\,\mathrm{d}\mathbf{S}_2}{r_{12}^3}\right\},$$

which is obviously unchanged by interchange of subscripts. Hence

$$m_{12} = m_{21} \quad \text{and} \quad M_{12} = M_{21}.$$

Electrical circuits

Up to now we have neglected to consider the very important class of problems presented by electrical circuits, such as can be represented by a collection of two-terminal or four-terminal elements, connected by wires whose resistive and inductive effects may be neglected in comparison with the effects produced by the individual elements. Typical two-terminal elements are a resistance, a capacitance and an inductance, while a mutual inductance is a four-terminal element. The reader will be assumed familiar enough with the treatment of resistive circuits by means of Kirchhoff's laws for the briefest synopsis to suffice. In a d.c. circuit, conservation of charge demands that the net current entering or leaving any point shall vanish, $\sum i = 0$; further, the electric field at all points is irrotational so that a potential can be defined. These two statements comprise the essential content of Kirchhoff's laws. The two two-terminal elements of special importance in elementary d.c. circuits are the resistor and the ideal cell, shown here as 'black boxes', but with a hint of what they contain. The resistor exhibits a field between its terminals when a current passes, and the potential difference $\int \mathbf{E}.d\mathbf{r}$ is proportional, for an ohmic resistor, to the current i; $V = Ri$ as conventionally written. The ideal cell exhibits a similar field which, however, is independent of i; a real cell may often be represented as an ideal cell and an ohmic resistor in series. If we form $\oint \mathbf{E}.d\mathbf{r}$ round any closed curve made up of the wires joining two-terminal elements and short paths in free space between the terminals, the vanishing of $\oint \mathbf{E}.d\mathbf{r}$ implies that $\sum Ri - \sum V_c = 0$,* where V_c is the e.m.f. of any ideal cell in the circuit. This result, together with charge conservation, enables any ohmic d.c. network to be solved.

174

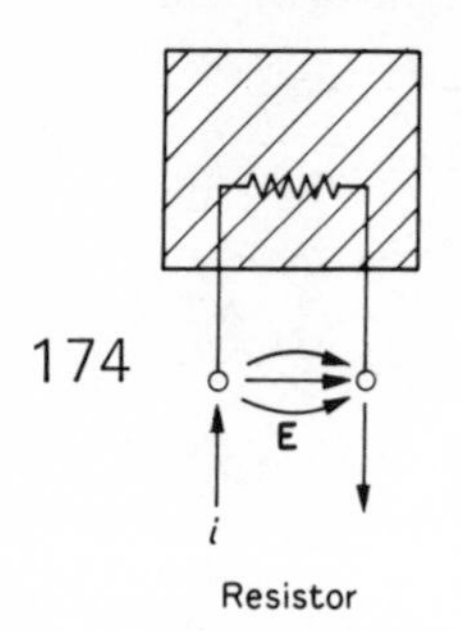

Resistor

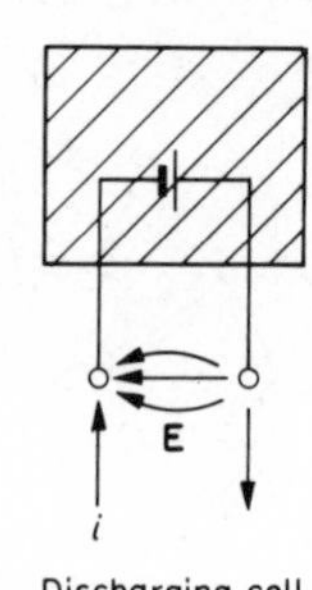

Discharging cell

To extend the argument to circuits containing capacitors and inductors, we must recognize, first, that the statement of Kirchhoff's laws needs justification when inductive effects are important and, secondly, that the potential difference between the terminals and the current through the element may be related by differential operations rather than simple algebraic constants. The first point arises from the result expressed in (12.4); when magnetic fields are changing, $\mathbf{E}$ ceases to be irrotational and a potential ϕ is no longer definable at all points. There is, however, a very important escape, for (12.3), when written in the form $\oint \mathbf{E}.d\mathbf{r} = -\dot{\Phi}$, shows that if the circuit round which the line integral is taken is free of changing magnetic fields, the line integral vanishes and a potential is definable. Provided the region outside the two-terminal inductive element is free from the magnetic field of the inductor, we may treat it as a circuit

* The diagram shows that a resistor passing a current and a cell discharging by passing the same current have oppositely directed fields across their terminals. This explains why V_c is given a negative sign in the equation, since by convention V_c is itself a positive number, as is R. In this sense the resistor may be said to exhibit a 'back e.m.f.' tending to resist the passage of current, but the concept of back e.m.f. is one that can easily mislead the student about the direction of the field, especially when applied to inductances, and it is probably safer to avoid it, and concentrate on the relation between current and field directions at the terminals.

element differing from a resistor only in the relation between V and i, but otherwise to be used in the application of Kirchhoff's laws in exactly the same way. This problem is peculiar to inductors; in a capacitor a potential may be defined everywhere, but it is still convenient to forget the details and take account only of the relationship between V and i at the terminals.

To take the inductor first, the electric field accompanying an increase in i points in the direction shown in the diagram, for according to Lenz's law if we were to join the terminals with a conducting link the resulting current should flow through the coil in the sense opposite to the additional increment due to $\mathrm{d}i/\mathrm{d}t$: the field shown would drive a current through the link and round the coil in an anticlockwise sense, while $\mathrm{d}i/\mathrm{d}t$ is clockwise. From the definition of L the magnitude of the potential difference is $L\,\mathrm{d}i/\mathrm{d}t$, and the sense for positive $\mathrm{d}i/\mathrm{d}t$ is the same as for a positive current flowing in a resistor. If, therefore, by convention we write the p.d. across the resistor as Ri, we must write the p.d. across the terminals of the inductor as $L\,\mathrm{d}i/\mathrm{d}t$.

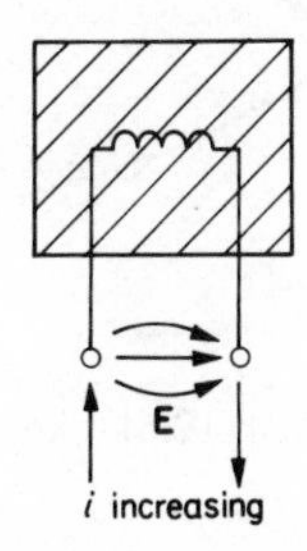

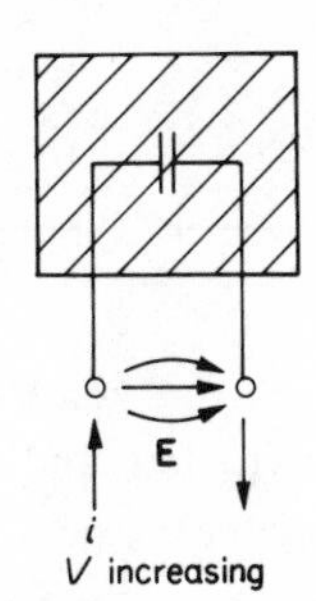

With a capacitor, the effect of the current i as shown is to increase the positive charge on the left-hand plate, and the negative charge on the right-hand plate, at a rate $\dot{Q} = i$. The field increases in the sense shown, and the potential difference is given by the equation $CV = Q$, or $C\,\mathrm{d}V/\mathrm{d}t = i$.

We may now apply these results to a generalization of Kirchhoff's laws. As with d.c. circuits, $\sum i$ at any junction must vanish to conserve charge, and $\oint \mathbf{E}.\mathrm{d}\mathbf{r}$ round any loop must also vanish. The latter condition implies that its time-derivative also vanishes, so that if we sum all the contributions to $\mathrm{d}V/\mathrm{d}t$ across the terminals of the circuit elements, we have for any closed loop in the circuit

$$\sum R\frac{\mathrm{d}i}{\mathrm{d}t} + \sum L\frac{\mathrm{d}^2 i}{\mathrm{d}t^2} + \sum \frac{i}{C} - \sum \frac{\mathrm{d}V_g}{\mathrm{d}t} = 0, \tag{12.6}$$

where V_g is the potential difference across the terminals of any ideal generator, i.e., any device producing a steady or time-varying p.d. whose behaviour is independent of the current passing through it. Remember that in a complicated network each element may be passing a different current, so that (12.6) is only one of a set of simultaneous linear differential equations involving all the different i's as variables: each independent loop in the circuit contributes an independent equation of the type (12.6). As we shall see later, if the generators produce sinusoidal p.d.s the solution of the problem resolves itself to the solution of simultaneous algebraic equations, nothing worse in fact than d.c. circuits. If, however, the generators are not sinusoidal, but pulse or square-wave generators for example, more powerful mathematical techniques must be brought into play to keep the analysis within manageable bounds.

EXAMPLES

(1) *Charging of a capacitor through a resistor.* When the switch is closed

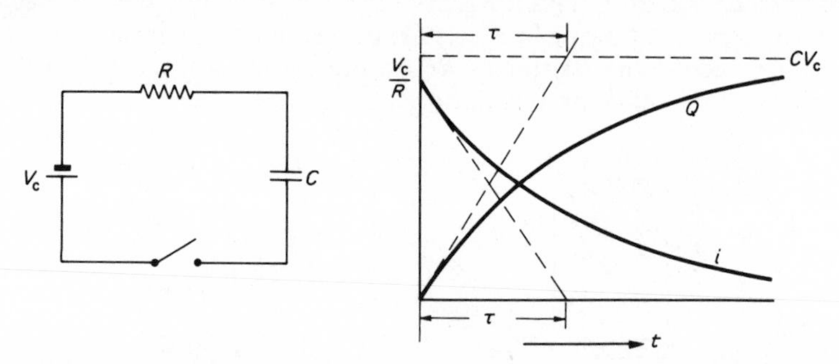

current begins to flow, and the cell p.d. is constant. Then (12.6) takes the form

$$R\,\mathrm{d}i/\mathrm{d}t + i/C = 0,$$

with solution $i = i_0\,\mathrm{e}^{-t/\tau}$, i_0 being an integration constant, and τ being written for RC. If at the moment $t = 0$, when the switch is closed, the capacitor is uncharged ($Q = 0$), the p.d.s across the resistor and the cell must be equal and opposite, so that $i_0 = V_c/R$. The total charge that flows before i falls to zero is found by integration, $\int_0^\infty i\,\mathrm{d}t = i_0\tau = CV_c$; this is of course just what is needed to bring the whole of V_c across the capacitor, leaving nothing to drive a current through the resistor.

Note the characteristic *time constant*, RC, for the combination of resistor and capacitor. Capacitors used in electronic circuits normally range between a few pF and many μF, and resistors between a few Ω and many MΩ; correspondingly, time constants may be adjusted from much less than a microsecond to many seconds, according to the need. It may also be noted that in a dry atmosphere the resistance between the open-circuited terminals of a good commercial capacitor may be many thousands of MΩ, so that a large capacitor can stay charged for a very long time. Since a capacitor charged to a voltage higher than 100 V is a real menace to life, great care should be taken not to expose the unwary to this hazard (by connecting a 'bleeder' resistor across the terminals).

(2) *Inductor and resistor.* In the absence of a capacitor there is no need to apply Kirchhoff's law in its time-derivative form, and we may write

$$Ri + L\,\mathrm{d}i/\mathrm{d}t - V_c = 0,$$

with solution $i = V_c/R + i_0\,\mathrm{e}^{-t/\tau}$, where τ is now L/R. At the moment of closing the switch, $i = 0$, so that $i_0 = -V_c/R$, and $i = V_c/R(1 - \mathrm{e}^{-t/\tau})$.

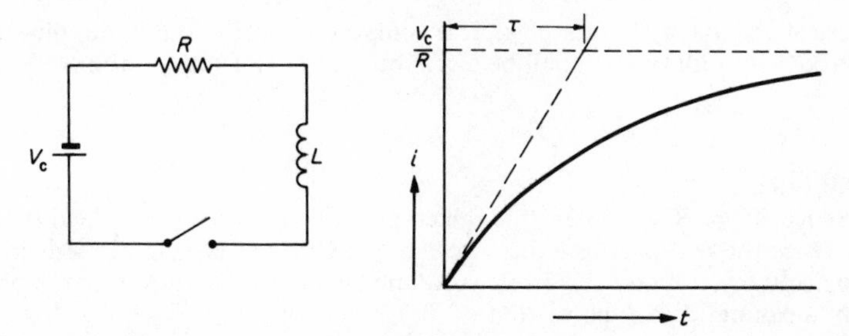

The initial rate of rise of current $(\mathrm{d}i/\mathrm{d}t)_0 = V_c/(R\tau) = V_c/L$; with no p.d. across the resistor, the whole of V_c must appear across the terminals of the inductor, giving V_c/L for the rate of change of current. Ultimately the current becomes steady and there is no p.d. across the inductor, so that in the end $i = V_c/R$.

(3) *Inductor, capacitor and resistor in series.* We now have, according to (12.6),

$$L\,\mathrm{d}^2i/\mathrm{d}t^2 + R\,\mathrm{d}i/\mathrm{d}t + i/C = 0.$$

This is the equation of damped simple harmonic motion, the damping being provided by the term $R\,\mathrm{d}i/\mathrm{d}t$. We shall first give a formal solution in complex

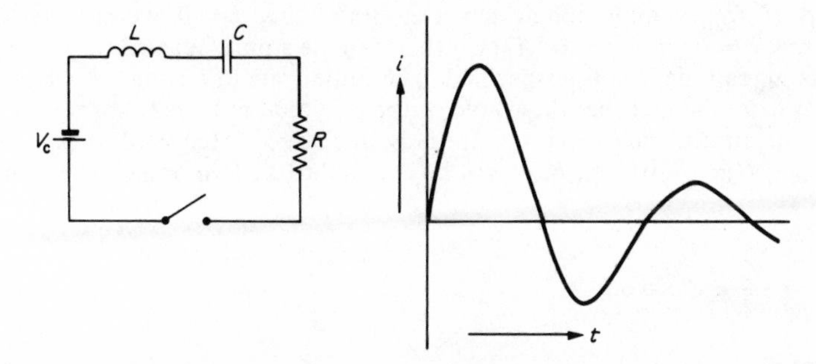

terms, and then discuss the physical interpretation. Since the equation is linear and of the second order, without any driving term of the form $F(t)$, the general solution can be written as the sum of arbitrary multiples of two independent functions, and in this case the functions are exponentials, as can be seen by substitution. Writing i as $A\,\mathrm{e}^{\mu_1 t} + B\,\mathrm{e}^{\mu_2 t}$, we find that μ_1 and μ_2 must be the solutions of the quadratic equation

$$L\mu^2 + R\mu + 1/C = 0,$$

so that

$$\mu = \frac{1}{2L}\{-R \pm \mathrm{j}(4L/C - R^2)^{1/2}\}. \tag{12.7}$$

We use j for $(-1)^{1/2}$ to avoid confusion with the current, i; this is standard notation in circuit theory. The reason for writing the solution in complex terms is that we are particularly interested in the oscillatory behaviour that occurs when $R^2 < 4L/C$. Putting $R/(2L) = \beta$ and $(4L/C - R^2)^{1/2}/(2L) = \omega$, the general solution is seen to take the form

$$i = (A\,\mathrm{e}^{\mathrm{j}\omega t} + B\,\mathrm{e}^{-\mathrm{j}\omega t})\,\mathrm{e}^{-\beta t},$$

where A and B may be complex. This is mathematically sound, but physically we ask that at all times i shall be a real quantity. This implies that $i^* = i$, or that

$$A^*\,\mathrm{e}^{-\mathrm{j}\omega t} + B^*\,\mathrm{e}^{\mathrm{j}\omega t} = A\,\mathrm{e}^{\mathrm{j}\omega t} + B\,\mathrm{e}^{-\mathrm{j}\omega t}$$

at all times.

Hence $A^* = B$, and $B\,\mathrm{e}^{-\mathrm{j}\omega t}$ is the complex conjugate of $A\,\mathrm{e}^{\mathrm{j}\omega t}$, so that i is just twice the real part of either $A\,\mathrm{e}^{\mathrm{j}\omega t}\,\mathrm{e}^{-\beta t}$ or $B\,\mathrm{e}^{-\mathrm{j}\omega t}\,\mathrm{e}^{-\beta t}$. Since A is in any case arbitrary, a factor 2 is irrelevant, and the general solution that is physically meaningful may be written

$$i = \mathrm{Re}(A\,\mathrm{e}^{\mu t}),$$

where μ is either solution of (12.7). If R^2 should be larger than $4L/C$, both solutions are real, and the most general solution that is physically meaningful is an arbitrary sum, with real coefficients, of two real exponentials, decreasing with time at different rates. We shall not discuss this case further here (see Problem below), but concentrate on the oscillatory solution occurring when $R^2 < 4L/C$.

Writing μ as $-\beta + \mathrm{j}\omega$, we have

$$i = \mathrm{e}^{-\beta t}\,\mathrm{Re}\,(A\,\mathrm{e}^{\mathrm{j}\omega t}),$$

in which the value of A is determined by the manner in which the current is set up. If, for example, the switch is closed at time $t = 0$, we must choose A so that $i = 0$ when $t = 0$; A must therefore be a pure imaginary, $-\mathrm{j}B$, say, but its magnitude is still unspecified. The initial rate of change of current is known from the fact that the whole of the potential must first appear across the inductance if the capacitor is initially uncharged. Hence $(\mathrm{d}i/\mathrm{d}t)_0$, which is $\mathrm{Re}\,[-\mathrm{j}B(\mathrm{j}\omega - \beta)]$, i.e., $B\omega$, must be equated to V_c/L in order to determine B.

Then

$$i = \frac{V_c}{\omega L}\,\mathrm{e}^{-\beta t}\sin\omega t.$$

PROBLEM

If $R^2 > 4L/C$ show that after closing the switch the current rises smoothly to a peak and dies away again without passing through zero at any time.

The use of complex numbers for a.c. circuits

We have just used complex algebra to solve a particular problem, but it is very much more widely valuable in analysing circuits excited by a constant frequency sinusoidal generator. The function $A\,\mathrm{e}^{\mathrm{j}\omega t}$ represents in an Argand diagram a line of length $|A|$ rotating at angular frequency ω, and the real part, being the projection of this motion on the real axis, is simple harmonic motion with amplitude $|A|$. We may conveniently represent the p.d. across the terminals of a generator by $V_g\,e^{j\omega t}$, the p.d. across the terminals of a circuit element by $V\,e^{j\omega t}$ and the current through it by $i\,e^{j\omega t}$, provided it is understood that only the real parts of these expressions refer to the measured quantities. The phase angles of what are in general complex numbers, V_g, V and i, then describe how the oscillations of the various quantities are related in phase; the examples that follow should make this clear.

Since the potential difference and current for an inductor are related by the equation $V = L\,di/dt$, it is clear by substitution that if we always understand V to be shorthand for $V\,e^{j\omega t}$, and i for $i\,e^{j\omega t}$, the amplitudes V and i must be related by the equation

$$V = j\omega Li.$$

Similarly for a capacitor C, since voltage and current are related by $C\,dV/dt = i$, the amplitudes at a given frequency must be related by

$$j\omega CV = i \quad \text{or} \quad V = -ji/\omega C.$$

In both cases one notes that V and i are proportional, as with an ohmic resistor, but the constants of proportionality are complex rather than real. On an Argand plot, the lines representing i and the p.d.s Ri, $j\omega Li$ and $-ji/(\omega C)$ are seen swinging round at constant angular velocity. The real

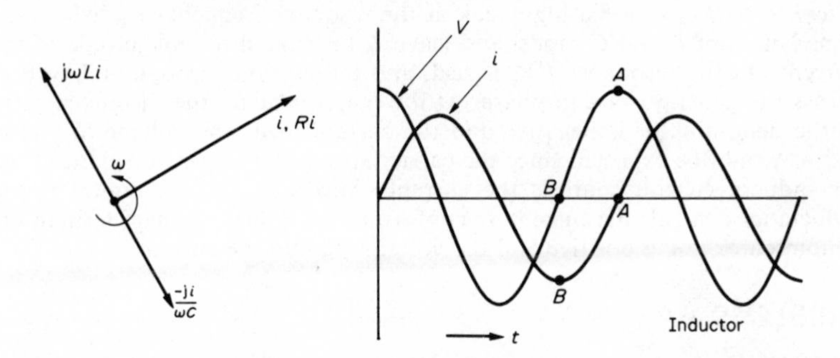

parts all execute simple harmonic oscillations, but while the p.d. across the resistor, Ri, oscillates in phase with the current, that across the inductor is a quarter of a cycle ahead and that across the capacitor a quarter of a cycle behind. The reason for the phase differences is easily appreciated; at a point in time like A, the p.d. takes a maximum value as the current through the inductor increases at its maximum rate. At B, where the current is momentarily not changing, no p.d. appears. Similar arguments will be found to account for the behaviour of a capacitor.

The practical application of this is that inductors and capacitors in a circuit supplied with a.c. of a single frequency may be treated exactly like resistors, but with complex rather than real *impedances* (i.e., V/i); the algebra of d.c. circuit analysis, using Kirchhoff's laws, applies unchanged and replaces the solution of differential equations.

EXAMPLE

The series resonant circuit, excited by an a.c. generator $V_g\, e^{j\omega t}$. If the circuit were composed of resistors in series we should add their resistances and state

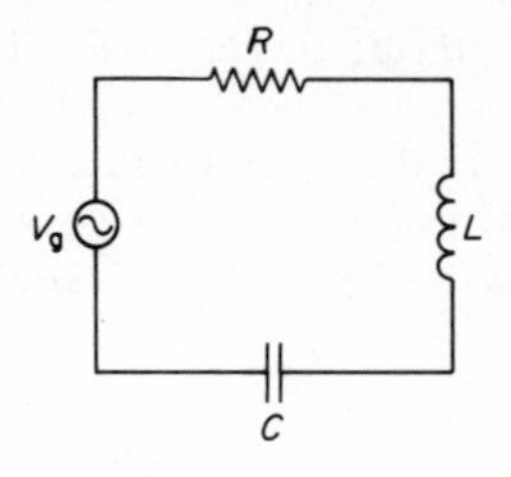

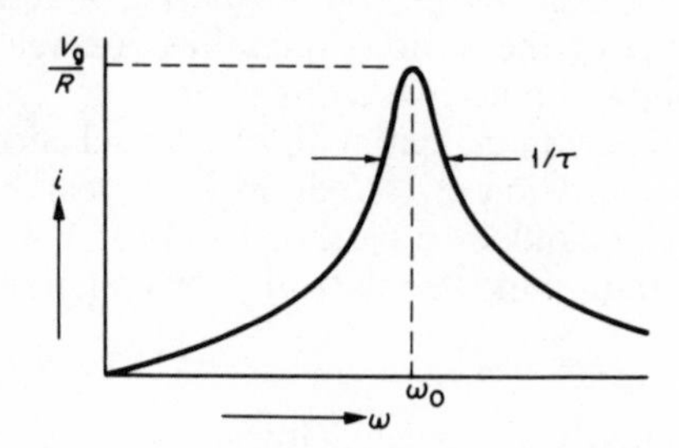

that the current flowing was $V_g/\sum R$. To generalize this result we add all impedances, and state that the current will be $V_g/\sum Z$, where in this case

$$\sum Z = R + j\omega L - j/(\omega C).$$

Writing $(LC)^{-1/2}$ as ω_0 and L/R as τ, we have that

$$i = \frac{V_g/R}{1 + j\omega\tau(1 - \omega_0^2/\omega^2)}.$$

If $\omega_0\tau \gg 1$, i/V_g shows a high peak at the resonant frequency ω_0, where the impedances of L and C cancel and leave R to control the magnitude of the current. At the resonance i/V_g is real, and the current through and voltage across the generator are in phase. At lower frequencies, the imaginary term in the denominator is negative and the current leads the voltage in phase, just as would be expected since the capacitance has a larger impedance than the inductance and controls the current. Above ω_0, on the contrary, the inductance controls the current, in conformity with the imaginary term in the denominator going positive and the voltage leading the current.

PROBLEMS

(1) If $\omega_0\tau \gg 1$, show that the two frequencies, one on either side of the resonance, at which $|i|^2$ takes half its peak value, are separated by $\Delta\omega = 1/\tau$. Thus the relative width of the resonant peak, defined as $\Delta\omega/\omega_0$, is $(\omega_0\tau)^{-1}$ or $R(C/L)^{1/2}$. Show further that at these frequencies the voltage and current have a relative phase angle of $\pm\pi/4$.

(2) A mass suspended from a spring and immersed in a viscous liquid is subjected to a sinusoidally varying vertical force. Show that the differential equation of motion of the mass can be written in the same way as that of the series resonant circuit, and identify the analogues (e.g., Mass $\equiv$ Inductance). Satisfy yourself that complex algebra can be usefully applied to mechanical systems also.

(3) A power line, carrying alternating current from a distant generator to a distant load, is monitored by displaying the p.d. between the conductors and across a small resistor in one of them on the two traces of a

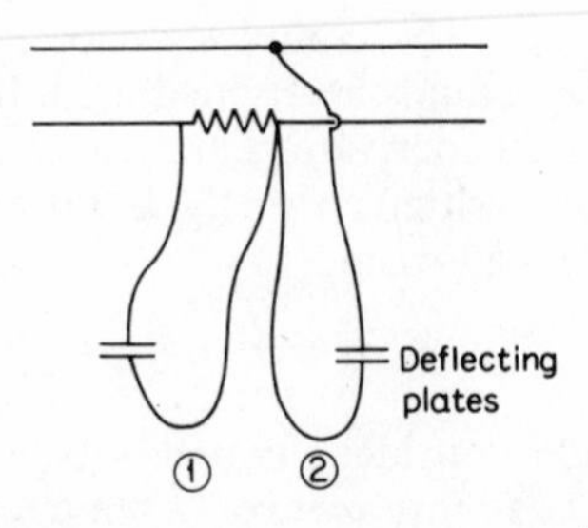

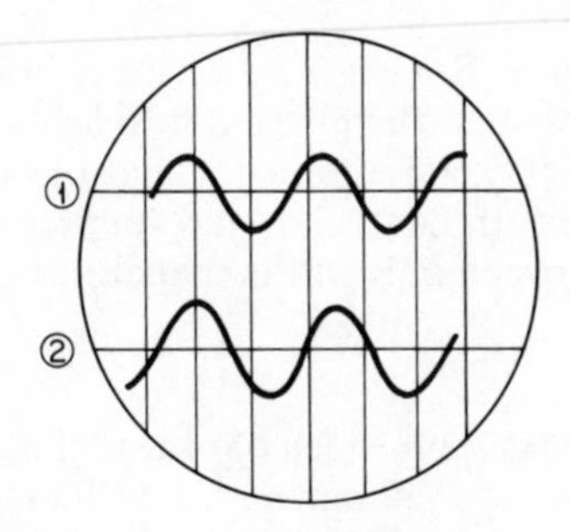

double-beam oscilloscope. The diagrams show the connections and the resulting traces (in which the spots move from left to right). On which end of the line is the generator, and can the load be described as a resistor in parallel with an inductor or with a capacitor?

Eddy currents and skin effect

Suppose a long hollow tube be placed in a solenoid carrying alternating current, so that the longitudinal field B_0 outside the tube oscillates with frequency ω. It is easily seen that the field inside will not be the same as B_0, for the oscillations of the internal field cause the flux through the tube to change, and by Faraday's law there must be produced an oscillating electric field circulating round the tube. This field excites a current in the walls, and so generates another longitudinal field which causes B and B_0 to take different values.

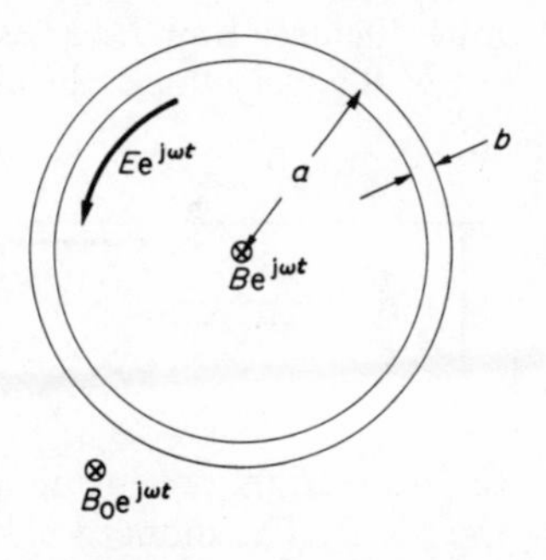

It is easy to work out the difference on the assumption that the current in the walls is uniformly spread. The flux in the tube is $\pi a^2 B$, and its rate of change $j\omega\pi a^2 B$ (from now onwards we assume all alternating fields to be represented by their amplitudes, and time differentiation to be effected by multiplication by $j\omega$). If the circulating field at the wall of the tube is E, (12.3) implies that

$$2\pi aE = j\omega\pi a^2 B, \quad \text{or} \quad E = \tfrac{1}{2}j\omega\pi aB.$$

For metal of conductivity σ, the current density J is σE and the current circulating in length L of the tube is LbJ or $\frac{1}{2}Lb\sigma\, j\omega\pi aB$. We now take a circuit which begins with a path L down the inside of the tube, goes through the wall and returns up the outside before passing back through the wall to the starting-point. The line-integral of **B** round this circuit must be μ_0 times the current encircled, so that

$$(B_0 - B)L = \mu_0 \times \tfrac{1}{2}Lb\sigma\, j\omega\pi a B,$$

or $B = B_0/(1 + jA)$, where $A = \frac{1}{2}\pi\omega\mu_0\sigma ab$. The complex denominator shows that the internal field lags in phase behind the applied field. It is a useful exercise to check that the signs in this derivation are correct, i.e., that both Lenz's law and Ampère's law are written correctly, but if we are interested only in the magnitude of B we may write

$$|B/B_0| = (1 + A^2)^{-1/2}.$$

To get some idea of orders of magnitude, consider a copper tube whose radius $a = 30$ mm and wall-thickness $b = 5$ mm; σ at room temperature is about $5 \times 10^7\ \Omega^{-1}\ \text{m}^{-1}$, so that $A \sim 1{\cdot}5 \times 10^{-2}\,\omega$. Screening begins to be significant when A is about unity, i.e., $\omega \sim 70$ or the frequency $\sim$10 Hz. Mains frequency oscillations (50 or 60 Hz) are screened by a factor 5 or 6, and higher frequencies correspondingly more. If the copper is cooled to increase σ, the screening is still more efficient.

When the screening is good, and A very large, we may begin to doubt the correctness of the calculation. For the assumption that E and J are constant in the wall depends on the flux contained within the material of the tube being small in comparison with the flux inside the tube. When $B \ll B_0$ it may well prove that it is the field falling from B_0 to B in the outside layers of the wall that provides the majority of the flux, and we must then see how E varies in the thickness of the wall. It is clear that if the wall is very thick, the currents cause B to fall steadily until at a great enough depth it may be considered to vanish completely. We shall examine this case in more detail and, to save physically irrelevant mathematical difficulties, we shall take the tube to be of so great a diameter that it may be considered to have a flat surface. We now have an applied field $\mathbf{B}_0$ in the z-direction, producing B_z in the metal which changes with depth x into the metal. The induced electric field and current flow in the y-direction. Consider first a circuit of unit length along y and of width dx. For this circuit

231

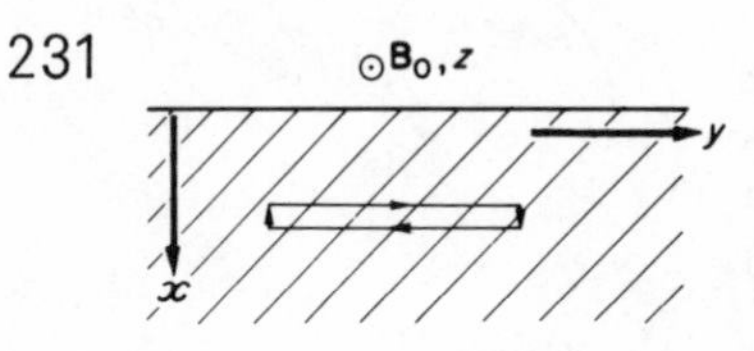

$$\oint \mathbf{E}.\mathrm{d}\mathbf{r} = E_y - \left(E_y + \frac{\partial E_y}{\partial x}\,\mathrm{d}x\right) = -\frac{\partial E_y}{\partial x}\,\mathrm{d}x.$$

This must be equated to the rate of change of flux in the circuit, i.e., to $j\omega B_z\,\mathrm{d}x$. According to Lenz's law, if B_z points out of the paper and is increasing, the induced current must set up a field in the circuit that points into the paper, and therefore E_y must decrease with depth, i.e.,

$$\partial E_y/\partial x = -j\omega B_z.$$

Since $\mathbf{J} = \sigma\mathbf{E}$, we have a relation between magnetic field and induced current,

$$\partial J_y/\partial x = -j\omega\sigma B_z. \tag{12.8}$$

A second relation is provided by taking a similar circuit in the x–z plane and equating the line-integral of $\mathbf{B}$ to μ_0 times the enclosed current: i.e.,

$$\partial B_z/\partial x = -\mu_0 J_y. \tag{12.9}$$

Differentiating (12.9) and substituting in (12.8) we obtain an equation* for B_z,

$$\partial^2 B_z/\partial x^2 = j\omega\mu_0\sigma B_z,$$

which has as a general solution, with two integration constants P and Q,

$$B_z = P\,\mathrm{e}^{kx} + Q\,\mathrm{e}^{-kx},$$

where

$$k^2 = j\omega\mu_0\sigma,$$

and therefore

$$k = (1 + j)(\tfrac{1}{2}\omega\mu_0\sigma)^{1/2}.$$

If the metal is thick, we must reject that part of the solution which increases exponentially with depth, putting $P = 0$; then Q must equal B_0, the field at the surface, and

$$B_z = B_0\,\mathrm{e}^{-kx}.$$

The field decays exponentially with distance into the metal, dropping by a factor e in every distance $(\tfrac{1}{2}\omega\mu_0\sigma)^{-1/2}$, which is called the *skin depth*. After a few skin depths the field is extremely small, and any shell thicker than this acts as a virtually perfect screen.

For copper at room temperature, the skin depth in millimetres is about $70/\nu^{1/2}$, where ν is the frequency in Hz. At very high frequencies, and still more at low temperatures because of the increased σ, the field may penetrate only a very small distance, possibly less than 1 μm. This is the fact that was referred to in the earlier discussion of cyclotron resonance. An

* This analysis has been presented in the old-fashioned language of Cartesian coordinates to give an opportunity to show in detail the physical processes at work. A much neater derivation starts from Ampère's and Faraday's laws in differential form:

$$\mathrm{curl}\,\mathbf{B} = \mu_0\mathbf{J} = \mu_0\sigma\mathbf{E}, \tag{10.11}$$

and

$$\mathrm{curl}\,\mathbf{E} = -\dot{\mathbf{B}} = -j\omega\mathbf{B}. \tag{12.4}$$

Then, taking the curl of (10.11) and applying XIII, bearing in mind that $\mathrm{div}\,\mathbf{B} = 0$, we have that $\mathrm{curl}\,\mathrm{curl}\,\mathbf{B} = -\nabla^2\mathbf{B}$, so that

$$\nabla^2\mathbf{B} = j\omega\mu_0\sigma\mathbf{B},$$

of which the equation derived in the text is a special case.

important technical consequence of the *skin effect* is that high-frequency currents carried by conductors are similarly confined to the skin depth, and the inner portions of the conductors might just as well be absent. The effective resistance of the wire is correspondingly higher, and this may prove a considerable nuisance, as it causes greater attenuation of signals on high-frequency transmission lines than one would otherwise expect.

A closely related phenomenon is the torque exerted on a conductor rotating in a magnetic field, as a result of the interaction of the field with the currents induced by motion in the field. Consider for example a sphere rotating with angular velocity $\boldsymbol{\omega}$ in a field $\mathbf{B}$ perpendicular to $\boldsymbol{\omega}$, and let us suppose the rotation to be so slow that the induced currents hardly affect the magnitude of $\mathbf{B}$ inside the sphere. We may think of the Lorentz force on the electrons carried with the sphere as equivalent to an electric field $\mathbf{v} \wedge \mathbf{B}$, or $(\boldsymbol{\omega} \wedge \mathbf{r}) \wedge \mathbf{B}$ if $\mathbf{r}$ is measured from the centre. This field is everywhere parallel to $\boldsymbol{\omega}$, and if it acted alone would drive current through the surface of the sphere. As a result, surface charges build up to provide an additional irrotational field which combines with the first field to give circulating field lines and hence circulating currents. One can see by inspection that such a circulating field pattern is obtained by interchanging the vectors in the double vector product, i.e., $(\boldsymbol{\omega} \wedge \mathbf{B}) \wedge \mathbf{r}$ represents a vector field which lies in the plane parallel to $\boldsymbol{\omega}$ and $\mathbf{B}$ and which, being normal to $\mathbf{r}$, has no radial component. We now demonstrate by direct calculation that we can supply a suitable constant so that the curl of both fields is the same everywhere. They thus differ only by an irrotational field which surface or space charges are able to supply. To show the required result, let us choose the x-axis parallel to $\mathbf{B}$ and the z-axis parallel to $\boldsymbol{\omega}$. Then

$$\begin{aligned}(\boldsymbol{\omega} \wedge \mathbf{r}) \wedge \mathbf{B} &= \boldsymbol{\omega}(\mathbf{r}.\mathbf{B}) - \mathbf{r}(\boldsymbol{\omega}.\mathbf{B}), \qquad \text{by VII,} \\ &= (0, 0, B\omega x);\end{aligned}$$

and

$$\begin{aligned}(\boldsymbol{\omega} \wedge \mathbf{B}) \wedge \mathbf{r} &= \boldsymbol{\omega}(\mathbf{r}.\mathbf{B}) - \mathbf{B}(\mathbf{r}.\boldsymbol{\omega}), \\ &= (-B\omega z, 0, B\omega x).\end{aligned}$$

The curl of the first vector is $(0, -B\omega, 0)$ and of the second $(0, -2B\omega, 0)$. Hence $\frac{1}{2}(\boldsymbol{\omega} \wedge \mathbf{B}) \wedge \mathbf{r}$ has the same curl as $(\boldsymbol{\omega} \wedge \mathbf{r}) \wedge \mathbf{B}$, and may be taken to express the electric field set up by rotation. The resulting current density is given by

$$\mathbf{J} = \tfrac{1}{2}\sigma(\boldsymbol{\omega} \wedge \mathbf{B}) \wedge \mathbf{r} = \tfrac{1}{2}\sigma B\omega(-z, 0, x). \tag{12.10}$$

To calculate the torque exerted by $\mathbf{B}$ on this current distribution we recollect that an elementary volume $\mathrm{d}V$ through which current density $\mathbf{J}$ is passing experiences a force $(\mathbf{J} \wedge \mathbf{B})\,\mathrm{d}V$. The moment of this force about the centre of the sphere is $\mathbf{r} \wedge (\mathbf{J} \wedge \mathbf{B})\,\mathrm{d}V$, so that the total torque on the sphere

$$\mathbf{G} = \int \mathbf{r} \wedge (\mathbf{J} \wedge \mathbf{B})\,\mathrm{d}V,$$

the integration being taken over the sphere. Expressed in Cartesian coordinates this has the form:

$$\mathbf{G} = \tfrac{1}{2}\sigma B^2\omega \int (-xz, 0, x^2)\,\mathrm{d}V.$$

The x-component vanishes identically, since for every volume element located at a given x and z there is another at x and $-z$; hence **G** is a torque about the z-axis, the axis of rotation, of magnitude

$$G_z = \tfrac{1}{2}\sigma B^2\omega \int x^2 \,\mathrm{d}V = \frac{2\pi}{15}\sigma B^2 \omega a^5 \tag{12.11}$$

for a sphere of radius a. The integration is most easily performed by sectioning the sphere by planes normal to the x-axis.

To gain some feel for the magnitude of this eddy-current braking effect, it is helpful to compare (12.11) with the torque acting on a sphere when it rotates in a liquid of viscosity η. In this case $G = 8\pi\eta\omega a^3$, so that we may imagine the eddy currents to be equivalent to a liquid for which $\eta = \sigma B^2 a^2/60$. For example, a copper sphere of radius 30 mm in a field of 1 T feels the equivalent of a viscosity of 750 kg s^{-1} m^{-1} (=7500 poise, the c.g.s. unit commonly used in books of tables). This is several times more viscous than treacle (molasses), and it is indeed an interesting experience to hold a large lump of copper between the poles of an electromagnet and try to turn it. If the same experiment were to be tried with pure copper at a very low temperature, σ might be 10^4 times as great, and the sphere would seem to be locked nearly solid to the magnet, so far as rotation about an axis normal to **B** was concerned; it would still rotate freely about **B** and could be translated without hindrance so long as **B** was uniform.

The calculation just performed is valid only for very slow rotation; at faster speeds the induced currents are large enough to reduce the field in the sphere substantially, and when the angular velocity ω is such that the corresponding skin depth $(\tfrac{1}{2}\omega\mu_0\sigma)^{-1/2}$ is much less than the sphere radius, the field is almost entirely excluded. The phenomenon is closely analogous to the skin effect, for if we look at it from the point of view of an observer rotating with the sphere, the field rotating with respect to him may be analysed into two fields at right angles in the plane normal to ω, oscillating harmonically with frequency ω and with a phase difference of $\pi/2$ between them—just as a conical pendulum may be thought of as executing two linear oscillations at right angles. The tendency for magnetic fields to be excluded from rotating conductors may play a considerable part in causing fluctuations of the Earth's magnetic field; eddies in the conducting core of the Earth are equivalent to rotations of large masses and distort the field in their vicinity. Because of the size of the eddies, which may be many kilometres in radius, they need rotate only very slowly to exclude the field almost completely.

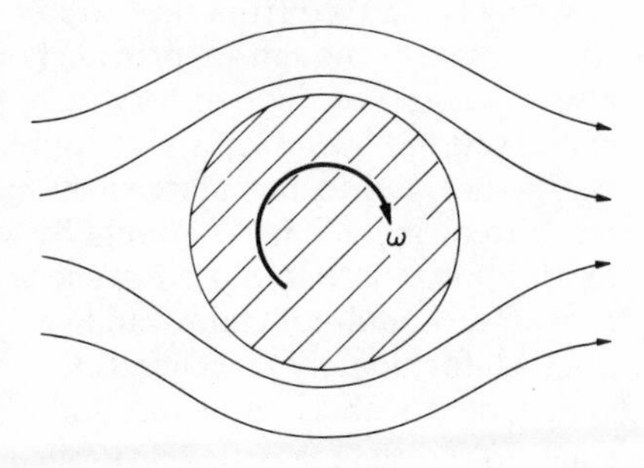

PROBLEM

Taking an eddy as equivalent to a rotating solid conductor 10 km in radius, and the conductivity to be $10^6\ \Omega^{-1}\ \mathrm{m}^{-1}$ (similar to that of liquid mercury, a poor conductor), find the period of rotation that will give a skin depth equal to the radius and thus cause substantial exclusion of field. [Answer: 13 years]

Energy in magnetic fields

For various reasons, we might expect energy conservation to hold when magnetic fields and induction effects are present. We know, for example, that permanent magnets behave as if the field obeyed a central inverse-square law, and if that were the complete picture of magnetism we would not hesitate to take over the results derived for electric fields and write the 166 energy density of a magnetic field as $\frac{1}{2}B^2/\mu_0$ (note that μ_0 appears in the denominator while ε_0 appears in the numerator in the electric case; the difference arises from the different relation of field to charge or pole). However, the basic origin of the magnetic field is not a pole but a moving charge, and we must not presume that the same result will apply, although in fact this turns out to be so. The close link between electromagnetic induction and the Lorentz force, which was brought out at the beginning of this chapter, suggests that as the latter does not violate energy conservation, neither will the former. But this is something that should be proved and not taken for granted. In some treatments, it is implied that Faraday's law may be derived from the law of conservation of energy; this is indeed true, or nearly enough so, provided one accepts the premise that energy is always conserved. The history of the development of the idea of energy conservation, however, shows only too clearly that the definition of energy has had to be enlarged on many occasions to allow new phenomena to be included. We should be wary, therefore, of assuming that the terms appropriate to old phenomena will remain valid in the presence of new, and should not claim validity for proofs based on energy conservation. Unfortunately, a general demonstration that the work done on a system of movable conductors and magnets by cells and external forces is equal to the change of $\int B^2\,\mathrm{d}V/(2\mu_0)$ requires rather elaborate development of the theory, beyond the point where we can take it. We must be content, therefore, to illustrate the result by specific examples and ask the reader to accept the assurance that a general proof is possible.

220 249 The first example is practical. We have already discussed the homopolar motor and generator, and have seen that the machine, considered as a motor, performs work at a rate $I\Phi\omega/(2\pi)$; at the same time the rotor, moving in the field B, induces an e.m.f. between hub and rim of magnitude $\omega\Phi/(2\pi)$. The work done by the source of the current I against this e.m.f. is clearly equal to the mechanical work performed by the motor.

The second example is artificial but instructive. Suppose a magnetic pole P be placed on the axis of a circular loop of wire whose resistance is R. When P is moved away from the loop a current is induced which creates a magnetic field, so that a force must be applied to P to keep it moving (this is a very elementary example of an eddy-current brake). The work done by this force may be compared with the heat generated in the wire. To make the argument quantitative, we note that if the solid angle subtended by the loop at P is Ω, the flux through the loop is $\mu_0 P\Omega/(4\pi)$, so that by (12.3) the e.m.f. round the circuit is $\mu_0 P\dot{\Omega}/(4\pi)$; this is to be equated to Ri to give the induced current (whose magnitude only we consider, not the sign);

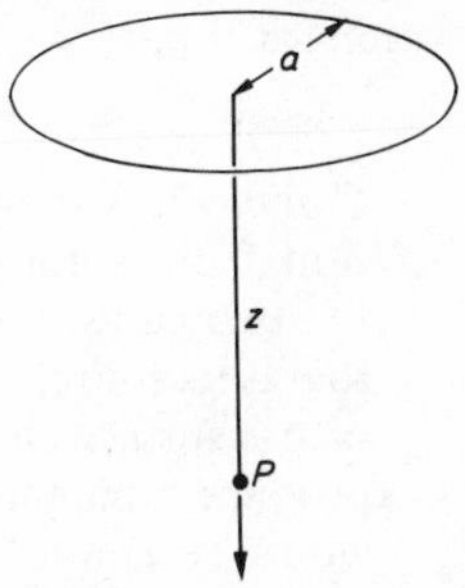

$$Ri = \mu_0 P\dot{\Omega}/(4\pi) = \frac{\mu_0 P}{4\pi}\frac{d\Omega}{dz}\frac{dz}{dt}.$$

Now by (10.4) the magnetic potential at P is $\mu_0 i\Omega/(4\pi)$, so that the value of B at P is $-\mu_0 i(d\Omega/dz)/(4\pi)$. Therefore

$$Ri^2 = PB\frac{dz}{dt},$$

and the heat Ri^2 exactly equals the rate at which work is done by the force PB.

The same answer will of course be obtained, without necessarily introducing the idea of induction, by imagining the pole at rest and the loop moving in the field. The Lorentz force then drives current round the loop and the argument proceeds to the same conclusion that the work done equals the heat generated. We have here an extension to electromagnetism of the principle which we saw hold in Newtonian mechanics, that if energy is conserved in one inertial frame it is conserved in all.

As a third example of energy conservation consider the build-up of current in a long solenoidal inductor. When a cell of e.m.f. V_c is applied to the solenoid, the rate of increase of current is given by the equation

$$L\,di/dt = V_c,$$

if the resistance is small enough to be neglected (the reader may care to develop the argument, and reach the same conclusion, for a resistive inductor). Now the cell does work at a rate $V_c i$, or $Li\,di/dt$. If we start with i equal to zero and continue until a certain current i_0 flows, the work W done by the cell is $\int_0^{i=i_0} Li(di/dt)\,dt$, or $\frac{1}{2}Li_0^2$. This may be expressed in terms of the field $B(=\mu_0 n i_0)$ in the solenoid, if we remember that $L = \mu_0 n^2 V$ for a solenoid of volume V and n turns per unit length. Then

$$W = \tfrac{1}{2}\mu_0 n^2 V i_0^2 = \frac{B^2}{2\mu_0}V.$$

As might have been expected, the work done by the cell may be regarded as stored in the magnetic field with an energy density $\frac{1}{2}B^2/\mu_0$.

PROBLEM

Show that the magnetic field inside a long solenoid exerts a pressure $\frac{1}{2}B^2/\mu_0$ on the current windings. It may be convenient to adopt the device

176 employed to analyse electrical forces on a conductor, and divide the field on the inside and outside of the windings into a symmetrical and an anti-symmetrical part.

Comment. A magnetic field exerts a pressure, an electric field a tension. This is not paradoxical, since there is no real analogy between the two cases, the magnetic field being parallel to the current sheet that is its source, and the electric field being normal to the conductor whose surface charge is its source. If there were free poles, one could produce a magnetic field normal to a 'magnetic conductor', and it would exert a tension.

As remarked earlier, the pressure that can be exerted by a magnetic field is very considerable, as is the energy density of a strong field. A superconducting solenoid with a field of 10 T stores energy at a density of about 4×10^7 J m^{-3}, about twice as much as the same volume of acetylene and oxygen at normal temperature and pressure. If the magnetic stress on the windings causes them to fracture, there is a chance that the current will die away rapidly, releasing this large amount of energy at an uncomfortably fast rate and destroying the coil and its container, if not much else besides. For this reason, in the design of large superconducting coils, alternative paths are provided for the current to take in the event of breakdown of the superconductor; these allow the field to die away slowly, so that the release of energy is at no greater rate than can be handled without severe damage.

The magnetic pressure can be used as a tool for forming metals into complicated shapes. If a metallic tube is placed in a solenoid through which a large capacitor is discharged to create a strong impulsive magnetic field, eddy currents in the tube hinder the penetration of field into its interior, so that the stronger external field exerts an uncompensated pressure and the tube collapses on to an internal insulating non-metallic mandrel. The whole process is rapid and does not, as hydraulic forming does, demand the sealing of the ends of the tube.

These examples should serve to illustrate the extension of the law of

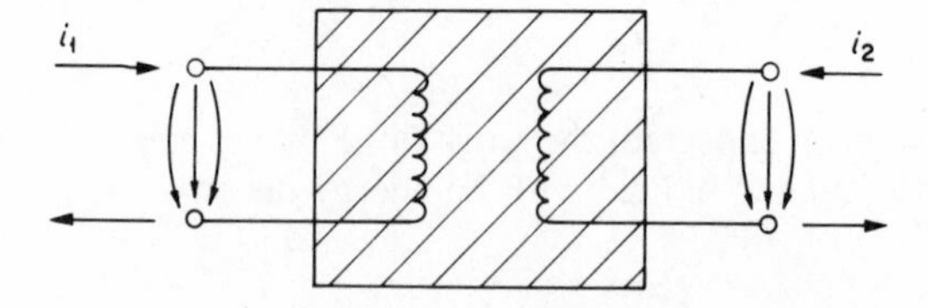

conservation of energy to magnetic fields in terms exactly analogous to what has already been proved for electric fields, and we shall assume that energy conservation is a valid tool to use in solving problems in electromagnetism. To illustrate this point, we consider the relation between two inductors L_1 and L_2 which are close enough together that they have a certain mutual inductance M, and we shall show that there is a limitation to the magnitude of M. A mutual inductor is a 4-terminal circuit element whose properties are represented by the equations

$$\left.\begin{aligned} V_1 &= L_1\,\mathrm{d}i_1/\mathrm{d}t + M\,\mathrm{d}i_2/\mathrm{d}t \\ V_2 &= L_2\,\mathrm{d}i_2/\mathrm{d}t + M\,\mathrm{d}i_1/\mathrm{d}t. \end{aligned}\right\} \tag{12.12}$$

These equations express the definitions of L_1, L_2 and M, whereby the flux linked with the first coil is $L_1i_1 + Mi_2$ and with the second coil $L_2i_2 + Mi_1$, and the e.m.f.s across the terminals reflect the rate of change of flux. The signs deserve consideration; L_1 and L_2 are by convention taken as positive, as discussed already, but M may be of either sign. For turning one of the coils back to front inside the box, without altering the external connections, reverses the sign of flux linkage with the other. The reader should satisfy himself, by detailed consideration of the signs of flux linkage when both coils are wound in the same sense, that in (12.12) the signs before M may either be both positive or both negative, but they must be the same. It is convenient to take them as positive and allow M to take either sign.

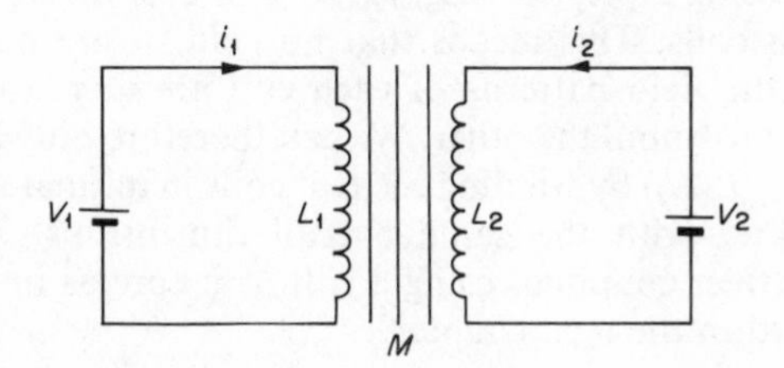

Let us now connect cells V_1 and V_2 to the circuits so that (12.12) represent the variation of i_1 and i_2. The work done by the cells in increasing the currents from zero to some final values i_1 and i_2 is given by

$$\begin{aligned} W &= \int V_1 i_1\,\mathrm{d}t + \int V_2 i_2\,\mathrm{d}t \\ &= \int \left(L_1 i_1 \frac{\mathrm{d}i_1}{\mathrm{d}t} + M i_1 \frac{\mathrm{d}i_2}{\mathrm{d}t} + M i_2 \frac{\mathrm{d}i_1}{\mathrm{d}t} + L_2 i_2 \frac{\mathrm{d}i_2}{\mathrm{d}t}\right)\mathrm{d}t; \end{aligned}$$

i.e.,

$$W = \tfrac{1}{2}L_1 i_1^2 + M i_1 i_2 + \tfrac{1}{2}L_2 i_2^2. \tag{12.13}$$

We expect this work to be stored as field energy with a density $B^2/(2\mu_0)$, and since this is essentially positive W also must be essentially positive. Now the first and third terms in (12.13) are positive, but the second can be made negative, whatever sign M takes, by appropriate choice of i_1 or i_2. And if M were too large it might even be possible to find i_1 and i_2 such as would make W negative. We know M cannot be as large as this, and proceed to discover what is the maximum value it can take so that no choice of i_1 and i_2 can ever make W negative. Treat (12.13) as a quadratic equation for i_1, with solution

$$L_1 i_1 = -M i_2 \pm \{(M^2 - L_1 L_2) i_2^2 + 2WL_1\}^{1/2}.$$

We can ensure that whenever W is taken as negative there shall be no solution by arranging that $(M^2 - L_1 L_2) i_2^2$ is negative; otherwise expressed, if this term should be positive, we could always find a negative value of W which still left the term in braces positive overall, giving a real solution for i_1, which we know to be impossible. Hence M must be such that $M^2 - L_1 L_2$ cannot be positive, or

$$M^2 \leqslant L_1 L_2. \tag{11.14}$$

The special case of an equality in (11.14), when M takes its maximum value, has a simple interpretation for, as (11.13) shows, the energy can then be written in the form

$$W = \tfrac{1}{2}(L_1^{1/2} i_1 + L_2^{1/2} i_2)^2,$$

so that by choosing $L_1^{1/2} i_1 = -L_2^{1/2} i_2$, $W = 0$ even though currents are flowing in both coils. This means that no field is set up anywhere, and therefore that the field patterns of each coil are identical, so that it is possible for one to annul the other. We can therefore only achieve perfect coupling ($M^2 = L_1 L_2$) by winding the two coils in intimate contact on the same former and with the same overall dimensions. An alternative approach to perfect coupling, using a soft iron core as in a transformer, will be discussed in the next chapter.

PROBLEMS

(1) Two solenoidal coils are intimately wound so as to achieve perfect coupling. If one has n_1 turns and the other n_2, show that

$$L_1 : L_2 : M :: n_1^2 : n_2^2 : n_1 n_2.$$

(2) A resistor R is connected across the terminals of the inductor L_2 in the preceding problem, and a generator of frequency ω across the terminals of L_1. Show that if ωL_1 and ωL_2 are both much larger than R, the current through the generator is the same as if a resistor of magnitude $(n_1/n_2)^2 R$ were connected straight across its output. This trick of using a transformer to modify the effective magnitude of a resistance is of great value in electronic circuits.

Cavendish Problems: 216, 236–240, 243–246, 249, 253–258, 260–262.

READING LIST

Circuit Analysis: E. A. FAULKNER, *Principles of Linear Circuits*, Chapman and Hall.

Measurement: A. CAMPBELL and E. C. CHILDS, *The Measurement of Inductance, Capacitance and Frequency*, Macmillan.

Geomagnetic Anomalies: E. C. BULLARD, *Monthly Notices of the Royal Astronomical Society, Geophysics Supplement* **5**, 248 (1948).

Generators: A. DRAPER, *Electrical Machines*, Longmans.

Electromagnetic Forming: *Metals Handbook*, 8th edn., vol. 4, p. 256, American Society for Metals.

13

Magnetic materials

This chapter covers ground similar to Chapter 9 and illustrates how the observed magnetic properties of materials can be understood in terms of their atomic structure. This is a very important branch of solid-state physics, not only because of the variety of interesting magnetic phenomena exhibited by solids but because of the technical importance of a small number of magnetic solids, chief among which is iron, without which the electrical industry would not have developed. In the form of permanent magnets in meters, synchronous clocks and a host of household appliances, and as cores in transformers and in the field coils of generators, iron is ubiquitous as a magnetic metal. It is humbling to the physicist to be forced to recognize that the problem of why iron has its peculiar properties (*ferromagnetism*), for all that they have been forced on his attention for centuries, is still only imperfectly resolved. To be sure, the atomic description of a ferromagnet has been developed in great detail, and if one assumes that iron is such a thing one can describe its properties thoroughly and consistently in terms of a very small number of assumptions; but why the phenomenon should be found in iron, cobalt and nickel and not in their neighbours in the Periodic Table appears to depend on such a nice balance of contributory and opposing causes that even the most advanced techniques of the theoretical solid-state physicist are still too imprecise to provide a convincing explanation.

In contrast to dielectrics, whose properties are governed by the existence of charges and the ability of molecules to become polarized by displacement of those charges, the magnetic properties of materials do not depend on the existence of poles, but rather on the possibility of setting up or rearranging current loops (somewhat more precisely, electron trajectories) in the molecules which will then act as elementary dipoles. Matter consists largely of free space, in which is to be found a swarm of charges. In so far as we neglect the motion of these charges, and concentrate on their mean positions, we can describe the state of electric polarization and, as in Chapter 9, relate it to laboratory observations; on the other hand, if we concentrate on the motion of the charges, representing their mean behaviour by microscopic currents, we are enabled to describe the state of magnetic polarization and relate it similarly to laboratory observations. Both descriptions are abstracts of the full description, which can only properly be made in quantum-mechanical terms, of the behaviour of the elementary charged particles constituting the atoms and molecules. But they are very adequate abstracts, and for magnetic purposes we may visualize the assembly of atoms in the same way as Ampère, as little current loops in otherwise empty space. These current loops may already exist before an external magnetic field is applied, or they may be induced by the very act of applying the field. In the former case, the field causes a degree of alignment and gives rise to paramagnetic polarization as discussed in Chapter 11; in the latter case, diamagnetic polarization results, in accordance with Larmor's theorem. 238

PROBLEM

An electron describes a circular orbit about an attractive centre (not necessarily inverse-square law). When a magnetic field **B** is slowly applied, normal to the plane of the orbit, show that the orbit remains the same size but, because of electromagnetic induction, the electron speed changes, and that the resulting change of centripetal acceleration is consistent with the presence of the Lorentz force in addition to the original central force. It is necessary to assume that the field is not so strong as to change the speed by more than a small fraction.

Comment. The change of speed, Δv, is $eBr/(2m)$, corresponding to a change in angular velocity $\Delta\omega = eB/(2m)$, in accordance with Larmor's theorem. This example illustrates how induction provides the mechanism for the change in the parameters of the orbit.

The contrast between dielectrics and magnetic materials, arising from the absence of poles, show up when we consider how the mean magnetic field **B** in a material, defined as an average over all regions inside and outside of molecules, is related to the field in a cavity. For a dielectric, we found the mean field **E** to be the same as that inside a pencil-shaped cavity cut parallel to **E**; in a magnetic material the mean field **B** is that inside a disc-shaped cavity with normal along **B**. The argument is simple; because there are no free poles, div $\mathbf{B} = 0$ and the lines of magnetic field are everywhere continuous on an atomic scale, though they may be highly contorted by the atomic currents and even form closed loops. But if we ask how many lines cross a given area drawn normal to the mean direction, such as the plane A, continuity demands that the number shall be the same whether A lies in the material or, like A', in the disc-shaped cavity. Since the mean field is the average number cutting unit area, we conclude that the field in the cavity is the same as $\bar{\mathbf{B}}$ in the neighbouring material. By the same argument we know that when a magnetic field lies normal to a boundary, it is **B** which is continuous across the boundary.

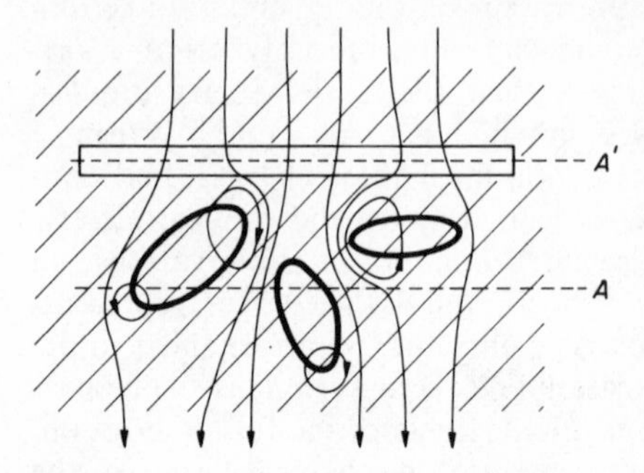

The difference between the electric and magnetic cases, revealed by the different cavities required to determine the mean fields, arises from the presence or otherwise of free charges or poles. Clearly the above argument could not be applied to a dielectric because the free charges create field lines, and continuity of the lines cannot therefore be assumed. Instead, we made use in the dielectric case of the vanishing of the line integral of **E** round a circuit partly in the material and partly in a pencil cavity. This argument fails to apply in its simplest form to magnetic materials since the circuit may well enclose currents and the line integral need not vanish. Suppose the magnetic properties are due to n current loops per unit volume, each having on the average a projected area a normal to the field direction and carrying on the average a current i. Then, by A1, the component parallel to **B** of the average magnetic moment of each is ai, and the parallel component of the magnetic moment per unit volume **M** is nai. For an isotropic material, **M** will be parallel to **B**, so that nai is the full magnitude of **M**. Now if these loops are randomly arranged in space, a line drawn along **B** will cut through na loops in going unit distance (if you do not find this statement obvious, devise a proof), so that the total current encircled by the path shown in $naiL$ or ML. If B' is the magnitude

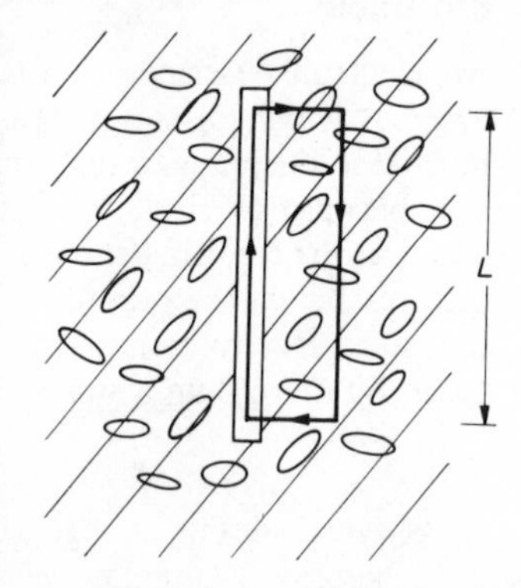

of the field in the cavity, the line integral $L(B - B')$ must be equated to $\mu_0 ML$, so that

$$B' = B - \mu_0 M. \tag{13.1}$$

For historical reasons, it is conventional to write $\mathbf{B}'$ as $\mu_0\mathbf{H}$, so that (13.1), expressed vectorially, takes the form

$$\mathbf{B} = \mu_0(\mathbf{H} + \mathbf{M}). \tag{13.2}$$

The reader may similarly demonstrate that in a dielectric

$$\mathbf{D} = \varepsilon_0\mathbf{E} + \mathbf{P}.$$

It is a pity that the definitions adopted for **M** and **P** do not allow the same form for what are, on the macroscopic scale, entirely analogous expressions, but the reader should not infer from the difference that the contrast in the atomic origins of the electric and magnetic properties leads to any discernible difference in effect between these two equations; the difference in form is purely accidental.

The derivation of (13.2) given above is of course not very rigorous, but the result is in fact of very general validity, and in no way depends on the magnetic medium being linear, in the sense of **M**, **B** and **H** being all in constant proportion. If the medium is linear we may define the *volume susceptibility* κ_m as the ratio M/H, and the *relative permeability* μ as B/B' or $B/(\mu_0 H)$. Then κ_m and μ are related, according to (13.2), by the analogue of (9.3),

$$\mu = 1 + \kappa_m. \tag{13.3}$$

An estimate of the diamagnetic susceptibility based on (11.10) shows that κ_m may be expected to be very much less than unity, in contrast to dielectric susceptibilities which we saw to be of the order of unity. Thus for n atoms per unit volume, each having n_e electrons, $\mathbf{M} = nn_e\mathfrak{m}$ and $\kappa_m = -\mu_0 ne^2\overline{r^2}n_e/(6m)$; for a solid in which the atoms are closely packed, nr^3 is about 1/6, so that κ_m is approximately $-\mu_0 n_e e^2/(40\ ma)$, if a is the atomic radius; taking n_e as 10, we find κ_m to be about -10^{-4}, so that the diamagnetic effect is very weak.

184

The paramagnetic susceptibility may be considerably stronger, especially at low temperatures where the attempts of the permanent moments of the atoms to align themselves along **B** are not disturbed so much by collisions. To calculate this effect properly demands the methods of statistical mechanics, but we can make a good estimate by a simple argument. If an atom has permanent moment $\mathfrak{m}$, it can have magnetic energy anywhere between $\pm\mathfrak{m}B$ according to its alignment. Now if the field is so strong that $\mathfrak{m}B$ is equal to the typical kinetic energy of a colliding molecule, kT say (k is Boltzmann's constant), the alignment will not be greatly disturbed by most collisions. We may guess therefore that in weak fields the degree of alignment increases linearly with B at such a rate as would lead, if it continued linearly, to total alignment in a field strength $B \sim kT/\mathfrak{m}$, or $H \sim kT/(\mu_0\mathfrak{m})$. But with total alignment the value of M is

$n\mathfrak{m}$, so that we may guess κ_m to be about $n\mathfrak{m} \div [kT/(\mu_0\mathfrak{m})]$, i.e., $\mu_0 n\mathfrak{m}^2/(kT)$. The complete calculation gives $\mu_0 n\mathfrak{m}^2/(3kT)$, differing only by a factor 3, and the diagram shows how the state of complete alignment (*saturation*) is approached. For a hydrogen atom in its lowest Bohr orbit, of radius a, an electron revolving with velocity v makes $v/(2\pi a)$ revolutions per second and therefore behaves as a current $ev/(2\pi a)$ in a loop of area πa^2; hence $\mathfrak{m} = \frac{1}{2}eva$. By using this result and (11.10), we may compare the paramagnetic and diamagnetic susceptibilities of the hydrogen atom, and find the simple result that they are roughly in the ratio of the kinetic energy of the electron to kT, which is about 500 at room temperature, and correspondingly more at lower temperatures. Even so, we are still some way from having $\kappa_m \sim 1$, and for most substances the permeability μ is only very little different from unity. Exceptions to this statement are soft iron and other ferromagnetic materials, to which we shall turn later.

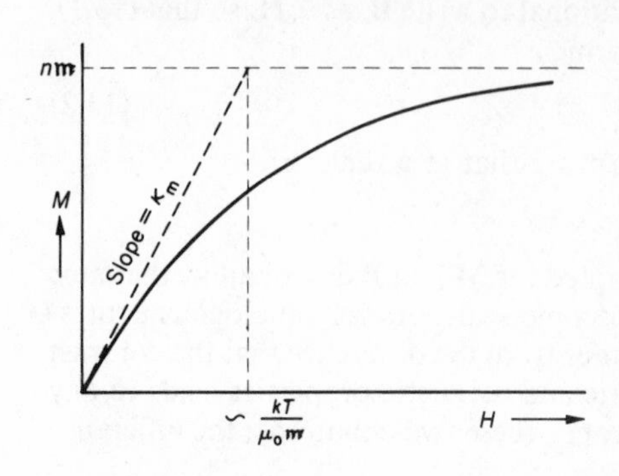

Before doing so, however, we must relate the fields **B** and **H** to their sources, e.g., poles and currents immersed in the magnetic material, and discuss the forces experienced by such poles and currents. First we note that in regions where there are no macroscopic currents the line-integral of **H** around any circuit vanishes; for we may cut a thin cavity to follow the circuit, and in it $\mu_0\mathbf{H}$ determines the component of field parallel to its length; the integral round the cavity vanishes if the circuit encloses no current (always remember that the cavity is to be thought of as a real cavity, obtained by removing atoms, not a geometrical abstraction obtained by drawing surfaces that may pass through atoms). By the same argument we infer that when there are currents present $\oint \mathbf{H} . d\mathbf{r}$ equals the current enclosed by the circuit, so that curl $\mathbf{H} = \mathbf{J}$. This result does not depend on the medium being present or not, and we may conclude that for a given current distribution **H** in a uniform linear magnetic medium shows exactly the same field pattern and magnitude as in free space, where **H** is just $\mathbf{B}/\mu_0$. It is necessary to specify that the medium shall be uniform, for a non-uniform medium (e.g., a mixture of regions of different properties) will in general distort the field pattern while still retaining locally the essential property that curl $\mathbf{H} = \mathbf{J}$. As a simple illustration, consider a current i

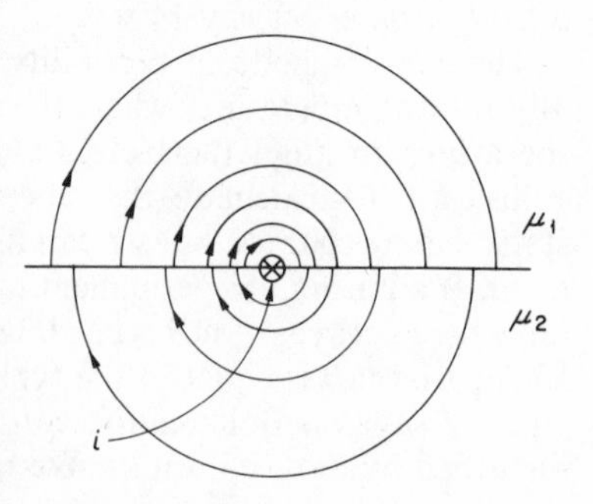

carried by a wire lying in the plane interface between two media of permeability μ_1 and μ_2. In each medium, the field drops off as $1/r$, but the magnitudes are different, since at the interface it is the normal component of **B**, i.e., $\mu\mu_0\mathbf{H}$ which is continuous. Thus in medium 1 we may have H varying as $A/(\mu_1 r)$ and in medium 2 as $A/(\mu_2 r)$, A being a constant, and the boundary condition is satisfied. Taking a circuit round the current at radius r we find

$$\oint \mathbf{H}.\mathrm{d}\mathbf{r} = \pi A(1/\mu_1 + 1/\mu_2),$$

and equating this to i we determine A. Hence, in medium 1, H is $\mu_2 i/[\pi r(\mu_1 + \mu_2)]$ and in medium 2, H is $\mu_1 i/[\pi r(\mu_1 + \mu_2)]$. It should be appreciated that this example has been chosen as one where the solution may be written down easily, but usually the solution of the boundary-value problem will be very troublesome, typically only to be solved by computer or other numerical means.

If a magnetic pole is embedded in the medium, we may assure ourselves, by cutting a spherical shell cavity round it, that **B** takes the same value as if the medium were absent. By extension to dipoles and permanent magnets, we see that **B** takes the same form and has the same magnitude in a uniform linear medium as in free space. To sum up these ideas, if we wish to express everything in terms of a single field variable **B**, we have the following set of rules, applicable to a linear medium:

(1) **B** is the field in a disc-shaped cavity cut normal to the field.

(2) At an interface the normal component of **B** and the parallel component of $\mathbf{B}/\mu$ are continuous.

(3) $\oint\mathbf{B}.\mathrm{d}\mathbf{r}$ equals $\mu\mu_0$ times the current enclosed, or curl $\mathbf{B} = \mu\mu_0\mathbf{J}$.

(4) For permanent magnets in a uniform medium, **B** is to be calculated as if the medium were absent.

If the medium is non-linear, both **H** and **B** must be retained to describe the field, and the medium characterized by the relationship between the two field vectors.

We may now consider the way in which the forces between poles and currents are modified by flooding the intervening space with a magnetic fluid (e.g., a solution of a transition metal salt, like $MnCl_3$, or liquid oxygen which is sufficiently paramagnetic to be picked up by a strong permanent magnet). For two poles placed in the fluid, the argument proceeds exactly as in Chapter 9, with pressure differences building up so as to reduce the force by a factor μ. Now in free space the force on a pole P in field **B** is $P\mathbf{B}$, and when the magnetic fluid is added the field **B** due to the other pole is unchanged. We therefore have that the force is $P\mathbf{B}/\mu$, which can also be written as $\mu_0 P\mathbf{H}$, in which form the permeability of the medium does not appear. This argument can clearly be extended to dipoles and 204 permanent magnets, with the conclusion that it is **H** that may be regarded as determining the force.

With this in mind consider the force acting on a pole in the vicinity of a current which generates a field **H**. When the fluid is added **H** remains unchanged, and the force on the pole must also stay the same. As discussed in Chapter 11, N3 applies to this system, so that the force exerted by a pole (or permanent magnet) on a wire carrying a current is unaffected by flooding the system with a magnetic fluid. Since it is the field **B** due to the pole which remains unchanged in this process, we may take it that Ampère's second law for the force, $i\,\mathbf{dl} \wedge \mathbf{B}$, on an element **dl** holds in the same form when a magnetic fluid is present.

From this point we proceed to the last case, the interaction of two wires carrying currents. When the fluid is added, the value of **B** due to one of the wires is increased by a factor μ, and therefore the force on the other wire must also increase by μ. We are led to the conclusion that the effect of the medium is to decrease by μ the force between poles and permanent magnets, to leave unchanged the force between a current and a pole or permanent magnet, and to increase by μ the force between currents.

This may at first seem rather unlikely, but a simple example will illustrate how the pressure distribution in the medium, as modified by magnetic effects, brings it about. Consider first two plane distributions of polarity, one North and one South, which attract each other. The field between them being stronger than it is outside, an excess pressure builds

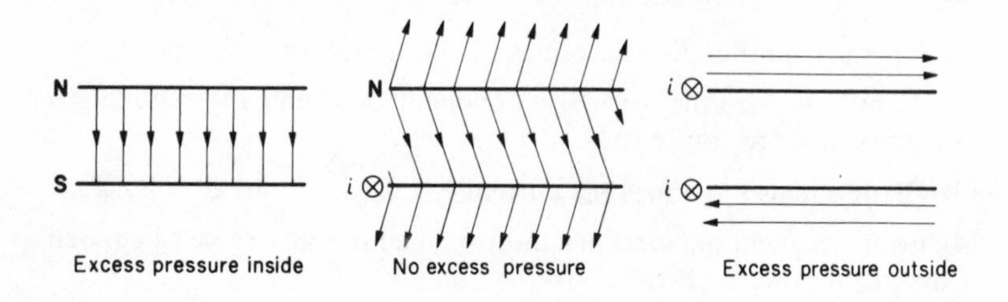

up inside if the fluid is paramagnetic ($\mu > 1$), and the force is consequently reduced; this exactly parallels the dielectric argument. Next, consider a plane of North polarity and a parallel plane current sheet. The diagram shows how the field patterns due to each combine into a zig-zag pattern, whose strength is the same everywhere but whose direction changes. Since the pressure differences are set up by variations of B^2, this source distribution does not disturb the uniform pressure and no force changes result. In any case, the force exerted by each plane on the other is parallel to the planes and could not be affected by pressure changes. Finally, consider the plane current sheets which, if they carry current in the same sense, attract each other. The field, however, in contrast to the attracting sheets of polarity, is strong outside and weak within, so that the pressure differential in the fluid increases the attractive force. Viewed in this way, the different effects of the medium do not seem so paradoxical.

185

However, before dismissing the subject, we must raise another paradox. If we believe that permanent magnets are nothing but an assemblage of current loops on an atomic scale, how is it that the force between them is affected differently from the force between macroscopic currents? The answer lies in the fact that the magnetic medium does not permeate the interstices of the elementary current loops in the permanent magnet, and the importance of this point is illustrated by a simple model of a permanent bar magnet, consisting of a long solenoid carrying a current i, and enclosed in a cylindrical box. Let us form $\oint \mathbf{H} . d\mathbf{r}$ round a path that passes along the axis of the solenoid. Since the value of the integral is not altered by the presence of the medium, and the greater contribution to it comes from inside the solenoid, **H** and therefore **B** inside the solenoid must be almost independent of the presence of the medium. Now at the interface it is **B** which is continuous, so that **B** in the medium is hardly affected by the presence of the medium. This is exactly how the permanent magnet behaves, and, for example, a current in the neighbourhood of the enclosed solenoid, being acted upon by **B**, will experience a force independent of the presence of the medium. But if we puncture the container and let the medium flow into the solenoid, the value of **B** inside must rise by a factor μ so as to maintain the same value of **H**, and continuity of **B** outside the solenoid will ensure that the force on a neighbouring current also increases by μ. Thus a simple example, while not giving a complete quantitative answer to the problem, shows clearly enough that there need be no major discrepancy in the picture of permanent magnets as assemblages of currents.

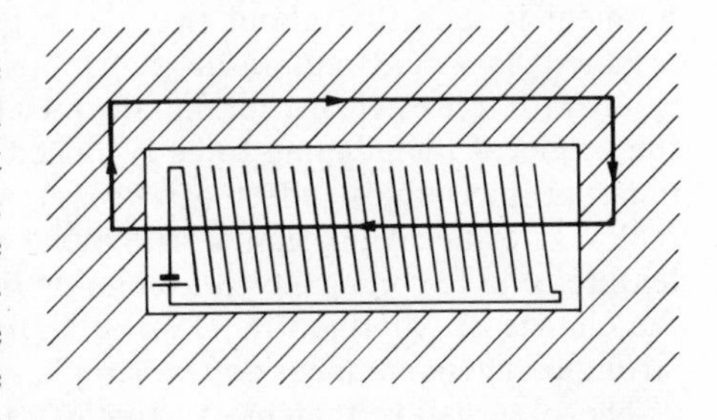

The foregoing account has given the impression that the agent responsible for the magnetic moment in an atom is an electron in orbit. This is normally true for diamagnetism and sometimes for paramagnetism (e.g., the oxygen molecule), but a commoner agent of paramagnetism is the intrinsic magnetic moment of the electron, which behaves rather as though it were a little spinning charged sphere. It is difficult to visualize this property of *spin* which, it must be emphasized, really has nothing to do with the structure of the electron. The quantum theory of the electron, due to Dirac, reveals the spin as a necessary property of a structureless, charged, point particle, and any picture we form of it must visualize an electron in motion more as a particle executing a finely coiled helical path than as a little magnetic dipole moving in a straight line. In other words, the magnetic effects of spin must still be regarded as essentially due to currents; there are experiments, which we need not discuss here, that demonstrate the mean field to be **B** (as is appropriate to currents)

rather than $\mathbf{B}/\mu$ or something in between these two (as would be appropriate to impenetrable dipoles). With this in mind, it usually will do no harm to think of the electrons as little magnets, provided we are not concerned with the mean field or, and this is far outside our domain, with phenomena that occur at very small distances, such as collisions of the electrons with other particles. Thus in iron, whose magnetic properties are due to the electron spin, we may visualize the phenomenon of permanent magnetization as caused by some process on the atomic scale that lines up a substantial fraction of the electrons with their moments all pointing in the same direction. In fully magnetized iron, for example, the moment is such that about two electrons in each atom must be firmly oriented. It would require an enormous magnetic field to achieve this result with free electrons continually misaligned by thermal agitation, but the origin of the aligning force is indeed not magnetic, but a peculiarly quantum-mechanical effect (*exchange*) which has no analogue in the realm of classical physics. Fortunately we can describe the large-scale features of a ferromagnet without going beyond the statement that, given the chance, it will align its electron spins in such a way as to confer a strong magnetic moment on the sample as a whole.

The observation that most samples of iron are unmagnetized or, at best, weakly magnetized, does not mean that there is no alignment, but only that the sample contains many small domains, each of which is fully magnetized, but which are randomly oriented so that the resultant magnetization is only weak. The domains can be revealed, in well-prepared samples, by covering the surface with a colloidal suspension of iron, a technique developed especially by Bitter and named after him. There is a tendency for the magnetic field on the surface to be stronger at the boundaries between domains, and the iron particles are attracted there, as can be seen in the photographs of a particularly simple pattern in a tiny single crystal of iron. The process of magnetization in a weak external field is achieved not by turning the domains round, but simply by those which are favourably oriented growing at the expense of the others. In a really soft magnetic material, which is one in which the movement of the boundaries between domains (*Bloch walls*) is not inhibited by structural defects or other factors, almost total alignment of the domains may be achieved with fields no stronger than the Earth's field ($\sim 10^{-4}$ T). These materials behave in such weak fields as substances of enormous paramagnetic permeability ($\mu > 10^3$). By contrast, if the domain boundaries are pinned, as in hard magnetic materials, a large field must be applied to overcome the pinning and magnetize the sample, but when the field is removed most of the magnetization remains. Such materials are suitable for permanent magnets, while soft ferromagnetics are used as cores in transformers and other components requiring high inductance.

When a solenoid or other coil is filled with a core of highly permeable ($\mu \gg 1$) material, its inductance is greatly increased for a reason which is readily understood from an atomic standpoint. Suppose for simplicity that the whole of the space in and around the coil is filled with a uniform

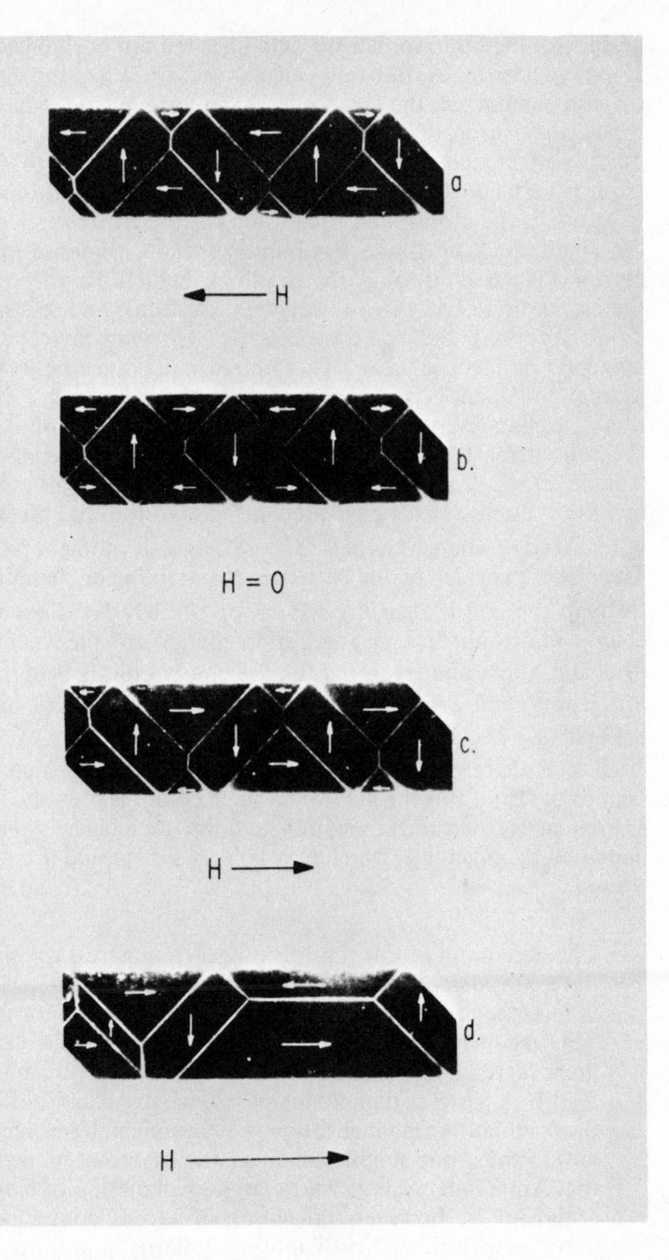

Domains in a very small (about 1/10 mm) crystal of iron, revealed by the Bitter technique. The arrows have been added to show the magnetization direction in each domain. Successive photographs show how favoured domains grow at the expense of the others.

permeable medium, so that the field lines remain unchanged in form, and **H** everywhere takes the same value as before. When the current through the coil is changed, the e.m.f. across the terminals may be thought of as the superposition of the inductive effect of the empty coil and any inductive effect due to the change in magnetization of the medium. The latter is very conveniently represented by changes of the currents flowing in loops in the atoms, and each atom can be treated independently and the resultant of all found by simple addition. Suppose then that when current i is passed through the windings the field $\mathbf{B}_0$ at a certain point in the empty coil is $\mathbf{A}i$ ($\mathbf{A}$ can vary with position), and let an atom at this point be represented as a loop of area $\mathbf{a}$ carrying current i_1, and having magnetic moment $\mathfrak{m} = \mathbf{a}i_1$. The mutual inductance between the atomic loop and the coil, being the flux through the loop when $i = 1$, is $\mathbf{A}.\mathbf{a}$, so that if i_1 changes at a rate $\mathrm{d}i_1/\mathrm{d}t$ (for example, in response to changes of the coil current) the e.m.f. appearing across the terminals of the coil is $\mathbf{A}.\mathbf{a}\ \mathrm{d}i_1/\mathrm{d}t$, or $\mathbf{A}.\dot{\mathfrak{m}}$. Adding up the contributions of all atoms in an element of volume $\mathrm{d}V$, we have an e.m.f. due to the changes in this volume of $(\mathbf{A}.\dot{\mathbf{M}})\ \mathrm{d}V$, where $\mathbf{M}$ as before is the magnetic moment per unit volume. The power provided by the battery in maintaining or changing the current through the coil is then $i(\mathbf{A}.\dot{\mathbf{M}})\ \mathrm{d}V$ or $(\mathbf{B}_0.\dot{\mathbf{M}})\ \mathrm{d}V$. Thus when a small change in i results in a magnetization change $\Delta\mathbf{M}$ the total work needed, over and above that required for the empty coil, is $\int(\mathbf{B}_0.\Delta\mathbf{M})\ \mathrm{d}V$. Now the empty coil requires work equal to the change in field energy $\Delta\int(\frac{1}{2}B_0^2/\mu_0)\ \mathrm{d}V$, or $(1/\mu_0)\int(\mathbf{B}_0.\Delta\mathbf{B}_0)\ \mathrm{d}V$. The effect of the magnetic medium is therefore to replace $\Delta\mathbf{B}_0$ by $\Delta(\mathbf{B}_0 + \mu_0\mathbf{M})$ which from (13.2) is seen to be $\Delta\mathbf{B}$, if $\mathbf{B}$ is the field when the medium is present. This is μ times as great as $\mathbf{B}_0$ for a linear medium, and the inductance is correspondingly increased by a factor μ, which may be very substantial if a really soft iron is used.

The derivation of this result is longer than necessary, for all we need have considered is that **B** is the mean field in the material, and if this is increased μ-fold so is the flux linkage with the turns of the coil. This argument is liable to lead to the belief that the observation of a great increase in inductance can be taken as implying that the mean field **B** is correspondingly increased, which would not be the case if the origin of the magnetization were atomic poles rather than current loops; thus, one might conclude, the macroscopic evidence shows that Ampèrian currents are the right explanation of magnetism. This argument is, however, fallacious. If atoms contained elementary particles with either North or South polarity, and it was the displacement of these that produced magnetic polarization, it is indeed true that the mean field, as in the dielectric case, would be **H** rather than **B**, and unchanged by the addition of the permeable medium. But just as moving electric charges produce magnetic fields, so we should find if we had free poles that when they were in motion they would produce electric fields; it would be this field, set up as the medium became magnetically polarized, that would lead to the appearance of an extra e.m.f. across the terminals of the coil. We can infer nothing from macroscopic observations of this sort about the microscopic structure of matter.

In the above analysis, we assumed the whole space to be uniformly filled with magnetic material. This is unnecessarily restrictive when the permeability is very high, for field lines tend to be attracted into the medium and are reluctant to emerge again. At an interface between soft iron and free space, the continuity of parallel **H** and normal **B** ensures that the field lines emerge normally into free space unless the field in the iron runs almost exactly parallel to the boundary. The effect is that when, for example, an iron anchor ring has a short coil wound on it, the field lines run round inside the iron with only very few escaping into free space. The inductance of the coil is therefore increased very nearly as much as if the whole of space were filled with iron.

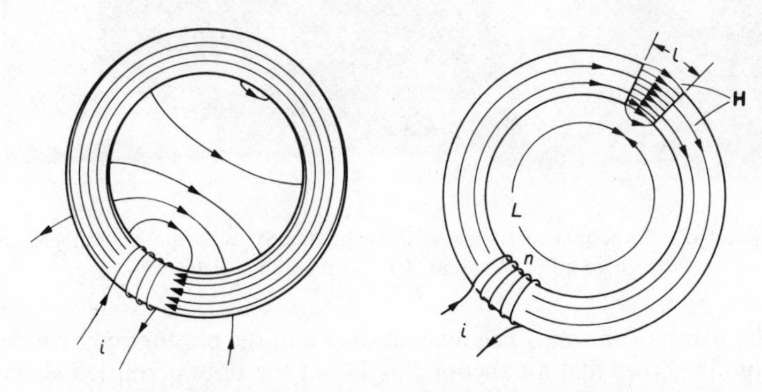

By taking a circuit of length L round the ring inside the material, one can equate the line integral LH to the current enclosed, ni for n turns carrying current i. Hence in the iron $B = \mu\mu_0 H = \mu\mu_0 ni/L$, and may be large if μ is high enough. It does not take much of a gap in the ring to change this result. For continuity of the normal component of **B** ensures that in the gap B is nearly the same (apart from fringing effects) as in the iron, and therefore H is μ times as great. We have then that $\oint \mathbf{H}.d\mathbf{r}$ is made up of $(L - l)H_1$ in the iron and $\mu l H_1$ in the gap, so that $H_1 = ni/[L + (\mu - 1)l]$. For $\mu = 10^3$, l need only be one thousandth of L to halve H, and hence B and the inductance of the coil. There are compensations, however; the value of B in the gap is the same as in the iron, i.e., $\mu\mu_0 H_1$, or approximately $\mu_0 ni/l$ if $\mu l \gg L$. This field is considerably greater, if l is much smaller than the coil dimensions, than what the coil would produce by itself; for alone it would give $\mu_0 ni$ divided by some characteristic length, say twice the length of the coil. This is the principle underlying the use of iron yokes for electromagnets; if they are large, they can carry large exciting coils, but the field is still mainly determined

The great electromagnet at Bellevue, Paris (c. 1935), with the pole pieces set to give a field of about 10 T in a volume of 1 cm^3.

by the width of the gap. The magnet shown in the photograph, which incidentally shows that an anchor ring is not the only permitted shape, is one of the largest ever made, and produces a field of 10 T in the tiny gap. The concentration of field by coning of the pole face helps to enlarge its magnitude; the tendency of field lines to run parallel to the surface inside the iron explains why this concentration occurs.

270 The use of an iron core in a transformer not only increases the self-inductance of the windings, but allows the mutual inductance to approach its maximum of $(L_1L_2)^{1/2}$. The condition we found for achieving this limit

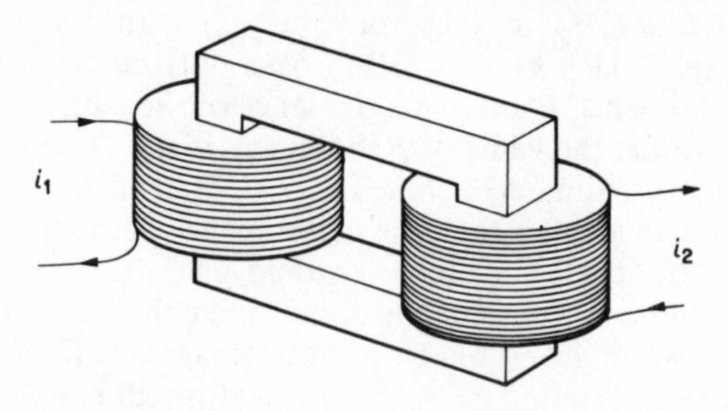

was that both coils must set up the same configuration of field, and this is very nearly possible, even with coils wound on different limbs of the core as in the diagram, if the field runs inside the core with very little leakage. Since transformers are used with alternating currents, the flux oscillates in magnitude and induces currents in the core. Unless the core is much smaller in cross-sectional dimensions than the skin depth at the frequency used, the inner parts are ineffective; moreover, the heat developed by the currents not only reduces the efficiency of the transformer but in large transformers poses a serious cooling problem. This is overcome in power frequency (50 or 60 Hz) transformers by laminating the core; the individual sheets are oxidized so as to be in only poor electrical contact, and the circulating eddy currents, which have to cross from one sheet to the next, are thereby greatly reduced. At higher frequencies, above 10^6 Hz say, transformers used in electronic circuits do not need such high inductances, since it is ωL that must be large compared with other circuit resistances etc., and the cores can be made of ferrites (mixed oxides of iron, nickel etc.) which have quite high permeability, though not as high as soft iron, and the advantage of being very poor conductors.

Cavendish Problems: 215, 220, 235, 248

READING LIST

Molecular Theory of Magnetism: C. Kittel, *Introduction to Solid State Physics*, Wiley.
Transformers: J. Hindmarsh, *Electrical Machines*, Pergamon.

14

Maxwell's displacement hypothesis and electromagnetic waves

Maxwell's equations

$$\text{curl}\,\mathbf{E} = -\dot{\mathbf{B}} \qquad \text{curl}\,\mathbf{B} = \mu\mu_0(\mathbf{J} + \varepsilon\varepsilon_0\dot{\mathbf{E}})$$
$$\text{div}\,\mathbf{E} = \rho/(\varepsilon\varepsilon_0) \qquad \text{div}\,\mathbf{B} = 0$$

Chapter 12 developed the idea that magnetic fields in motion generated electric fields, in accordance with (12.1), $\mathbf{E} = \mathbf{B}\wedge\mathbf{u}$, and it is now time to enquire whether the reciprocal effect exists whereby an electric field in motion generates a magnetic field. It was the discovery of this complement to Faraday's law of induction that led Maxwell to his description of the electromagnetic wave, a landmark in the history of physics. There is no need to recapitulate his arguments, which provide a reason for the otherwise mysterious term *displacement*, nor to worry that his postulate has come to be known as a hypothesis rather than a law; it has all the validity of the other laws we have been discussing, being a plausible suggestion consolidated by verification of its predicted consequences. Let us rather assemble some simple arguments and models that will serve for us as his did for him.

First, consider the generation of a magnetic field by a current in a wire. Since the current is due to moving charges, each of which carries with it its own electric field, we may tentatively imagine that it is the motion of the electric field that is to be held responsible for the presence of the magnetic field. It may be objected that a neutral wire carrying a current produces no electric field, but we must remember that observed effects are attributed to the superposition of the effects due to individual particles; if a moving negative charge produces different effects from a stationary positive charge the difference should be observable, and in this case we suggest that it may be the magnetic field. To take the specific case of a straight wire in which n particles per unit length, each carrying charge e, are drifting at a velocity v to produce a current nev, the magnetic field at a distance r follows from (10.1),

$$B(r) = \mu_0 nev/(2\pi r).$$

Now the moving charges, if they were present alone, would produce a radial electric field $E(r)$ which is easily calculated by application of Gauss' theorem,

$$E(r) = ne/(2\pi r\varepsilon_0).$$

Comparing these two results we see that, as far as magnitudes are concerned, $B = \mu_0\varepsilon_0 vE$. The diagram, which is drawn for moving electrons, shows that at every point **B**, **E** and **v** are mutually perpendicular, so that this particular result may be written in vector form

$$\mathbf{B} = \mu_0\varepsilon_0 \mathbf{v}\wedge\mathbf{E}. \qquad (14.1)$$

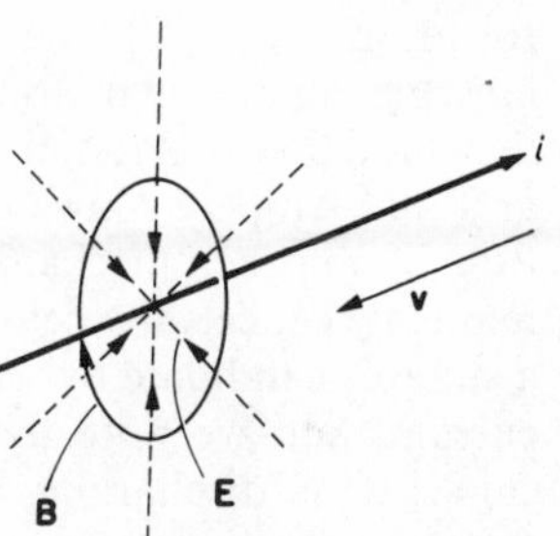

This is similar in form to (12.1), but has the components of the vector product in reverse order and contains in addition the dimensional constant $\mu_0\varepsilon_0$.

For the second argument, we turn back to the simple case, treated in Chapter 12, of a charge moving past a stationary pole. Already, in the first discussion of this case, we have noted the possibility of difficulties, but we still prefer to turn a blind eye to them and explore the further plausible implications of the analysis. If the moving charge experiences a Lorentz force due to the magnetic field of the pole, we may expect, from N3, that the pole will experience an equal and opposite force. If **r** is the radius vector from the pole P to the charge e, the magnetic field at the charge is, by (10.5) and (10.6),

244

$\mu_0 P\mathbf{r}/(4\pi r^3)$; the charge, moving at velocity $\mathbf{v}$, thus experiences a Lorentz force $\mu_0 eP(\mathbf{v}\wedge\mathbf{r})/(4\pi r^3)$, and we suppose the pole to experience an opposite force:

$$\mathbf{F} = \mu_0 eP(\mathbf{v}\wedge\mathbf{r}')/(4\pi r'^3),$$

where $\mathbf{r}'$ is $-\mathbf{r}$, the vector from e to P. Now the charge produces a field $\mathbf{E} = e\mathbf{r}'/(4\pi r'^3\varepsilon_0)$ at P, so that the force takes the form

$$\mathbf{F} = \mu_0\varepsilon_0 P\mathbf{v}\wedge\mathbf{E},$$

as if the moving charge produced a magnetic field in accordance with (14.1). This argument is clearly of the same nature as the first, but applied to an individual moving charge rather than an assembly of moving charges in a wire. One would be surprised if they gave a different answer.

These two arguments suggest a law complementary to Faraday's in the form of a moving field inducing an additional field. The alternative form of Faraday's law, in terms of the change of flux through a circuit being related to the line-integral of $\mathbf{E}$, as in (12.3), may be paralleled by application of a type of argument somewhat different from those just discussed. A special case of the argument will make its physical nature apparent, and we shall then state it in more general mathematical terms.

Suppose we charge a concentric spherical capacitor whose interspace is a slightly conducting dielectric; after charging there is a slow discharge during which a uniform radial current flows between the spheres. Now if we take Ampère's laws at their face value, we expect a magnetic field to accompany this current, for it is easy to make a circuit in the interspace that encloses current and we expect $\oint \mathbf{B}.d\mathbf{r}$ round this circuit to equal μ_0 times the current enclosed. The field must have a tangential component that satisfies the requirement of spherical symmetry, and this is impossible; for at any point all tangential directions are equally valid, and to choose any one in particular would violate spherical symmetry. We are bound to conclude that this radial current is not accompanied by a magnetic field, and that Ampère's laws require some change, for example supplementation by another mechanism for generating a magnetic field that automatically cancels the field of the current in this case. It is easy to discover a suitable candidate by seeing what other field quantities are linked to the current, and we note that the existence of the current means that the capacitor is discharging and the radial electric field is changing. If the charge on the inner sphere is Q, the electric field at a point $\mathbf{r}$ in the interspace is given by

$$\mathbf{E} = Q\mathbf{r}/(4\pi\varepsilon\varepsilon_0 r^3),$$

so that

$$\dot{\mathbf{E}} = \dot{Q}\mathbf{r}/(4\pi\varepsilon\varepsilon_0 r^3).$$

Now the total current is $-\dot{Q}$, so that the current density $\mathbf{J}$ may be written

$$\mathbf{J} = -\dot{Q}\mathbf{r}/(4\pi r^3) = -\varepsilon\varepsilon_0\dot{\mathbf{E}}.$$

In this particular case, then, $\mathbf{J} + \varepsilon\varepsilon_0\dot{\mathbf{E}}$ vanishes at all times, so that if we suppose the generator of magnetic field to be not simply the current (or current density) but current density supplemented by the term $\varepsilon\varepsilon_0\dot{\mathbf{E}}$, which only matters when **E** is changing, and therefore would naturally have been overlooked in the stationary experiments of Ampère and others, we do not worry ourselves looking for **B** where by symmetry no **B** can exist.

The argument can be expressed more generally by noting the differential form of Ampère's law,

$$\text{curl } \mathbf{B} = \mu\mu_0 \mathbf{J}. \tag{14.2}$$

The dilemma discussed in connection with the spherical condenser reappears here when we take the divergence of both sides of the equation. For according to XIII, div curl **B** is identically zero for any vector field **B**; on the other hand div **J** does not necessarily vanish, though when it is non-zero there must be charge densities accumulating in accordance with the equation of charge conservation,

$$\text{div } \mathbf{J} + \dot{\rho} = 0. \tag{14.3}$$

Thus Ampère's law as stated in (14.2) presents no problems when **J** is non-divergent, but only when charges are accumulating and electric fields changing, as in the discharging capacitor. Now we can relate ρ and **E** by means of (8.4), modified in a dielectric medium to the form

$$\text{div } \mathbf{E} = \rho/(\varepsilon\varepsilon_0).$$

With the help of this equation we write (14.3) in the form

$$\text{div } (\mathbf{J} + \varepsilon\varepsilon_0\dot{\mathbf{E}}) = 0,$$

which shows that if we take $\mathbf{J} + \varepsilon\varepsilon_0\dot{\mathbf{E}}$ to be the magnetic field generator instead of merely **J**, we shall never run into difficulties from the vanishing of div curl **B**. We propose therefore to replace (14.2) by the equation

$$\text{curl } \mathbf{B} = \mu\mu_0(\mathbf{J} + \varepsilon\varepsilon_0\dot{\mathbf{E}}). \tag{14.4}$$

The term $\varepsilon\varepsilon_0\dot{\mathbf{E}}$ is the *displacement current* discovered by Maxwell.*

In a situation where no real current **J** is flowing, in an insulator for example, it may still be possible to generate a magnetic field by changing the electric field,

$$\text{curl } \mathbf{B} = \mu\mu_0\varepsilon\varepsilon_0\dot{\mathbf{E}}. \tag{14.5}$$

This equation should be compared with the differential equation expressing Faraday's law,

$$\text{curl } \mathbf{E} = -\dot{\mathbf{B}}. \tag{14.6}$$

250

The similarity of form shows that just as we could picture the movement of magnetic lines generating an electric field according to the rule

* It should be clear that (14.4) implies (14.3), the equation of charge conservation, which need not therefore be explicitly included among the equations governing electromagnetism.

$\mathbf{E} = \mathbf{B} \wedge \mathbf{v}$, so we may adopt a parallel view, without violating (14.4), in which the movement of charges and their associated electric fields generates a magnetic field according to the rule

$$\mathbf{B} = -\mu\mu_0\varepsilon\varepsilon_0\mathbf{E} \wedge \mathbf{v}. \tag{14.7}$$

In writing this down, we have noted the sign difference between (14.5) and (14.6), and supplied the appropriate constants. This result is the same as (14.1), to which we were led by the first examples, extended to material media by including μ and ε as factors. On the other hand, we may choose to interpret (14.5) and (14.6) in parallel ways in terms of the flux through a circuit. Just as (14.5) expresses in differential form the fact that the integral $\oint \mathbf{E}.d\mathbf{r}$ equals the rate of change of flux of $\mathbf{B}$ through the circuit, so (14.5) tells us that $\oint \mathbf{B}.d\mathbf{r}$ is the rate of change of the flux of $\mu\mu_0\varepsilon\varepsilon_0\mathbf{E}$ through the circuit. In the spherical condenser, for example, the decrease of $\mathbf{E}$ as it discharges changes the flux of $\mu\mu_0\varepsilon\varepsilon_0\mathbf{E}$ through any circuit we choose to draw, in such a way as to give a contribution to $\oint \mathbf{B}.d\mathbf{r}$ that exactly annihilates the contribution of the discharge current. Usually, of course, the two do not annul one another, and magnetic fields are produced by one or the other, or both together.

If the movement of an electric field produces a magnetic effect, we must also expect that moving through a stationary electric field will produce on the moving body effects equivalent to applying to it a magnetic field of magnitude given by (14.7). Such an effect is significant in atoms, where the motion of the electron through the electric field of the nucleus makes it behave as if there were an additional magnetic field acting on its intrinsic 279 magnetic moment (spin). The energy of the electron now has an extra contribution depending on the orientation of the spin relative to the apparent magnetic field direction $\mathbf{v} \wedge \mathbf{E}$. The quantum theory of the electron allows only two orientations, parallel and anti-parallel, and the resulting two slightly different energies for the electron reveal themselves as splitting of the radiated spectral lines into *doublets*. The best-known of these is the sodium doublet with wavelengths 5895·93 and 5889·96 Å, differing by one part in 1000. Any moderately good spectrograph (prism or grating) will reveal the characteristic orange light of a sodium lamp as made up of these two lines.

It will be appreciated that the electron in the sodium atom is moving very fast, say 10^7 m s^{-1}, in an extremely strong electric field, say 10^{11} V m^{-1}; yet the magnetic field so generated is only enough to modify the energy very little. In fact, the observed difference is such as would be produced by a magnetic field of about 10 T, the sort of field that a good superconducting coil produces in the laboratory. The relative weakness of this effect arises from the constant $\mu_0\varepsilon_0$ in (14.7); since μ_0 is defined in SI units to be $4\pi \times 10^{-7}$, and ε_0 is $8{\cdot}9 \times 10^{-12}$ in this system, $\mu_0\varepsilon_0$ is about 10^{-17}. It will be seen that the orders of magnitude quoted (and they are not intended to be more than rough estimates) are in agreement with (14.7). We cannot therefore expect that laboratory experiments on macroscopic bodies will readily exhibit displacement effects, although the corresponding induction effects are easily seen. This is another example of what was discussed in Chapter 10, that the constitution of matter out of electrically rather than magnetically charged particles favours the production in the laboratory of strong magnetic fields and also the detection of weak electric fields; it is therefore much easier to detect the electric field produced by a moving magnetic field than the reverse effect. It would be quite simple (as for example in the Gaussian system of units still favoured by many physicists interested in atomic and nuclear problems) to redefine the units of **E** and **B** so as to give a more symmetrical appearance to (14.5) and (14.6). But the new units would be far removed from the magnitudes normally found in everyday applications, which dictated the choice of practical units.

197

The magnitude of $\mu_0\varepsilon_0$ has just been quoted as about 10^{-17} in SI without stating the dimensions; these are easily discovered by referring back to the laws of force between charges (4.2) and currents (10.14), which show that, in terms of charge $[Q]$, length $[L]$, force $[F]$, and time $[T]$,

$$[\varepsilon_0] = [Q]^2[L]^{-2}[F]^{-1};$$

$$[\mu_0] = [Q]^{-2}[T]^2[F].$$

Hence

$$[\mu_0\varepsilon_0] = [L]^{-2}[T]^2;$$

that is $(\mu_0\varepsilon_0)^{-1/2}$ has the dimensions of velocity, and a magnitude of about 3×10^8 m s^{-1}, which is close to the velocity of light in free space. In fact, the better the measurements of $(\mu_0\varepsilon_0)^{-1/2}$ by electrical means and of the velocity of light, the nearer is the agreement. This was recognized by workers before Maxwell, but it was his triumph to discover in the displacement hypothesis the link in the argument that had been missing hitherto. Given the phenomena of induction and displacement one can show that waves can be propagated with the velocity $(\mu_0\varepsilon_0)^{-1/2}$ in free space, and it is natural then to believe that these waves are none other than light waves. We shall give several arguments leading to the notion of electromagnetic wave propagation and shall then describe briefly and in very general terms the enormous range of physical phenomena that is illuminated by an understanding of these waves and their properties.

217

First, imagine that by some means, using suitable sources, we have set up in space a field pattern such that in any plane normal to the z-axis **E** and **B** are uniform fields lying in the plane and mutually perpendicular. In different planes they may take different values and orientations, but they maintain the same simple pattern.* Let us now project the whole pattern bodily, sources and all, along the z-axis with velocity **v**. Then the

* The reader who recognizes, and is pedantically dismayed by, the fact that this prescription entails having available 'magnetic currents' as well as electric currents may prefer to ignore this argument and proceed to the following two which are not so objectionable.

motion of **B** generates an extra electric field $\mathbf{B} \wedge \mathbf{v}$ parallel to and enhancing the already existing field **E**; similarly the motion of **E** generates an extra magnetic field $\mu_0 \varepsilon_0 \mathbf{v} \wedge \mathbf{E}$, parallel to and enhancing the already existing field **B**. Note that the sign difference between (12.1) and (14.1) allows both fields to be simultaneously enhanced by the motion. As **v** is increased, the field strength for given sources becomes greater, and there is a certain velocity at which it is possible to arrange that the sources are dispensed with entirely. For if we choose **E**, **B** and **v** to be such that, in any plane normal to **v**, $\mathbf{E} = \mathbf{B} \wedge \mathbf{v}$ and $\mathbf{B} = \mu_0 \varepsilon_0 \mathbf{v} \wedge \mathbf{E}$, each field maintains the other by its motion. We have therefore discovered a field pattern which requires no sources to maintain itself, and which propagates unchanged at a certain velocity which we shall designate by **c**. This is an electromagnetic wave, and since **E**, **B** and **c** are mutually perpendicular we can derive c from the requirements that $E = Bc$ and $B = \mu_0 \varepsilon_0 cE$, i.e.,

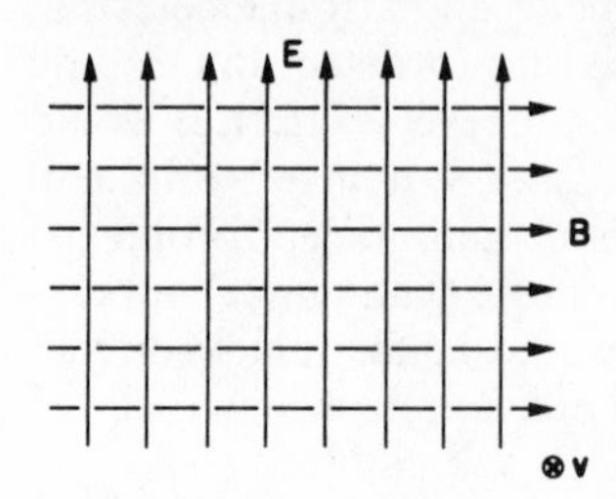

$$c = (\mu_0 \varepsilon_0)^{-1/2}. \tag{14.8}$$

The principal characteristics of the wave are immediately clear from this argument. A plane electromagnetic wave is transverse, i.e, **E** and **B** lie in the plane normal to the direction of propagation; **E** and **B** are mutually perpendicular, and their relative magnitudes are such that $\varepsilon_0 E^2 = B^2/\mu_0$, i.e., the energy density associated with each field is the same; the magnitude and direction of **E** may change in any way from one plane to another (**B** of course follows the variations of **E**), without the velocity needing to be altered, i.e., the wave is non-dispersive and in particular simple harmonic waves of all frequencies propagate at the same velocity.

The foregoing argument may be restated in analytical rather than pictorial terms by manipulating the equations (14.5) and (14.6) that relate the fields. If we take the curl of (14.6), we have by XIII,

$$\text{curl curl } \mathbf{E} \equiv \text{grad div } \mathbf{E} - \nabla^2 \mathbf{E} = -\text{curl } \dot{\mathbf{B}}.$$

Now in empty space, where there are no free charges, div **E** vanishes, so that

$$\nabla^2 \mathbf{E} = \text{curl } \dot{\mathbf{B}} = \mu_0 \varepsilon_0 \ddot{\mathbf{E}} \quad \text{from (14.5), if } \mu = \varepsilon = 1.$$

Hence the space and time variations of **E** must be related by the equation,

$$\nabla^2 \mathbf{E} - \ddot{\mathbf{E}}/c^2 = 0, \tag{14.9}$$

which is the non-dispersive wave equation in three dimensions with a characteristic wave velocity c. This equation is capable of a great variety of solutions, which we shall not attempt to discuss since we are here concerned only to show the existence and display the properties of the simplest form, the plane wave. It is sufficient for this purpose to start by assuming a plane wave propagating along the z-direction; in other words, we suppose that in any x–y plane the fields are constant, so that $\partial/\partial x$ and $\partial/\partial y$ are both zero. If we now write out (14.5) and (14.6) in coordinate form, retaining only derivatives with respect to z, we obtain the following equations:

$$-\partial B_y/\partial z = \dot{E}_x/c^2 \tag{14.5a}$$
$$\partial B_x/\partial z = \dot{E}_y/c^2 \tag{14.5b}$$
$$0 = \dot{E}_z/c^2 \tag{14.5c}$$
$$\partial E_y/\partial z = \dot{B}_x \tag{14.6a}$$
$$-\partial E_x/\partial z = \dot{B}_y \tag{14.6b}$$
$$0 = \dot{B}_z \tag{14.6c}$$

(14.5c) and (14.6c) show that the wave is transverse. Moreover, E_x and B_y alone appear in (14.5a) and (14.6b), while E_y and B_x alone appear in the other pair of equations. This allows us, if we wish, to construct two independent solutions based on the presence of one or other of these pairs; in each solution the fields **E** and **B** are mutually perpendicular, and the two independent solutions correspond to two transverse *plane polarized* waves, with field directions* as shown in the diagram. The ability to be plane polarized, by a Nicol prism or sheet of Polaroid, is the characteristic of light that demonstrates it to be a transverse wave, and the electromagnetic wave enables this property to be interpreted in terms of the directions taken by the field vectors.

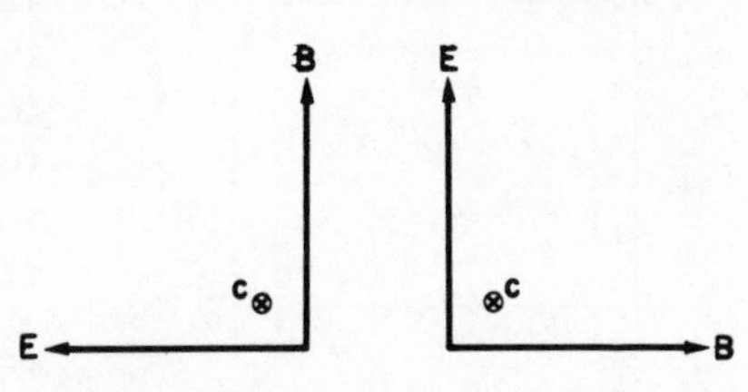

If we take (14.5a) and (14.6b) as typical, we can, by differentiation of one with respect to z and substitution of the other, write down the one-dimensional equation for either field, e.g.,

$$\frac{\partial^2 E_x}{\partial z^2} - \frac{1}{c^2}\frac{\partial^2 E_x}{\partial t^2} = 0, \tag{14.10}$$

which is, of course, the special case of (14.9) appropriate to a plane wave. The general solution, as discussed earlier for a wave on a string, describes the superposition of any two waveforms, $f_1(z - ct)$ and $f_2(z + ct)$, propagating unchanged in opposite directions. Taking only that one which moves in the positive direction towards larger z, if we write $E_x = f_1(z - ct)$, we have from (14.5a) and (14.6b) that

$$\partial B_y/\partial z = f_1'/c \quad \text{and} \quad \partial B_y/\partial t = f_1',$$

* The two independent solutions have **E** and **B** lying along x and y only by a mathematical accident, quite irrelevant to the physics. Superposition of the two solutions with their crests coincident (i.e., in phase) yields a plane-polarized wave with **E** and **B**, still perpendicular, pointing in intermediate directions. All orientations are equally good solutions.

both of which equations are satisfied if $B_y = f_1/c = E_x/c$, as we found in our pictorial demonstration.

A third approach to electromagnetic waves is by way of transmission lines, such as the coaxial cable used to connect a television aerial to the receiver. The easiest type of transmission line to understand is a pair of parallel wires. If they are carrying equal and opposite currents, they will produce a magnetic field pattern like that shown in the diagram on the left, and if the current is alternating, the change with time of the magnetic field will induce an electric field pattern. Looking down on the plane containing the wires, one sees that the flux contained within the rectangular circuit $ABCD$ changes, and that therefore $\oint \mathbf{E}.\mathbf{dr}$ round this circuit does not vanish. With good conductors for the lines, the only parts of the circuit which can support a substantial electric field are AB and CD, and it follows that there must be an electric field between A and B different from that between D and C. The requirement that changes in time of the field pattern demand changes in space as well leads us to suspect that the fields propagate as a wave.

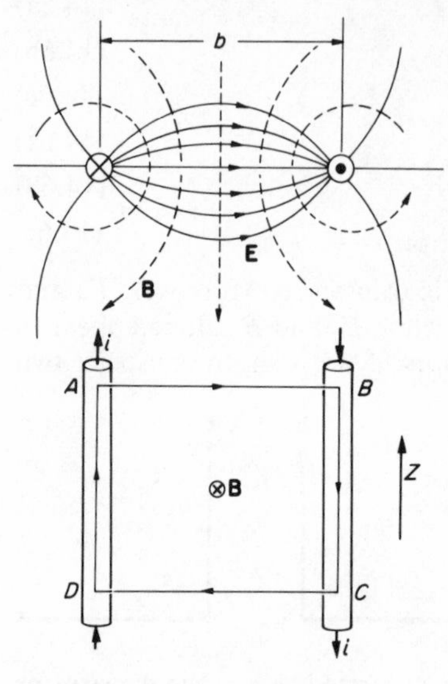

To make the ideas more exact, let us write the 'potential difference' between the lines, $\int_B^A \mathbf{E}.\mathbf{dr}$ as $V(z)$; then the e.m.f. round the shaded circuit, of length δz, is $V(z + \delta z) - V(z)$, or $(\partial V/\partial z)\, \delta z$. This must be equated to the rate of change of magnetic flux within the shaded area. Since the magnetic field strength is proportional to the current, i, we can write the flux as $Li\, \delta z$; L is called the inductance per unit length of the transmission line, since the flux linkage with a rectangular area of unit length is just Li. We shall return in due course to the evaluation of L, but for the present we need only note the form taken by Faraday's law, as applied to the elementary rectangle,

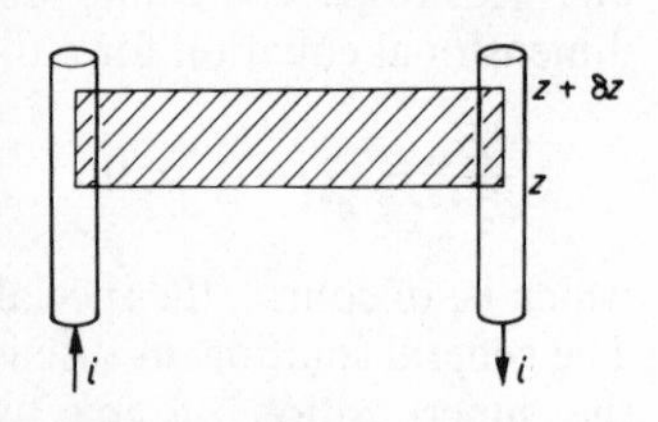

$$\partial V/\partial z = -L\, \partial i/\partial t. \tag{14.11}$$

Next we observe that if the current i, along with the field strengths, varies with z there must be charges built up on the lines. For the current entering an element δz of one of the lines is $i(z)$ and that leaving is $i(z + \delta z)$ or $i(z) + (\partial i/\partial z)\, \delta z$; thus the rate at which charge is being withdrawn from the element is $(\partial i/\partial z)\,\delta z$, so that if the excess charge on unit length of the line is ρ, we may write the condition for charge conservation in the form

$$\partial\rho/\partial t = -\partial i/\partial z. \tag{14.12}$$

The charge pattern on the surface of one wire is matched by an equal and opposite charge pattern on the other, and these surface charges match up with the electric field which we have seen to exist in the space surrounding the wires. The wires, in fact, act as the two plates of a capacitor, and the field between them takes the form shown in the first diagram, running from one wire to the other as we already know is required to satisfy Faraday's law. Just as we introduced formally the inductance per unit length, so we can introduce the capacitance per unit length, C, by requiring that CV shall be the charge ρ on unit length of each wire when there is a potential difference V between them. We may then write (14.12) in the form

$$C\,\partial V/\partial t = -\partial i/\partial z. \tag{14.13}$$

Hence

$$CL\,\partial^2 V/\partial t^2 = -L\,\partial^2 i/\partial z\,\partial t = \partial^2 V/\partial z^2, \quad \text{from (14.11),}$$

or

$$\frac{\partial^2 V}{\partial z^2} - \frac{1}{v^2}\frac{\partial^2 V}{\partial t^2} = 0, \tag{14.14}$$

where $v = (LC)^{-1/2}$.

Thus V (and i also, as is easily shown) is propagated as a wave along the line, and to find the velocity v we must evaluate L and C. This is a simple matter if the wires have a radius a, which is so much less than their separation b, that we may assume the charge and current to be uniformly spread around their perimeter*, and not attracted towards those parts that face one another, as happens with closely spaced wires. Given this assumption, the reader will easily verify by use of Gauss' theorem that the potential difference between wires carrying charges $\pm\rho$ per unit length† is

$$\rho \ln\left(\frac{b-a}{a}\right)\Big/(2\pi\varepsilon_0),$$

so that

$$C = 2\pi\varepsilon_0/\ln\left(\frac{b-a}{a}\right);$$

* The charge always finds its way to the surface at great speed (about 10^{-15} s), and if the wires have high conductivity and the frequency is high, the skin effect keeps the current also near the surface.

† It will be observed that we are assuming constant ρ along the line, as if the wavelength were infinite; this is reasonable so long as the wavelength is much greater than b, and in fact the displacement current ensures that the result derived holds even when the wavelength is short. But we shall not pursue this point.

furthermore, by use of the formula $\mu_0 i/(2\pi r)$ for the field at a distance r from a wire carrying current i, the flux linkage per unit length is found to be $\mu_0 i \ln[(b - a)/a]/(2\pi)$, so that

$$L = \mu_0 \ln\left(\frac{b-a}{a}\right)\bigg/(2\pi).$$

Hence

$$LC = \mu_0\varepsilon_0 \quad \text{and} \quad v = (\mu_0\varepsilon_0)^{-1/2} = c.$$

The wave that propagates along the transmission line is an electromagnetic wave in which **E** and **B** are wholly transverse and are in fact (as will be proved immediately) at every point perpendicular to each other. It is no accidental consequence of choosing thin, well-spaced wires that the velocity is exactly c—cylindrical conductors of any cross-section produce the same result if only the conductivity is high enough that we can suppose the current to be carried entirely on the surface. To show this, consider any two parallel cylindrical conductors carrying constant charge $\pm\rho$ per unit length, so that the field between them is described by a potential ϕ which depends only on the transverse coordinates x and y:

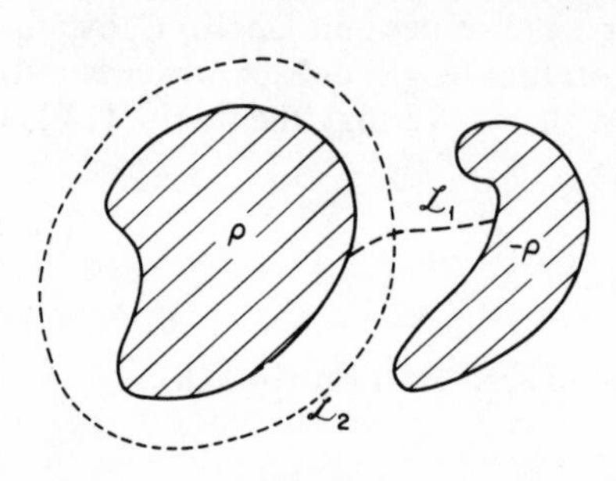

$$\left.\begin{aligned} &\mathbf{E} = (-\partial\phi/\partial x, \quad -\partial\phi/\partial y, \quad 0) \\ &\text{and} \\ &\nabla^2\phi = 0. \end{aligned}\right\} \qquad (14.15)$$

We can now define another field in terms of ϕ, having the correct properties to describe the magnetic field **B** produced by equal and opposite currents on the surfaces of the wires. We write

$$\mathbf{B} = (\partial\phi/\partial y, \quad -\partial\phi/\partial x, \quad 0), \qquad (14.16)$$

and note by direct calculation of the components (remembering that $\partial/\partial z = 0$) that curl $\mathbf{B} = 0$ and div $\mathbf{B} = 0$. The field **B** thus defined possesses the required properties of a magnetic field in a current-free region, but to complete the demonstration we must show that it is consistent with currents carried on the surface of the conductors. If the skin effect prevents a magnetic field from entering the conductor, the field lines immediately outside the conductors must run parallel to the surface, whatever its shape; since the electric field is normal to the surface, the real **B** and **E** must be mutually perpendicular at the surface. In fact the fields defined by (14.15) and (14.16) are mutually perpendicular everywhere, as can be proved by evaluating $\mathbf{B}.\mathbf{E}$ as identically zero. We have shown, then, that **B** as here defined has the properties needed to describe the magnetic field, and, incidentally, that **B** lies on the equipotentials of **E** (this is only strictly true when the conductors have no resistance and the skin effect confines the current to the surface).

To determine the velocity of a wave we need to evaluate LC rather than each of L and C separately, and we can achieve the required result by considering certain integrals along two lines $\mathscr{L}_1$ and $\mathscr{L}_2$, the former being any line joining the conductors and the latter any line surrounding one. Then we can write

$$V = \int_{\mathscr{L}_1} \mathbf{E}.\mathrm{d}\mathbf{r} = \int_{\mathscr{L}_1} \left(\frac{\partial \phi}{\partial x}\mathrm{d}x + \frac{\partial \phi}{\partial y}\mathrm{d}y\right),^*$$

and

$$i = \frac{1}{\mu_0}\oint_{\mathscr{L}_2} \mathbf{B}.\mathrm{d}\mathbf{r} = \frac{1}{\mu_0}\oint_{\mathscr{L}_2} \left(\frac{\partial \phi}{\partial y}\mathrm{d}x - \frac{\partial \phi}{\partial x}\mathrm{d}y\right).$$

Further, we can apply Gauss' theorem to a cylinder of unit length bounded by $\mathscr{L}_2$ to determine ρ, making use of the vector product to express the normal flux through the surface,

$$\rho = \varepsilon_0 \oint_{\mathscr{L}_2} \mathbf{E}\wedge \mathrm{d}\mathbf{r} = \varepsilon_0 \oint_{\mathscr{L}_2} \left(\frac{\partial \phi}{\partial y}\mathrm{d}x - \frac{\partial \phi}{\partial x}\mathrm{d}y\right).$$

Finally, we can also use the vector product device to evaluate the magnetic flux linking unit lengths of the two conductors,

$$\Phi = \int_{\mathscr{L}_1} \mathbf{B}\wedge \mathrm{d}\mathbf{r} = \int_{\mathscr{L}_1} \left(\frac{\partial \phi}{\partial x}\mathrm{d}x + \frac{\partial \phi}{\partial y}\mathrm{d}y\right).$$

When we use these four results to determine LC, i.e., $\Phi/i \times \rho/V$, we see that the value is $\mu_0\varepsilon_0$ irrespective of the actual magnitudes of the integrals.

Once more we have shown the existence of a wave travelling in free space with velocity $c = (\mu_0\varepsilon_0)^{-1/2}$, though in this case the wave is confined to the vicinity of the conductors. There are many different configurations used in practice for transmission of power or information by electromagnetic waves, of which the two shown here, the coaxial cable and the screened twin, are the most important. Both confine the wave within a conducting sheath and therefore prevent interference of one cable with another when many are stacked close together. If the interspace were empty, the wave velocity would be c, but the conductors need support, and one must not neglect the influence of the dielectric constant. If we had kept ε in all our calculations, we would have found it everywhere

* We can afford to ignore the overall sign of the integrals in what follows since the sign of the final result is not in doubt. Thus we do not put a minus sign in front of this integral.

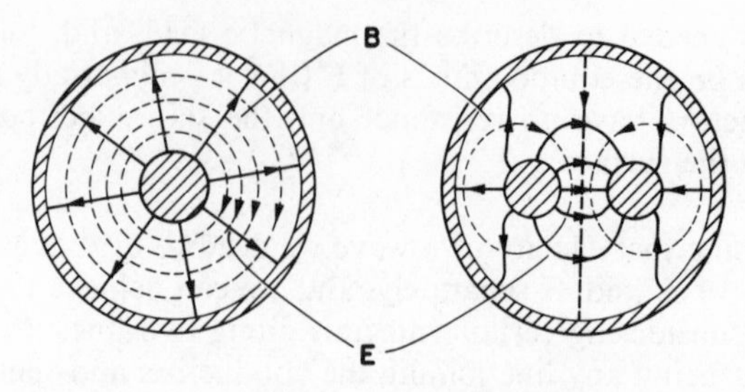

(e.g., in (14.4) and (14.7), and as a factor increasing the capacitance of a transmission line by ε) as a multiplier of ε_0; the velocity of a wave on a dielectric-filled line is therefore $c/\sqrt{\varepsilon}$, as also is the velocity of a plane wave in a uniform dielectric. Similarly if the medium has linear magnetic properties we must multiply μ_0 by μ, and the velocity will be $c/(\varepsilon\mu)^{1/2}$.

It is worth commenting briefly on the derivation of the wave velocity on a transmission line, where at no point have we introduced the idea of displacement (see however the footnote on p. 295). Faraday's law enters when we write (14.11), but instead of completing the argument by the complementary displacement law, we write (14.12), which expresses charge conservation. This should be enough to show how very intimately the various ideas and laws in electromagnetism are related, and is wholly consistent with the arguments used in the last few chapters to introduce and make plausible each new law by showing how it might *almost* be predicted from the more elementary laws already analysed. However, one must not try to eliminate that word 'almost'; wisdom after the event tells us that the fundamental laws are very closely interlocked, and indicate a great simplicity in the basic physical structure of the universe—we should not pretend that this is obvious until we have seen the interlocking of the ideas to be justified by the most searching experiments.

For this reason, it is a matter of great importance to discover how exactly the velocity of light (if we believe light to be an electromagnetic wave) agrees with independent measurements of the electrical constant $(\mu_0\varepsilon_0)^{-1/2}$. This point was carefully studied by Maxwell, who devised an ingenious technique for determining $\mu_0\varepsilon_0$, and his method was refined still further by Rosa and Dorsey in 1907. Their full account of the determination, and the precautions taken to reduce errors as far as possible, take 170 pages—by no means untypical of what is needed to describe a precise measurement in sufficient detail for the reader to be able to form a sound judgement on the competence with which the work has been carried out. If we here confine our account to the bare principles, there should be no misunderstanding of the gulf that divides the schematic description and the actual realization; every task of this nature is the labour of several years at least, and demands of the investigator a scrupulously critical attention to even the most unlikely sources of error, and such patience that it can only be well performed by someone whose personality is diametrically opposed to that of the brilliant scientist conventionally portrayed in fiction.

220

The determination proceeds in two stages, of which only the second was carried out by Rosa and Dorsey themselves, the first being already at that time a fairly routine matter for the various laboratories equipped to do it. This first stage involved calibrating a resistor in absolute terms, and depends essentially on the laws of Ampère and Faraday. For a given coil of wire, the constant of proportionality between the current in the coil and the magnetic field produced is determined by the dimensions, and we assume α to be a measurable constant when we write

$$B = \mu_0 \alpha i.$$

If we now move a conductor in this field, say by rotating a metal disc at angular velocity ω about an axis parallel to $\mathbf{B}$, the potential difference, V, induced between any two points on the disc is proportional to the magnetic field and the speed of rotation,

$$V = \beta B \omega = \mu_0 \alpha \beta \omega i,$$

where β is, like α, a constant determined by the geometrical configuration. The ratio V/i is $\mu_0 \alpha \beta \omega$ and is known absolutely from the configuration and the rotational speed. Strictly speaking, of course, α and β depend on position, but this does not alter the basic fact that the induced voltage V is proportional to ωi with a proportionality constant fixed by μ_0 and the configuration of the apparatus. Let us now pass the same current i through a resistor R and compare by means of a potentiometer the potential difference across the resistor and that induced in the rotating disc; if the former is measured to be n times the latter, we know that R is $n\mu_0\alpha\beta\omega$, so that R has been calibrated in terms of μ_0 and the standards of length and time which enter into $\alpha\beta$ and ω respectively.

One form of the apparatus used in this measurement (F. E. Smith's modification of Lorenz's original conception) is illustrated in the diagram.

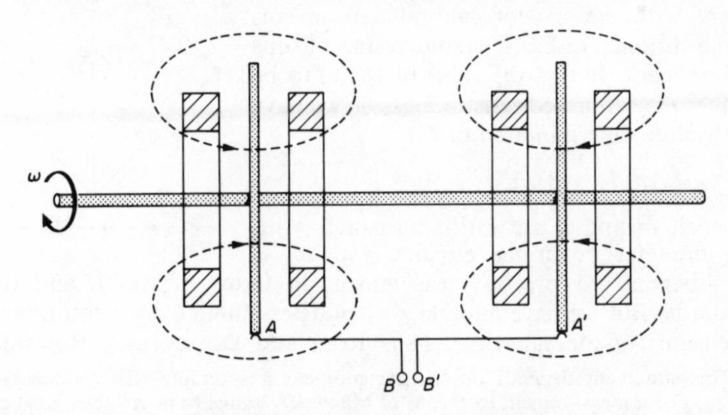

Two pairs of coils produce oppositely directed fields in which the copper discs spin on a common metallic shaft. Sliding contacts A, A' bring the induced potential difference to the terminals B, B'. The merit of the double disc arrangement over Lorenz's single disc is that because the field

249

is oppositely directed inside and outside the coils, there is a certain diameter at which the field is zero, and if this is chosen for the disc diameter, both sliding contacts are at points where the induced voltage is not appreciably affected by small changes of diameter; put another way, at this diameter the constant $\alpha\beta$, which depends on the total flux through the disc, is independent of the diameter, which therefore need not be measured with the same accuracy as, say, the geometrical form of the coils.

To proceed now to the second part of the determination, Maxwell suggested an ingenious bridge circuit which would enable a capacitor to be compared with a resistor. Consider first the circuit containing a cell, a capacitor and a switch which oscillates between the two positions, A and B, ν times a second. When it is at A the capacitor is discharged, but, as soon as it contacts B, charge CV flows through the cell to bring the capacitor to a potential difference V; with the return of the switch to A this charge is removed from the capacitor, but no charge flows through the cell. Thus the cell passes $CV\nu$ of charge per second in the form of a succession of pulses, and if a galvanometer is placed in the circuit it will experience the same mean deflexion as if a current $CV\nu$ were passing through it. If its response time is much longer than the time $1/\nu$ between pulses it will not respond to the variation of current but only record the mean.*

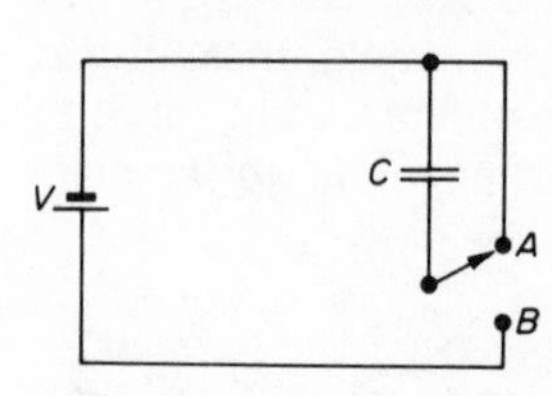

In this arrangement, then, the combination of capacitor and switch behaves like a resistor of magnitude $1/(C\nu)$. By incorporating it in a Wheatstone bridge the resistance of the capacitor-switch combination can be compared with a genuine resistor, and ultimately with the resistor calibrated by means of the Lorenz disc. If as the result of this measurement we find the value of $1/(C\nu)$ to be m times the calibrated resistor $n\mu_0\alpha\beta\omega$, we have an absolute determination of C,

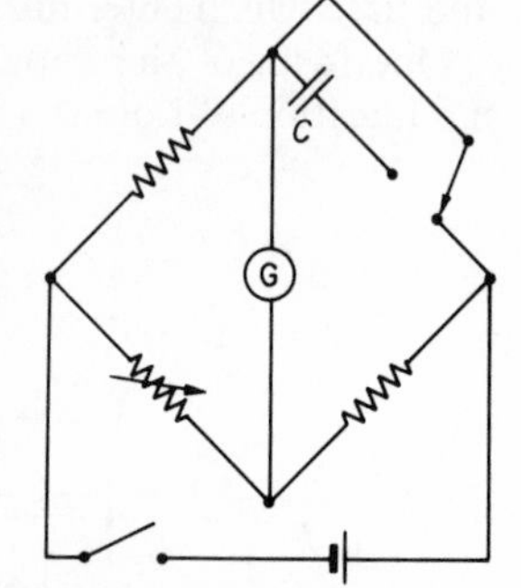

$$C = 1/(mn\mu_0\alpha\beta\omega\nu),$$

in which m and n are ratios measured with potentiometer or bridge circuits, and $\alpha\beta$, ω and ν depend on precise measurement in terms of length and time standards. But we have another way of determining C by measuring the dimensions of the capacitor. Thus Rosa and Dorsey used Rowland's

217

* This statement depends on the galvanometer responding strictly linearly to current. For a measurement by means of Maxwell's bridge to be reliable, great care must be taken to avoid even the smallest hint of non-linearity, such as might arise from imperfect joints which sometimes exhibit rectifying action. This sort of trouble is, however, so common a hazard in precise measurements that it need not surprise us not to find it mentioned by Rosa and Dorsey; their procedures were almost instinctively such as to show it up if it should have occurred.

finely constructed concentric spherical air-spaced capacitors whose capacitance is $4\pi\varepsilon\varepsilon_0(1/a - 1/b)^{-1}$, i.e, $\gamma\varepsilon_0$, where γ is a geometrical constant slightly corrected for the dielectric constant of air, 1·000 58. Equating these two measurements of C, we have that

$$\mu_0\varepsilon_0 = 1/(mn\alpha\beta\gamma\omega\nu),$$

and $\mu_0\varepsilon_0$ is therefore experimentally determined.

The most difficult aspect of this experiment is the switch, since frictional effects at make and break can produce electrostatic charging and thermoelectric voltages. Much careful design and testing therefore was needed before the rotating commutator used by Rosa and Dorsey fulfilled its purpose. Ultimately the accuracy of the determination was fixed by the bridge measurement, and they believed their final value for $(\mu_0\varepsilon_0)^{-1/2}$ to be no more than 1 part in 10^4 in error. Their value $(2{\cdot}997\,84 \pm 0{\cdot}000\,30) \times 10^8$ m s^{-1} agrees within their estimated error with the best modern determinations of the velocity of light in free space $(2{\cdot}997\,930 \pm 000\,005) \times 10^8$ m s^{-1}.

It helps one to appreciate the problems involved in the determination of $\mu_0\varepsilon_0$ if one considers a difficulty that only becomes apparent when the magnitudes involved are calculated. The Lorenz discs can be said to generate a standard resistance of magnitude $\mu_0\alpha\beta\omega$ which must be compared with the resistance $1/(\nu C)$ generated by the capacitor and switch. If these two are very different in size, the comparison will be less accurate than if they are similar. It is difficult to construct coils of enough turns (yet still capable of being measured accurately) and to spin the discs fast enough to generate a resistance larger than 10^{-2} Ω; on the other hand, the spherical capacitors used by Rosa and Dorsey had capacitance of about 50 pF, so that with a switching frequency of 400 times a second the resistance generated was 50 MΩ. The comparison involves therefore two resistances differing by a factor of about 10^9. It should be recognized that this is not accidental. The quantity being determined is a velocity of 3×10^8 m s^{-1}, and the method employs apparatus whose dimensions are less than a metre, and time intervals (e.g., $1/\nu$) of the order of 10 ms,which together are incapable of yielding anything as large as 3×10^8 m s^{-1} unless a very large number is involved in the apparatus or the measurement; there are opportunities for numbers of the order of 10^3 in the equipment itself, as for example in the number of turns on the Lorenz field coils, but these do little to bridge the gap and the really large number appears as a ratio of resistances. It is here that accuracy may be lost, as may be seen by considering the following hypothetical scheme of comparing two resistors. Let us construct a standard unit of resistance (not necessarily 1 Ω) and then make another equal to it (e.g., by substituting one for the other in a Wheatstone bridge and verifying that the balance of the bridge is unchanged); we may then put the two together in series and construct two 2-unit resistors, each equal to the sum of the original units, and proceed in this way to 2^2, 2^3, 2^4 etc., until in the end we have a resistor 10^9, or 2^{33} times the original. Clearly about 33 successive processes are involved, and the errors introduced in each will accumulate; they will not amount to 33 times the individual error, since some will be positive and some negative, and under these conditions the likely final error will be about $\sqrt{33}$, say 6, times the individual error. In the light of this, Rosa and Dorsey's accuracy can be recognized as very impressive.

If we look back over the chain of argument involved, and note how this experiment depends for its interpretation on Coulomb's, Ampère's and Faraday's laws, all of which we have seen to fall in a consistent manner into a logically connected argument whose final, and equally consistent, step is Maxwell's hypothesis, we must agree that the measurement of Rosa and Dorsey puts a seal on the whole process and leaves little room for scepticism about the essential qualitative and quantitative correctness of the construction. This may be said to be the culmination of classical physics, and so magnificent an achievement indeed that one would be tempted to regard it as perfect if one did not know that its sphere of application was limited to things that were neither too large nor too small, nor yet moving at too great a speed—larger than an atom, smaller than a galaxy, and not rivalling the speed of light.

Before proceeding to discuss the limitations of the classical scheme, let us pause to note that the association of light and electromagnetic waves does not rest merely on a measurement of velocity, but that the properties of electromagnetic waves, as deduced from the theory, agree in so many and diverse ways with observation that there can be no doubt left. It is possible, indeed, to use electronic circuits to generate and study the propagation of undeniably electromagnetic waves with frequencies ranging from as little as a few cycles per second (Hz) up to 10^{12} Hz or more, at which frequency they merge into, and are seen to be identical with, the far end of the infra-red spectrum of radiation which at 10^{15} Hz we know as light. As Hertz showed from Maxwell's equations, and demonstrated experimentally in 1888, oscillations or accelerations of charge provide sources for electromagnetic waves, a fact that is made use of every time a dipole aerial is connected to a radio transmitter or, since the process is reciprocal, used to pick up the transmitted waves and convey the information they carry to a radio receiver. To transmit measurable amounts of power at very low frequencies is hard, unless the oscillating system of charges is extended over a great distance; this condition is fulfilled by lightning strokes which generate a wide band of frequencies, some of which are guided by a peculiar mechanism (the *magneto-ionic wave*) along the Earth's magnetic lines of force high in the ionized regions of the atmosphere, and appear many thousands of miles away as signals that can be picked up, amplified and fed into loud-speakers to give a descending whistle, audibly exhibiting frequencies lower than 100 Hz. At higher frequencies, it gets progressively easier to radiate power, and indeed in everyday life, with so many oscillatory devices in use, such as sparking-plugs and electrical machinery, the problem often is to subdue the continuous background of radiated electromagnetic noise that interferes badly enough with television reception but still more disastrously with the sensitive measurements that form the observational basis of radio-astronomy.

Oscillatory circuits to generate alternating currents, and excite suitable aerials or other radiators, have been devised to operate from the lowest frequencies up to over 10^{11} Hz (giving waves with a wavelength of about 3 mm), and this can be exceeded by the use of frequency multipliers which, when fed with a considerable amount of power at this frequency, emit tiny quantities of power at harmonics of the exciting frequency. In this way waves have been generated by purely macroscopic electromagnetic means with wavelengths as small as 0·4 mm. Such waves can be handled like light waves, focused by mirrors and lenses, and spectrally analysed by diffraction gratings.

To move to another point of vantage, we may note that visible radiation lies in the wavelength range 0·38 μm (violet) to 0·78 μm (red), some 1000 times less than the shortest waves generated artificially. But the visible spectrum only represents that very limited range of wavelengths that happen not to be significantly absorbed by our atmosphere; both longer and shorter waves are heavily absorbed, and if our eyes were capable of responding to them, it would do us little good, for it would be no better than living in a dense fog. It is a striking example of a successful evolutionary process that so large a proportion of the 'optical window' is used, and so small a potentially useful band of wavelengths is wasted. The reciprocity between radiation and reception, already commented on for radio aerials, holds here too—molecules, suitable excited, are capable of emitting the same wavelengths as they absorb. The absorption processes that render air opaque in the ultra-violet (wavelengths shorter than visible) are electronic excitations analogous to the transitions between different states of the Bohr atom, while the infra-red absorptions (wavelengths longer than visible) are associated with rotations and vibrations of whole molecules. These can be studied as band-spectra, a multitude of closely-spaced spectral lines, extending from the visible spectrum out as far as 100 μm and even further. A particularly interesting example is the series of spectral lines for ammonia, NH_3, in the range of wavelengths between 10 and 30 mm, for these can be detected as absorption lines when electromagnetic waves from an oscillator are passed through the gas, and in this way a linkage between the oscillator end and the molecular end of the spectrum is achieved. It may further be noted that the ammonia molecule was the first to be used, by Townes in 1951, in a *maser*—an oscillator which derives its power from the radiation emitted by excited atoms or molecules. It can fairly be said now that there is no hiatus in the electromagnetic spectrum between the radiation by man-made oscillations and the natural radiations of atoms and molecules. Moreover, the mathematical description in quantum mechanics of an atom radiating energy as it makes a transition from one state to another resembles very closely the description of charge oscillations in a radio aerial—only the scale is different.

The electromagnetic spectrum does not end in the ultra-violet, but can be extended without any theoretical limit, that is known at present. At wavelengths below about 10 μm the far ultra-violet region may be considered to merge into the X-ray region. The production of X-rays is achieved by acceleration of charge, either during atomic transitions of the inner electronic shells of heavy atoms, or by allowing energetic electrons to strike a target and be suddenly brought to rest. The latter process results in a broad spectrum of wavelengths rather than sharp spectral lines, and this spectrum has one very significant feature, that it contains no frequencies higher than a certain critical frequency determined by the energy of the electron that is brought to rest. The critical frequency ω_c is related to E, the energy of the electron, by the simple equation $\hbar\omega_c = E$, so closely indeed that a determination of ω_c for electrons of known energy is one of the best methods for measuring Planck's constant precisely. This behaviour shows very clearly the quantization of the emitted radiation; at a frequency ω the radiation is emitted only in quanta of $\hbar\omega$, and no electron can produce radiation of a frequency higher than ω_c, for which more than its total energy would be needed to produce a single quantum.

X-ray generators employing accelerating voltages of many million volts have been constructed, and the radiation thus generated is of the same nature as the γ-rays spontaneously emitted by radioactive nuclei when they decay. Still more energetic are the γ-rays produced when particles of enormous energy enter our atmosphere from outer space. Particles with energies of more than 1 J($\sim 10^{19}$ eV) are occasionally observed directly, by the dense trail of ionization they leave in photographic plates flown to high altitudes in balloons, and indirectly by the appearance from time to time of extended showers of electrons which simultaneously cause the discharge of Geiger counters displayed over an area of several acres. *Cosmic-ray showers* arise from an extremely important property possessed by γ-ray quanta, that they can generate particles when they pass close to an atomic nucleus. In this process of *pair production*, the γ-ray quantum disappears, and in its place appear an electron and a positron, which has the same mass as the electron but is positively charged. The energy of the pair of particles equals that of the γ-ray, provided we allow that each particle of mass m possesses, by virtue of its mass, energy mc^2 even at rest. Because of this mass-energy relation, no γ-ray of quantum energy less than 10^6 eV can create an electron-positron pair, since for electrons $2mc^2 \sim 2 \times (9 \times 10^{-31}) \times (9 \times 10^{16})\text{J} \sim 1{\cdot}6 \times 10^{-13}\text{J} \sim 10^6$ eV. In cosmic rays, the energy of the primary particle as it enters the atmosphere is usually very much larger than this threshold, and it generates pairs itself or, being accelerated as it passes by a nucleus, emits energetic γ-rays which generate pairs, and these in their turn may generate more pairs or γ-rays, so that the original energy may be shared ultimately among a large number of electrons and positrons; showers of 10^9 electrons reaching the ground are not unusual, and all derive from the entry into the atmosphere of a single charged particle.

166

We have clearly entered a realm of observation far removed from the genteel world of the laboratory and the telescope, which led our thoughts to the classical description of fairly readily observable phenomena, and we have recorded in the last pages examples of two different types of failure of classical concepts—quantization and the mass-energy equivalence, which is one of the spectacular triumphs of relativity theory. We shall say no more about quantum effects, but introduce some of the leading ideas in relativity by pointing out how the logical processes that led us to Maxwell's equations contain in themselves an inconsistency which is not to be resolved except by changing our fundamental ideas. We should not, however, regard the development of relativity theory as a blow to the classical scheme; it represents yet one more step in the process of refinement which does not falsify anything that has gone before, but rather extends it and delineates the sphere of application within which the approximations of the earlier treatment were justified.

An important part was played at several stages of the development by the idea of Galilean invariance: different inertial observers moving relative to one another at a constant speed agree on their observations of acceleration and therefore can all assent to the same Newtonian dynamical laws. By looking at the same phenomena from the standpoint of different observers, we were able, in Chapter 12, to link Faraday's law of induction with the Lorentz force experienced by a moving charged particle in a magnetic field. But even there we suspected, while agreeing to ignore for the time being, the existence of a difficulty. Now that we have developed and confirmed Maxwell's displacement hypothesis, we must return to this difficulty, to show how serious it really is. We found Maxwell's idea to be strictly analogous to Faraday's; e.g., if a moving magnetic field generates an electric field, then a moving electric field generates a magnetic field. This symmetry reflects itself in the symmetry of Maxwell's equations for the fields. We cannot therefore accept the asymmetry that we were prepared to put up with temporarily when we wrote for the relation between the fields as seen by two observers with relative velocity $\mathbf{u}$,

30

245

$$\mathbf{E}' = \mathbf{E} + \mathbf{u} \wedge \mathbf{B} \quad \text{and} \quad \mathbf{B}' = \mathbf{B}.$$

If we are to incorporate the displacement concept, we must write something more symmetrical than this; for example

$$\mathbf{E}' = \mathbf{E} + \mathbf{u} \wedge \mathbf{B} \quad \text{and} \quad \mathbf{B}' = \mathbf{B} - \mathbf{u} \wedge \mathbf{E}/c^2. \tag{14.17}$$

The new term reflects the generation of a magnetic field by a moving electric field. Let us then see what these equations imply for the force on a charge moving at velocity $\mathbf{v}$ relative to the first (unprimed) observer. What he sees is a force $\mathbf{F}$ given by

$$\mathbf{F} = e(\mathbf{E} + \mathbf{v} \wedge \mathbf{B}),$$

while the other observer sees

$$\mathbf{F}' = e[\mathbf{E}' + (\mathbf{v} - \mathbf{u}) \wedge \mathbf{B}'].$$

Substituting (14.17), we find that **F** and **F**′ are now unequal:

$$\mathbf{F}' = \mathbf{F} - e\mathbf{v}\wedge(\mathbf{u}\wedge\mathbf{E})/c^2. \tag{14.18}$$

This result strikes at the very heart of our assumptions. For if two observers see a particle subjected to a different force they will also, if the mass is seen to be the same by both, see different accelerations, and this means that they do not agree on the basic issue of standards of length and/or time. We might discuss at great length, though with little profit, how to modify our concepts in not too drastic a fashion so as to salvage the theory, but it is more profitable to note that the difficulty implied by the second term on the right of (14.18) is serious only when v and u are both comparable with the velocity of light, c. Under normal circumstances of relatively slow motion, the symmetrical transformation (14.17) will cause no embarrassment by its inconsistency with Newtonian mechanics. When, however, we consider phenomena involving velocities close to c, there are much more glaring difficulties than this. Consider, for example, the propagation of light itself, which puts us in a dilemma over the existence or otherwise of an absolute frame of reference. The strength of our application of Galilean transformations lay in the recognition that dynamical observations and, by inference, electromagnetic observations would not reveal our motion relative to any such absolute frame. Now, however, we have discovered a wave that travels through empty space with a certain velocity c, and we may ask: relative to what is this velocity to be measured? Since the signal does not require any material medium for its propagation, we may send a light signal past two observers moving at different speeds, and ask them to measure its speed to a high enough precision to ascertain whether they get the same value or values differing by, say, their relative speed. If the answer is the latter alternative, we may accept that such a fundamental idea as the law of composition of velocities is sound, but we have to admit that the laws of physics, as exemplified by the velocity of light, differ for the two observers, and that different inertial observers are not equivalent; we may then allow ourselves the luxury of an absolute frame of reference and modify all our subsequent arguments as necessary to make them self-consistent, without having to ensure that the laws take exactly the same form for all observers. In fact, however, experiment has ruled this solution out—light signals pass all inertial observers at the same speed, and the law of composition of velocities breaks down when speeds approaching c are involved. Granted this result, we need not flog our brains to discover some trivial change that will put everything right—we know that a fundamental assumption is wrong and must be righted by a reformulation of the fundamental principles.

The experimental result, due originally to Michelson and Morley in 1887 (the idea for the experiment was Maxwell's), is so important as to deserve some account of how it was obtained. The Michelson–Morley experiment itself was an optical experiment but it is easier to appreciate a later version by Essen in 1957 in which radio waves of about 3 cm wavelength were used. Suppose we allow plane waves to bounce to and fro between parallel metal surfaces; steady behaviour is only possible if the separation of the planes is an integral multiple of one-half the wavelength so that standing waves are set up with the electric field vanishing at the metal surfaces (this boundary condition is necessary if the metal is taken to be a perfect conductor, and is a very good approximation for real metals). The system is therefore capable of sustaining electromagnetic oscillations only at those frequencies ω for which $n\pi c/\omega = l$. This result depends on the assumption that the wave is carried by a medium with respect to which the reflectors are at rest, so that the velocity relative to the reflectors is the same in both directions. If, however, we suppose the medium to be flowing past the reflectors at velocity u, the wave travels at speeds $c \pm u$ in the two directions. Since the number of wave crests leaving a reflector in a given time must equal the number arriving, the frequencies of the two travelling waves that bounce backwards and forwards must be the same, and they therefore have different wavelengths, $2\pi(c \pm u)/\omega$. Let us write an expression for two such travelling waves, adjusted in phase and amplitude so as to have automatically zero resultant amplitude at all times at the first plate, where x is taken as zero:

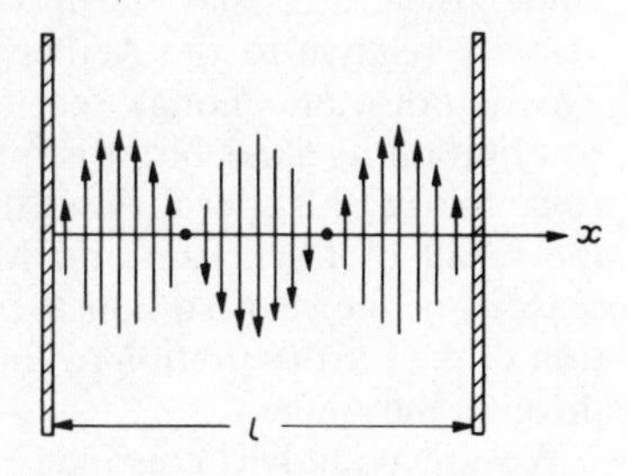

$$E(x, t) = A\left[\sin \omega\left(t - \frac{x}{c+u}\right) - \sin \omega\left(t + \frac{x}{c-u}\right)\right].$$

If E also vanishes at all times when $x = l$, $-\omega l/(c+u)$ and $\omega l/(c-u)$ must differ by an integral multiple of 2π, i.e.,

$$\omega = \frac{n\pi c}{l}(1 - u^2/c^2). \tag{14.19}$$

This result may be obtained more economically by noting that the permitted resonances are such that a point on the wave takes an integral number of periods of oscillation to traverse the resonator once in each direction; this condition ensures that after each double traverse it reinforces the wave already there. It is easy to see that if the speed is $c - u$ in one direction and $c + u$ in the other, the time taken is increased by a factor $(1 - u^2/c^2)^{-1}$ from that taken when $u = 0$, so that the resonant frequency is reduced as in (14.19).

In Essen's experiment, the wave was not contained between two infinite parallel plates but in a cylindrical box; the general principle is, however, unaltered. If the velocity of light is constant in some hypothetical frame, the Aether, with respect to which the laboratory is moving with velocity $\mathbf{u}$ (because of the Earth's rotation on its axis, or round the Sun, or with the whole solar system through space), the natural frequency of the resonator will depend on how it is oriented with respect to this movement, and turning the resonator on a turntable should result in a sinusoidal variation of frequency with orientation, having two maxima and two minima per revolution. The relative amplitude of this variation measures u^2/c^2 directly. Essen found the variation to be no more than one part in 10^{11}, and even this could be accounted for as the effect of the Earth's magnetic field on the dimensions of the resonator. We may conclude, therefore, that during the course of the experiment the drift velocity relative to the Aether was less than $c \div 10^{11/2}$, i.e., 1 km s^{-1}. This is not quite enough sensitivity to detect effects due to the Earth's rotation on its axis* but it is quite enough to detect motion in the orbit, since the Earth's speed round the Sun is about 30 km s^{-1}. This confirms the result of Michelson and Morley, who found no effect at different seasons of the year, so that it cannot be explained as a chance cancellation of the Earth's motion round the Sun and the drift of the solar system through the Aether.

We are bound to conclude, then, that this experiment has failed to demonstrate the existence of an absolute framework, and at the same time, by its criticism of the law of composition of velocities, has pointed to a deficiency in our assumptions about the measurement of distance and time by different observers, of practical importance only when very high speeds or very precise measurements are involved. The *theory of relativity*, in the form developed by Poincaré, Einstein and many others, does not attempt to patch up the classical assumptions but instead asserts boldly that no absolute framework need be assumed since no experiment will reveal its existence. In particular, all inertial observers will agree on the same formulation of the laws of physics, even though this implies that various 'commonsense' assumptions must be abandoned. The theory systematically explores the problem of devising new rules which meet this requirement, and are at the same time consistent with the undeniably successful classical picture under not too extreme conditions. The law of composition of velocities, for instance, must remain very nearly true, yet must be modified so that light passes all observers at the same speed; this in its turn means that two observers in relative motion will not be able to agree on standards of length and time. To illustrate the implications of a constant speed of light, let us note how two observers will disagree about whether two events occur simultaneously or not. Imagine a long train carrying an observer A at the centre and flash tubes X and Y at each end, moving past an observer B sitting beside the track. As A passes B, he presses a switch which causes the tubes to flash once, and he is pleased to observe that the two flashes reach him simultaneously and at the very moment when he is opposite B (as indicated, for example, by his head being hit by a stick held by B), so that B agrees that the flashes also reached him simultaneously. But B, seeing the train rushing past, and knowing that

* Townes and his co-workers have used an ammonia maser to achieve still better sensitivity, enough to show up effects due to the Earth's rotation, and have obtained a null result also.

light travels at a finite speed, perceives that when the flashes were made, X was further away from him than Y, and he congratulates A on timing his flashes so nicely as to allow for this and give X the required start over Y; A brusquely rejects the compliment, remarking that since X and Y are equidistant from him at all times there was no difficulty in arranging that they flashed simultaneously. B's counter to this is that the velocity of the

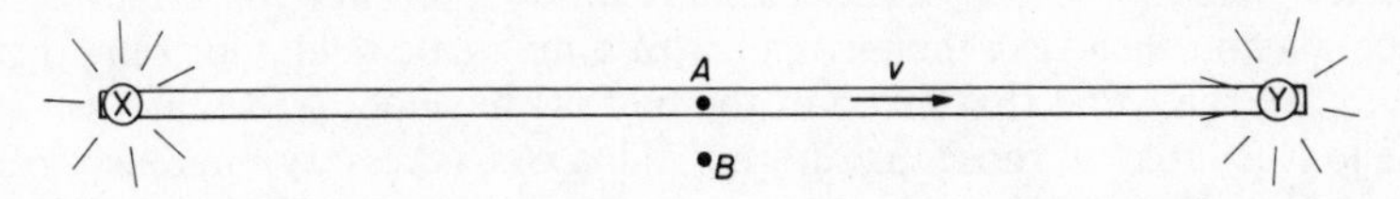

light signals relative to A is different in the two directions because of the motion of the train, but A, who has an Essen experiment set up, assures him that this is not so. And the argument proceeds without resolution so that ultimately the train recedes with A and B agreeing to differ, which in this case is the right resolution of their argument—they will never agree on whether the flashes were simultaneous or not.

One could invent similar demonstrations and dialogues to illustrate the other paradoxes of relativity, but since they are better studied quantitatively and as a special topic in their own right, we shall do no more than list some of the most important ways in which classical conceptions have to be modified.

(1) Two inertial observers with relative velocity u cannot agree on standards of length, even if they define them in relation to something readily reproducible, e.g., the wavelength of a certain spectral line. Each thinks the other's standard is short by a factor $(1 - u^2/c^2)^{-1/2}$ (*Lorentz contraction*).

(2) Similarly each thinks the other's standard clock is running slow by the same factor (*time dilatation*).

These two are obviously closely related, and can be illustrated by a striking phenomenon of cosmic-ray physics. Primary cosmic rays reaching the upper atmosphere generate unstable particles, μ-mesons, by collision with atomic nuclei, and many of these mesons reach ground level. Now observations on fairly slow-moving μ-mesons show that they spontaneously decay according to a simple exponential law—if we start with n particles then after time t there are only $ne^{-t/\tau}$ left, with τ equal to 2·2 μs. However, when they are in motion with a velocity approaching c, they survive for a longer time. In an experiment in which the number at a height of 1900 m was compared with the number at sea-level it was found that 73% survived the journey even though, moving as they did with a velocity very near to c, they must have taken at least 6·4 μs and only 5% should have survived. In fact the 'clock' of the mesons appears to observers on the Earth to be running slow by a factor of about 9, in reasonable agreement with the factor $(1 - u^2/c^2)^{-1/2}$ if u was, as estimated for the particular mesons selected for detection, only $\frac{1}{2}$% less than the velocity of light. From the point of view of a hypothetical observer moving with the mesons, this result is interpreted differently; he measures their lifetime as 2·2 μs and is not surprised that so many survive to ground level, since he sees the mountain as only about 200 m high, on account of Lorentz contraction.

If we generate a group of fairly slow mesons, and keep half of them by us (e.g., orbiting in a magnetic field) while we accelerate the rest with an electric field to a velocity near c, send them on a journey for a few microseconds and then turn them round with a magnetic field and bring them back, we shall find that fewer of the mesons have decayed that went on the journey than of those that did not. This is a necessary inference from the cosmic-ray experiment and, expressed in these terms, it does not seem more disturbing than any other relativistic effect; but for some reason, when it is told as a story in which space travellers return to find themselves less aged than their stay-at-home friends, it meets a strong emotional rejection from some otherwise rational thinkers. The literature is full of quarrels about this so-called 'clock-paradox', and one must conclude that commonsense (i.e., classical) views can be set aside by physicists so long as the problem is posed in the neutral terms of an experiment with particles, but that they are very deeply embedded in our thinking about people and everyday events. This is not surprising, nor need one be upset by the abandonment of logic in favour of passion when such a matter is discussed—the dedicated scientist is not a disembodied logical thinker, but one whose passions are engaged alongside his reasoning powers and may easily take control. Nevertheless, where logic can lead to a unique answer, passion is a poor substitute, and in this case, if one accepts the basic assumptions of relativity, one must accept also its paradoxical consequences. It is not the clock-paradox that should especially disturb our commonsense, but the whole world-picture implied by relativity, that time and space are not independent absolutes but are part of a unity which our limited senses do not permit us to appreciate as such. It is the strength of the experimental method and the use of logical analysis that they reveal such a unity in spite of the misgivings of our everyday senses.

(3) If the momentum, P, of a body, whose mass at rest is m_0, is written as $m_0v/(1 - v^2/c^2)^{1/2}$ instead of m_0v, the Newtonian equation $\dot{P} = F$ still holds. The mass m may be treated as velocity-dependent, though not appreciably so until v approaches c; $m = m_0/(1 - v^2/c^2)^{1/2}$.

(4) The total energy of a body of mass m may be written as mc^2. Taking account of the variation of mass with velocity, we may expand the energy as a power series in v/c:

$$E \Rightarrow mc^2 = m_0c^2/(1 - v^2/c^2)^{1/2} = m_0c^2 + \tfrac{1}{2}m_0v^2 + \tfrac{3}{8}m_0v^4/c^2 + \cdots.$$

The second term is the kinetic energy as derived in Newtonian mechanics, and the higher terms are the expected corrections when $v \sim c$. The first term is attributed to the body at rest and is the distinctive novelty of relativistic dynamics. In Einstein's interpretation, mass and energy are different manifestations of the same property, and the conservation laws of mass and energy are only valid when conflated into a single law of conservation of mass or energy (it is a matter of taste which one prefers, since they are related by a constant c^2). All forms of energy have mass, and all masses exhibit identical inertial and gravitational properties, as shown by the Eötvös experiment. When pairs of particles, e.g., electrons and positrons, are produced by a γ-ray quantum, there is conservation of mass provided one ascribes mass to the γ-ray, or conservation of energy provided one takes account of the rest-energy, $2m_0c^2$, of the two new particles. The γ-rays, like all quanta of electromagnetic radiation (*photons*) are peculiar objects in that they have finite energy while moving at the velocity of light; they may be considered to have zero rest-mass so that unless they move with velocity c their energy is strictly zero and they have no existence.

No body can be accelerated to the speed of light, since its mass increases without limit and it can acquire any momentum and energy, however large, without moving quite as fast as light.

(5) All inertial observers agree on the form of Maxwell's equations relating **B** and **E**, even though they measure different values for the field strengths, which are related for different observers by equations such as (14.17). Unlike mass, charge is not dependent on velocity, and all observers can agree on the measurement of a given charge. Charge is conserved; for example, when particles are produced by a γ-ray quantum they always appear as positive and negative pairs.

These few notes must suffice to indicate how the classical formulation of the laws of mechanics and electromagnetism needs to be adjusted to include effects occurring at high velocities. To a considerable degree, the relativistic rules may be treated in practice as well-established tricks that can be applied automatically to otherwise classical reasoning, and this is a sound view to take if the object is to get the right answer. But one should not forget that relativity has jolted the philosophical framework very severely, as must any discovery that reveals uncompromisingly a flaw in the beliefs that were developed through trusting in the evidence of the senses. It is not the least of the triumphs of physics in the present century to have penetrated so deeply behind the veil of our everyday perceptions as to reveal beyond doubt that our first-hand experience of the universe is at best a narrow and distorted view of whatever structure it is of which we are a part; yet we have this assurance also, that our universe, in so far as we have been able to probe it, is a marvellously ordered creation whose fuller understanding is a continual source of joy, in which intellectual satisfaction is mingled with wonder and humility.

Cavendish Problems: 222, 228.

READING LIST

Measurement of $\mu_0\varepsilon_0$: A. GRAY, *Absolute Measurements in Electricity and Magnetism*, Dover.

Whistlers: L. R. O. STOREY, *Phil. Trans. Roy. Soc.*, **A 246**, 113 (1953).
Microwave Generators: S. F. ADAM, *Microwave Theory and Applications*, Prentice-Hall.
Infra-red, Optical and X-ray Spectra: M. BORN, *Atomic Physics*, Blackie.
Relativity: J. H. SMITH, *Introduction to Special Relativity*, Benjamin.

APPENDIX

Vector formulae

The following notes summarize the principal results of vector algebra used in this book. The simplest example of a vector is the displacement of a point in Euclidean space, which illustrates the fundamental property of a vector—the rule by which vectors are added. A particle which moves in a straight line from A to B and then in a straight line from B to C, suffers the same resultant displacement as if it had moved straight from A to C; if we represent the two displacements by arrows whose length and direction define the displacement, summation of the two vectors is achieved by drawing the arrows head to tail. This process may be repeated for any number of vectors. Any physical quantity whose behaviour can be represented by a vector is called a vector quantity, and is denoted by a bold-face letter; thus **v** may denote the velocity of a particle, both in magnitude and direction, while v is used for the magnitude alone. Unless otherwise stated, vectors will be taken to be unlocated, i.e., all arrows of a given length and pointing in a given direction represent the same vector. The following properties of vectors are used in the text:

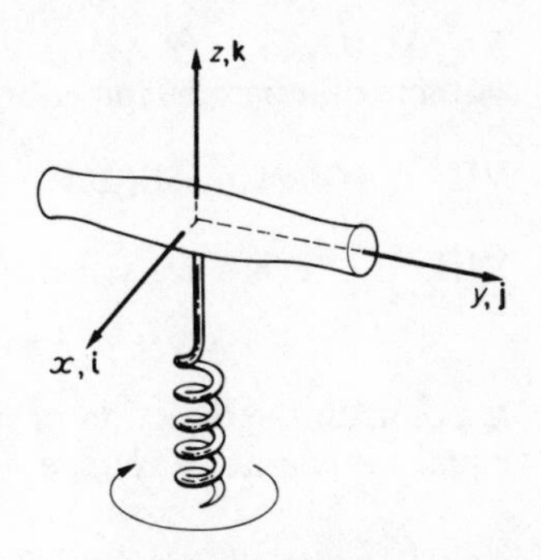

I. The unit vectors **i**, **j**, **k** point along the x, y, z axes respectively. Right-handed axes are used; a corkscrew pointing along **k** and twisted from **i** to **j**, as shown by the arrow, moves in the direction of **k**.

II. If ϕ is a scalar and **A** a vector, ϕ**A** points in the same direction as **A** and is ϕ times greater in magnitude. Hence for any vector **A**, **A**/A is a unit vector pointing in the same direction as **A**.

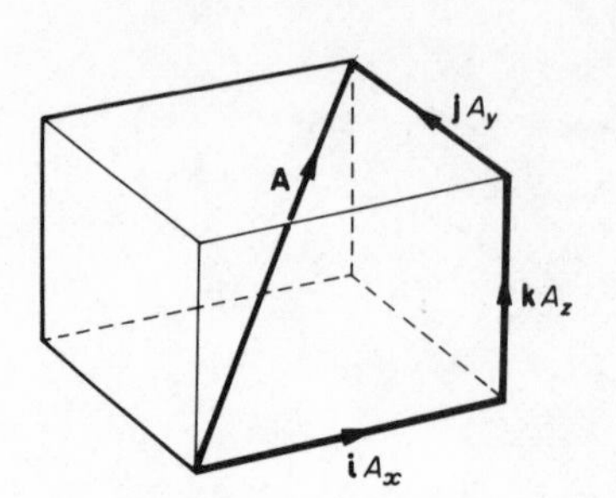

III. Any vector **A** may be decomposed into three vectors

$$\mathbf{A} = \mathbf{i}A_x + \mathbf{j}A_y + \mathbf{k}A_z,$$

which is also written (A_x, A_y, A_z); A_x, A_y and A_z are the components of the vector. Hence

$$\phi\mathbf{A} = (\phi A_x, \phi A_y, \phi A_z).$$

IV. The scalar product **A**.**B** is a scalar of magnitude $AB \cos \theta$, where θ is the angle between the vectors.

$$\mathbf{i}.\mathbf{i} = \mathbf{j}.\mathbf{j} = \mathbf{k}.\mathbf{k} = 1; \quad \mathbf{i}.\mathbf{j} = 0, \text{ etc.}$$

By writing **A** and **B** in Cartesian coordinates and multiplying out the brackets, we obtain a useful expression for **A**.**B**:

$$\mathbf{A}.\mathbf{B} = (\mathbf{i}A_x + \mathbf{j}A_y + \mathbf{k}A_z).(\mathbf{i}B_x + \mathbf{j}B_y + \mathbf{k}B_z) = A_xB_x + A_yB_y + A_zB_z.$$

Note also that $\mathbf{A}.\mathbf{A} = A^2$.

V. The vector product $\mathbf{A}\wedge\mathbf{B}$ is a vector of magnitude $AB \sin \theta$ directed normal to the place containing **A** and **B**. A corkscrew pointing along the vector $\mathbf{A}\wedge\mathbf{B}$, and turned in the sense that swings **A** towards **B** by the shortest route, travels along $\mathbf{A}\wedge\mathbf{B}$. Hence $\mathbf{B}\wedge\mathbf{A} = -\mathbf{A}\wedge\mathbf{B}$, and $\mathbf{A}\wedge\mathbf{A} = 0$.

$$\mathbf{i}\wedge\mathbf{j} = \mathbf{k}, \quad \mathbf{j}\wedge\mathbf{k} = \mathbf{i}, \quad \mathbf{k}\wedge\mathbf{i} = \mathbf{j}; \quad \mathbf{i}\wedge\mathbf{i} = \mathbf{j}\wedge\mathbf{j} = \mathbf{k}\wedge\mathbf{k} = 0.$$

$$\mathbf{A}\wedge\mathbf{B} = \begin{vmatrix} \mathbf{i} & \mathbf{j} & \mathbf{k} \\ A_x & A_y & A_z \\ B_x & B_y & B_z \end{vmatrix} = \mathbf{i}(A_yB_z - A_zB_y) + \mathbf{j}(A_zB_x - A_xB_z) + \mathbf{k}(A_xB_y - A_yB_x).$$

VI. $\mathbf{A}.(\mathbf{B}\wedge\mathbf{C}) = (\mathbf{A}\wedge\mathbf{B}).\mathbf{C} = (\mathbf{C}\wedge\mathbf{A}).\mathbf{B}$, and each expression gives the volume of a parallelepiped whose sides are **A**, **B** and **C**.

VII. $\mathbf{A}\wedge(\mathbf{B}\wedge\mathbf{C}) = \mathbf{B}(\mathbf{A}.\mathbf{C}) - \mathbf{C}(\mathbf{A}.\mathbf{B})$.

VIII. If ϕ (**r**) is a scalar field, grad ϕ is defined as the vector

$$(\partial\phi/\partial x, \quad \partial\phi/\partial y, \quad \partial\phi/\partial z).$$

It points in the direction of most rapid increase of ϕ and has magnitude equal to the rate of change of ϕ along this line.

$$\int_A^B (\text{grad}\, \phi).\mathrm{d}\mathbf{r} = \phi_B - \phi_A.$$

IX. If **A**(**r**) is a vector field, div **A** is defined as the scalar

$$\partial A_x/\partial x + \partial A_y/\partial_y + \partial A_z/\partial z.$$

If a volume V is bounded by a surface S, Gauss' theorem states that

$$\int_S \mathbf{A}.\mathrm{d}\mathbf{S} = \int_V (\mathrm{div}\,\mathbf{A})\,\mathrm{d}V.$$

X. If ϕ (**r**) is a scalar field,

$$\mathrm{div}\ \mathrm{grad}\ \phi = \partial^2\phi/\partial x^2 + \partial^2\phi/\partial y^2 + \partial^2\phi/\partial z^2,$$

which is written as $\nabla^2\phi$. The value of $\nabla^2\phi$ for a given ϕ (**r**) is independent of the choice of axes.

$$\nabla^2\mathbf{A} = (\nabla^2 A_x,\ \nabla^2 A_y,\ \nabla^2 A_z).$$

XI. grad (**r**.**A**) = **r**.div **A** + **A**.
In particular, if **A** is a constant vector, grad(**r**.**A**) = **A**.

XII. If **A**(**r**) is a vector field, curl **A** is defined as the vector

$$\begin{vmatrix} \mathbf{i} & \mathbf{j} & \mathbf{k} \\ \partial/\partial x & \partial/\partial y & \partial/\partial z \\ A_x & A_y & A_z \end{vmatrix},$$

i.e.,

$$\mathbf{i}(\partial A_z/\partial y - \partial A_y/\partial z) + \mathbf{j}(\partial A_x/\partial z - \partial A_z/\partial x) + \mathbf{k}(\partial A_y/\partial x - \partial A_x/\partial y).$$

XIII. curl grad ϕ = 0; div curl **A** = 0; curl curl **A** = grad div **A** − ∇^2**A**.

The results V, VI, VIII, XI, XIII are conveniently demonstrated, like IV, by the use of unit vectors and Cartesian components.

READING LIST

R. Gans, *Vector Analysis*, Blackie.

Index